SEISMIC WAVE THEORY

Perfect for senior undergraduates and first-year graduate students in geophysics, physics, mathematics, geology, and engineering, this book is devoted exclusively to seismic wave theory. The result is an invaluable teaching tool, with its detailed derivations of formulas, clear explanations of topics, exercises along with selected answers, and an additional set of exercises with derived answers on the book's website. Some highlights of the text include a review of vector calculus and Fourier transforms and an introduction to tensors, which prepare readers for the chapters to come; a detailed discussion of computing reflection and transmission coefficients, a topic of wide interest in the field; and a discussion in later chapters of plane waves in anisotropic and anelastic media, which serves as a useful introduction to these two areas of current research in geophysics. Students will learn to understand seismic wave theory through the book's clear and concise pedagogy.

EDWARD S. KREBES is a professor emeritus in geophysics in the Department of Geoscience at the University of Calgary. He has published many peer-reviewed papers on seismic wave propagation theory in leading journals such as the *Journal of Geophysical Research, Geophysics, Geophysical Journal International,* and the *Bulletin of the Seismological Society of America*. He recently coauthored and published another book, *Seismic Forward Modeling of Fractures and Fractured Medium Inversion* (2018). He was an associate editor of the journal *Geophysics*. His primary research interests are in theoretical and computational seismology, and seismic wave propagation in particular.

"Based on my many years of teaching university seismology courses, I must give a full and enthusiastic endorsement for the book *Seismic Wave Theory* by Professor Edward Krebes. This book is ideally suited for both senior undergraduate and graduate school courses on theoretical seismology. The book gives lucid mathematical descriptions of seismic wave theory accompanied by numerical examples and physical descriptions of seismic waves. It contains several descriptive illustrations, enlightening exercises, and useful references to seismic papers. The book ranks at the top of my list for recommended seismology textbooks, and it provides a valuable research reference for seismologists."

– Larry Lines,
University of Calgary

"A valuable textbook for students and researchers in geophysics and related disciplines . . . [including] detailed demonstrations of the equations [and] exercises with derived answers. The wave reflection-transmission problem is treated in great detail and [there's] a useful introduction to the complex problem of wave propagation in anelastic and anisotropic media. The emphasis is on seismic waves, but students in the fields of rock acoustics and material science – including many branches of acoustics of fluids and solids . . . – may also find this text useful. . . [T]his book is a perfect guide to learn how to analyze problems related to wave propagation, explaining in detail the mathematical tools and models that describe the different phenomena."

– José Carcione,
Istituto Nazionale di Oceanografia e di Geofisica Sperimentale

"This timely volume is an elegant and clear treatise on the theory of stress waves in fluids and solids. A unique feature is that it is one of the few texts in this area that has the needs of the exploration seismologist explicitly in mind. Coverage includes seismic imaging and (tau,p) analysis, both rarely found in other books of its genre."

– Sven Treitel,
TriDekon Inc.

". . . an excellent overview of the basic theoretical concepts that all seismologists should understand . . . The text covers the essential mathematical background leading to the concise and clear analytic derivation of many key problems related to seismic reflectivity or ray tracing through layered media. These theoretical developments are made tangible by being placed in a broad range of contexts from earthquake seismology to borehole sonic logging. [With] lucid discussions of more advanced topics related to seismic migration, anisotropy, and attenuation, [this] is an ideal text to lead senior undergraduate and junior graduate students to a fuller appreciation of the science of seismology."

– Doug Schmitt,
Purdue University

". . . *Seismic Wave Theory* is highly recommended for both students and lecturers of seismology. . . This book is a result of many years of fruitful lecturing and educating many seismologists for both academia and industry. Hence, its style is exceptionally clear and guides the readers in a most comfortable manner. . . Each chapter ends with a series of excellent exercises, which allow the readers to ensure their understanding. . . Thanks to the balanced choices of subjects and their careful presentation, *Seismic Wave Theory* is to become a classic textbook for quantitative seismology, a subject that – *sensu stricto* – cannot be but quantitative."

– Michael A. Slawinski,
Memorial University of Newfoundland

"After reading the table of contents, I felt like I was approaching a detective story – how is this is all going to turn out? I was not disappointed! . . . The clarity of development of the basic ideas, suitable for senior undergraduates and new graduate students, enables the easy grasp of the fundamentals. . .The text is replete with problems at the end of each chapter, [and] throughout the text, there are practical examples which illuminate principles that were presented previously. . . Its singular defining feature, often absent in most texts, is the conversational style of presentation. . . I strongly recommend this text for instructors and students who wish to obtain a fundamental understanding of seismic wave theory."

– Matthew Yedlin,
University of British Columbia

SEISMIC WAVE THEORY

EDWARD S. KREBES

University of Calgary

CAMBRIDGE
UNIVERSITY PRESS

CAMBRIDGE
UNIVERSITY PRESS

University Printing House, Cambridge CB2 8BS, United Kingdom

One Liberty Plaza, 20th Floor, New York, NY 10006, USA

477 Williamstown Road, Port Melbourne, VIC 3207, Australia

314–321, 3rd Floor, Plot 3, Splendor Forum, Jasola District Centre, New Delhi – 110025, India

79 Anson Road, #06–04/06, Singapore 079906

Cambridge University Press is part of the University of Cambridge.

It furthers the University's mission by disseminating knowledge in the pursuit of education, learning, and research at the highest international levels of excellence.

www.cambridge.org
Information on this title: www.cambridge.org/9781108474863
DOI: 10.1017/9781108601740

First published 2019

Printed in the United Kingdom by TJ International Ltd. Padstow Cornwall

A catalogue record for this publication is available from the British Library.

Library of Congress Cataloging-in-Publication Data
Names: Krebes, Edward Stephen, author.
Title: Seismic wave theory / Edward S. Krebes (University of Calgary).
Description: Cambridge ; New York, NY : Cambridge University Press, [2019] | Includes bibliographical references and index.
Identifiers: LCCN 2018041296 | ISBN 9781108474863 (hardback)
Subjects: LCSH: Seismic waves–Problems, exercises, etc.
Classification: LCC QE538.5 .K74 2019 | DDC 551.22–dc23
LC record available at https://lccn.loc.gov/2018041296

ISBN 978-1-108-47486-3 Hardback

Additional resources for this publication at www.cambridge.org/krebes

If geophysics requires mathematics for its treatment, it is the Earth
that is responsible, not the geophysicist.

– Sir Harold Jeffreys, Cambridge University

Contents

Preface

This book is based mostly on a set of notes that I prepared for a senior-level undergraduate geophysics course in seismic wave propagation theory with some basic applications that I taught in the Department of Geoscience at the University of Calgary for many years. The course was required for undergraduate geophysics majors.

The background knowledge required of readers, to get the most out of the book, is single- and multi-variable calculus, vector algebra and calculus, linear algebra, complex variables, Fourier series and transforms, basic concepts from ordinary and partial differential equations, and basic junior-level physics (mechanics, thermal physics, electricity and magnetism, optics). Some of the mathematical knowledge is reviewed in Chapter 1, which also includes some topics that may not be covered in an undergraduate curriculum, e.g., an introduction to tensors. Readers should also have some familiarity with basic concepts from reflection and refraction seismology as used in exploration geophysics, e.g., shot records, normal moveout, common-midpoint (CMP) gathers, and stacking; but even if such familiarity is lacking, readers should still be able to acquire an understanding of these concepts from the discussions in the text.

This book is intended to be pedagogical, i.e., its purpose is to teach seismic wave theory. It is intended to be an introduction, at the advanced undergraduate level, to seismic wave theory and some of its basic applications. The basic applications presented generally come from the field of exploration seismology, which is a reflection of my own area of specialization. The book is not intended to be an up-to-date, all-encompassing reference book on modern methods in seismology, and hence many modern topics of interest are not covered in the book, and many important and interesting books and papers have not been cited. The applications of seismic wave theory covered in the book are examples that were chosen to teach and show the reader how seismic wave theory can be applied to problems of interest. For instance, Chapter 8, on seismic migration, covers only some basic topics of seismic wave equation migration such as frequency-wavenumber migration and finite-difference migration (topics that were originally developed many years ago) and not other topics such as reverse time migration or modern developments in seismic migration and imaging or seismic inversion. In other words, Chapter 8 is not so much a chapter on seismic migration as it is a chapter showing an example of how to apply seismic wave theory to an important topic, i.e., seismic migration, in order to develop an important method, i.e., wave equation migration.

Exercises are included at the end of each chapter. Readers are encouraged to work through the exercises in order to acquire a deeper understanding of seismic wave theory. Answers to selected exercises are provided in a section at the back of the book. Fully derived answers to all the exercises are available, to instructors only, on the book's website.

I am grateful to my colleagues Pedro Enrique Martinez Fernandez, Pat Daley, and Larry Lines for reading the text. Pedro also prepared most of the figures, including those that required some numerical computation of mathematical curves. Pat also assisted in checking and preparing a set of additional problems with derived answers that I developed, which is available on the book's website.

It is hoped that students of geophysics, including those from other backgrounds such as mathematics, physics, engineering, and geology, will find the book a useful text for learning about seismic wave theory.

Symbols

The Greek Alphabet

Name	Lower case	Upper case
alpha	α	A
beta	β	B
gamma	γ	Γ
delta	δ	Δ
epsilon	ϵ , ε	E
zeta	ζ	Z
eta	η	H
theta	θ , ϑ	Θ
iota	ι	I
kappa	κ	K
lambda	λ	Λ
mu	μ	M
nu	ν	N
xi	ξ	Ξ
omicron	o	O
pi	π	Π
rho	ρ , ϱ	P
sigma	σ , ς	Σ
tau	τ	T
upsilon	υ	Υ
phi	ϕ , φ	Φ
chi	χ	X
psi	ψ	Ψ
omega	ω	Ω

1

Vectors, Tensors, and Fourier Transforms

This chapter reviews vectors and Fourier transforms, and introduces tensors. There are a number of classic texts that cover these topics. See, for example, Arfken (1985), Mathews and Walker (1970), and Spiegel (1959, 1971).

1.1 Basic Definitions

A vector $\mathbf{A}$ is expressed as

$$\mathbf{A} = A_x \mathbf{e}_x + A_y \mathbf{e}_y + A_z \mathbf{e}_z = (A_x, A_y, A_z), \tag{1.1}$$

where A_x, A_y, and A_z are the x, y, and z components of $\mathbf{A}$ and $\mathbf{e}_x$, $\mathbf{e}_y$, and $\mathbf{e}_z$ are unit vectors in the x, y, and z directions. The unit vectors are often written as $\mathbf{i}$, $\mathbf{j}$, and $\mathbf{k}$. The position vector $\mathbf{r}$ of a point with Cartesian coordinates x, y, and z is $\mathbf{r} = x\mathbf{e}_x + y\mathbf{e}_y + z\mathbf{e}_z$. See Figure 1.1.

Addition and subtraction for two vectors $\mathbf{A}$ and $\mathbf{B}$ (see Figure 1.2) are defined algebraically by

$$\mathbf{A} \pm \mathbf{B} = (A_x \pm B_x)\mathbf{e}_x + (A_y \pm B_y)\mathbf{e}_y + (A_z \pm B_z)\mathbf{e}_z. \tag{1.2}$$

The **dot product**, or **scalar product**, of two vectors $\mathbf{A}$ and $\mathbf{B}$ is given by

$$\mathbf{A} \cdot \mathbf{B} = A_x B_x + A_y B_y + A_z B_z. \tag{1.3}$$

The scalar product is also given by

$$\mathbf{A} \cdot \mathbf{B} = AB \cos \theta, \tag{1.4}$$

where

$$A = |\mathbf{A}| = \sqrt{A_x^2 + A_y^2 + A_z^2} = \sqrt{\mathbf{A} \cdot \mathbf{A}} \tag{1.5}$$

1

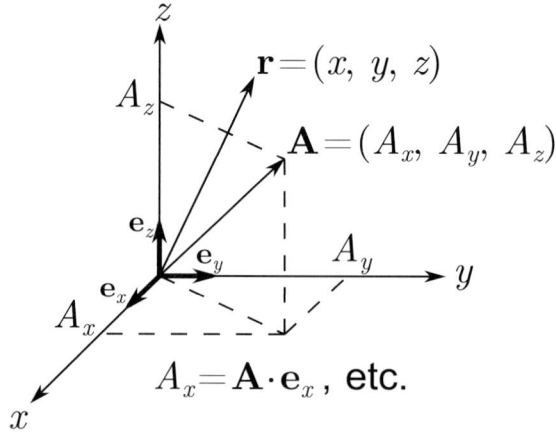

Figure 1.1 Vectors in a Cartesian coordinate system.

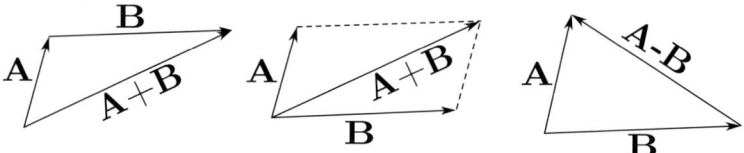

Figure 1.2 Vector addition and subtraction.

is the length, or magnitude, of **A** (with a similar formula holding for B), and θ is the angle between **A** and **B**. $\mathbf{A} \cdot \mathbf{B} = 0$ if and only if **A** and **B** are perpendicular to each other. $\mathbf{A} \cdot \mathbf{B} = AB$ if and only if **A** and **B** are parallel. The unit vectors satisfy

$$\mathbf{e}_x \cdot \mathbf{e}_x = \mathbf{e}_y \cdot \mathbf{e}_y = \mathbf{e}_z \cdot \mathbf{e}_z = 1 \tag{1.6}$$

and

$$\mathbf{e}_x \cdot \mathbf{e}_y = \mathbf{e}_y \cdot \mathbf{e}_z = \mathbf{e}_x \cdot \mathbf{e}_z = 0 . \tag{1.7}$$

A_x is the projection of **A** onto the x axis and can be written as $A_x = \mathbf{A} \cdot \mathbf{e}_x$. Similar formulas hold for A_y and A_z (see Figure 1.1).

The **cross product**, or **vector product**, of **A** and **B** is given by

$$\mathbf{A} \times \mathbf{B} = (A_y B_z - A_z B_y)\mathbf{e}_x + (A_z B_x - A_x B_z)\mathbf{e}_y + (A_x B_y - A_y B_x)\mathbf{e}_z$$
$$= (AB \sin \theta)\mathbf{n}, \tag{1.8}$$

where θ is the angle between **A** and **B**, and **n** is a unit vector perpendicular to both **A** and **B** · **n** gives the direction of $\mathbf{A} \times \mathbf{B}$ and can be determined from the "right-hand rule": if the fingers of the right hand are pointing from **A** to **B**, then the thumb is pointing in the direction of **n**. See Figure 1.3. Note also that $\mathbf{A} \times \mathbf{A} = \mathbf{0}$.

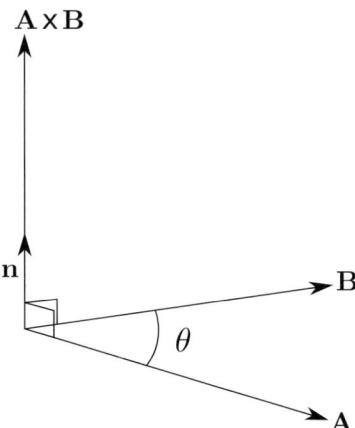

Figure 1.3 The cross product.

The first of the preceding formulas for the vector product can also be obtained by expanding a 3×3 determinant whose first, second, and third rows are given by $(\mathbf{e}_x, \mathbf{e}_y, \mathbf{e}_z)$, (A_x, A_y, A_z), and (B_x, B_y, B_z), respectively. $\mathbf{A} \times \mathbf{B} = \mathbf{0}$ if and only if $\mathbf{A}$ and $\mathbf{B}$ are parallel. The unit vectors satisfy

$$\mathbf{e}_x \times \mathbf{e}_x = \mathbf{e}_y \times \mathbf{e}_y = \mathbf{e}_z \times \mathbf{e}_z = \mathbf{0} \qquad (1.9)$$

and

$$\mathbf{e}_x \times \mathbf{e}_y = \mathbf{e}_z, \quad \mathbf{e}_y \times \mathbf{e}_z = \mathbf{e}_x, \quad \mathbf{e}_z \times \mathbf{e}_x = \mathbf{e}_y. \qquad (1.10)$$

The area of a parallelogram with sides $\mathbf{A}$ and $\mathbf{B}$ is given by $|\mathbf{A} \times \mathbf{B}|$. The volume of a parallelopiped with sides $\mathbf{A}$, $\mathbf{B}$, and $\mathbf{C}$ is given by $|\mathbf{A} \cdot (\mathbf{B} \times \mathbf{C})|$. Scalar and vector products satisfy the rules $\mathbf{A} \cdot \mathbf{B} = \mathbf{B} \cdot \mathbf{A}$ and $\mathbf{A} \times \mathbf{B} = -\mathbf{B} \times \mathbf{A}$. Many other algebraic rules among vectors exist and can be found in texts on vectors.

For a familiar application, consider a particle of mass m. Let $\mathbf{r}$ be the position vector of the particle, $\mathbf{p} = m\mathbf{v}$ be its linear momentum, $\mathbf{L}$ be its angular momentum, $\mathbf{F} = d\mathbf{p}/dt$ be the force on the particle, and $\mathbf{T}$ be the torque on the particle. Then, using the rule in (1.18) for the time derivative of a vector product, the following hold:

$$\mathbf{L} = \mathbf{r} \times \mathbf{p}, \qquad \mathbf{T} = \mathbf{r} \times \mathbf{F} \qquad (1.11)$$

$$\frac{d\mathbf{L}}{dt} = \frac{d\mathbf{r}}{dt} \times \mathbf{p} + \mathbf{r} \times \frac{d\mathbf{p}}{dt} = \mathbf{v} \times m\mathbf{v} + \mathbf{r} \times \mathbf{F} = \mathbf{T} \qquad (1.12)$$

since $\mathbf{v} \times m\mathbf{v} = \mathbf{0}$. Just as the force is the time derivative of the linear momentum ($\mathbf{F} = d\mathbf{p}/dt$), the torque is the time derivative of the angular momentum ($\mathbf{T} = d\mathbf{L}/dt$).

A scalar function

$$\psi = \psi(x, y, z) \tag{1.13}$$

defines a **scalar field**. Every point (x, y, z) has a scalar (a number) $\psi(x, y, z)$ associated with it. For example, $T(x, y, z)$ might represent the temperature at the point (x, y, z). A vector function

$$\mathbf{A} = \mathbf{A}(x, y, z) = A_x(x, y, z)\mathbf{e}_x + A_y(x, y, z)\mathbf{e}_y + A_z(x, y, z)\mathbf{e}_z \tag{1.14}$$

defines a **vector field**. Every point (x, y, z) has a vector $\mathbf{A}(x, y, z)$ associated with it. For example, $\mathbf{E}(x, y, z)$ might represent the electric field at the point (x, y, z).

Suppose that $\mathbf{A}$ is a function of one variable, such as the arc length s along some path in space, or the time t:

$$\mathbf{A}(t) = A_x(t)\mathbf{e}_x + A_y(t)\mathbf{e}_y + A_z(t)\mathbf{e}_z. \tag{1.15}$$

Then, since $\mathbf{e}_x$, $\mathbf{e}_y$, and $\mathbf{e}_z$ are independent of time,

$$\frac{d\mathbf{A}}{dt} = \frac{dA_x}{dt}\mathbf{e}_x + \frac{dA_y}{dt}\mathbf{e}_y + \frac{dA_z}{dt}\mathbf{e}_z. \tag{1.16}$$

For example, if $\mathbf{r} = x\mathbf{e}_x + y\mathbf{e}_y + z\mathbf{e}_z$ is the position vector of a particle moving in space, then

$$\mathbf{v} \equiv \frac{d\mathbf{r}}{dt} = \frac{dx}{dt}\mathbf{e}_x + \frac{dy}{dt}\mathbf{e}_y + \frac{dz}{dt}\mathbf{e}_z \tag{1.17a}$$

is the particle's velocity $\mathbf{v}$. See Figure 1.4.

Note that if the unit vectors actually depend on time t, then they must be differentiated as well, using the product rule. For example, if $\mathbf{r}$ is the position vector of a

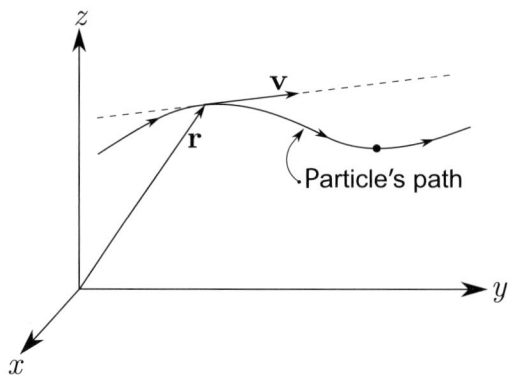

Figure 1.4 $\mathbf{r}$ and $\mathbf{v}$ ($= d\mathbf{r}/dt$) for a particle.

particle moving in space, i.e., $\mathbf{r} = r\mathbf{e}_r$, where $r = |\mathbf{r}|$, and $\mathbf{e}_r$ is the unit vector along $\mathbf{r}$ (see Figure 1.1), then $\mathbf{e}_r$ changes with time as the particle moves. Hence,

$$\mathbf{r} = \mathbf{e}_r \quad \Rightarrow \quad \mathbf{v} = \frac{d\mathbf{r}}{dt} = \frac{dr}{dt}\mathbf{e}_r + r\frac{d\mathbf{e}_r}{dt}. \tag{1.17b}$$

Various rules exist, which can be found in mathematics texts. For example,

$$\frac{d}{dt}(\mathbf{A} \times \mathbf{B}) = \frac{d\mathbf{A}}{dt} \times \mathbf{B} + \mathbf{A} \times \frac{d\mathbf{B}}{dt}, \quad \frac{d}{dt}(\mathbf{A} \cdot \mathbf{B}) = \frac{d\mathbf{A}}{dt} \cdot \mathbf{B} + \mathbf{A} \cdot \frac{d\mathbf{B}}{dt}. \tag{1.18}$$

Partial derivatives are similar. For example, for $\mathbf{A} = \mathbf{A}(x, y, z)$, we have

$$\frac{\partial \mathbf{A}}{\partial x} = \frac{\partial A_x}{\partial x}\mathbf{e}_x + \frac{\partial A_y}{\partial x}\mathbf{e}_y + \frac{\partial A_z}{\partial x}\mathbf{e}_z. \tag{1.19}$$

Differentials are often useful. If $\psi = \psi(x, y, z)$, then

$$d\psi = \frac{\partial \psi}{\partial x}dx + \frac{\partial \psi}{\partial y}dy + \frac{\partial \psi}{\partial z}dz. \tag{1.20}$$

If $\mathbf{A} = \mathbf{A}(x, y, z)$, then

$$d\mathbf{A} = (dA_x)\mathbf{e}_x + (dA_y)\mathbf{e}_y + (dA_z)\mathbf{e}_z, \tag{1.21}$$

where dA_x, dA_y, and dA_z can each be evaluated using (1.20).

A rule that is sometimes used is

$$\mathbf{A} \cdot \frac{d\mathbf{A}}{dt} = A\frac{dA}{dt}, \tag{1.22}$$

which can be seen from

$$2A\frac{dA}{dt} = \frac{d}{dt}(A^2) = \frac{d}{dt}(\mathbf{A} \cdot \mathbf{A}) = \mathbf{A} \cdot \frac{d\mathbf{A}}{dt} + \frac{d\mathbf{A}}{dt} \cdot \mathbf{A} = 2\mathbf{A} \cdot \frac{d\mathbf{A}}{dt}. \tag{1.23}$$

The **gradient** $\nabla\psi$ of a scalar field ψ is given by

$$\nabla\psi = \frac{\partial \psi}{\partial x}\mathbf{e}_x + \frac{\partial \psi}{\partial y}\mathbf{e}_y + \frac{\partial \psi}{\partial z}\mathbf{e}_z. \tag{1.24}$$

If the preceding equation is written without "ψ," it defines the gradient operator "∇," which can be thought of as a vector (in that case, the basis vectors are usually written in front of the partial derivative operators, i.e., $\nabla = \mathbf{e}_x \frac{\partial}{\partial x} + \cdots$.

The **divergence** $\nabla \cdot \mathbf{A}$ of a vector field $\mathbf{A}$ is given by

$$\nabla \cdot \mathbf{A} = \frac{\partial A_x}{\partial x} + \frac{\partial A_y}{\partial y} + \frac{\partial A_z}{\partial z}. \tag{1.25}$$

The **curl** $\nabla \times \mathbf{A}$ of a vector field $\mathbf{A}$ is given by

$$\nabla \times \mathbf{A} = \left(\frac{\partial A_z}{\partial y} - \frac{\partial A_y}{\partial z}\right)\mathbf{e}_x + \left(\frac{\partial A_x}{\partial z} - \frac{\partial A_z}{\partial x}\right)\mathbf{e}_y + \left(\frac{\partial A_y}{\partial x} - \frac{\partial A_x}{\partial y}\right)\mathbf{e}_z. \tag{1.26}$$

The **Laplacian** $\nabla^2 \psi$ of a scalar field ψ is defined by

$$\nabla^2 \psi = \nabla \cdot \nabla \psi = \frac{\partial^2 \psi}{\partial x^2} + \frac{\partial^2 \psi}{\partial y^2} + \frac{\partial^2 \psi}{\partial z^2}. \tag{1.27}$$

Various rules exist among these quantities, e.g.,

$$\nabla \cdot (\psi \mathbf{A}) = (\nabla \psi) \cdot \mathbf{A} + \psi (\nabla \cdot \mathbf{A}), \quad \nabla \times (\psi \mathbf{A}) = (\nabla \psi) \times \mathbf{A} + \psi (\nabla \times \mathbf{A}). \tag{1.28}$$

Three rules useful in seismic wave theory are

$$\nabla \times \nabla \psi = \mathbf{0}, \qquad \nabla \cdot (\nabla \times \mathbf{A}) = 0 \tag{1.29a}$$

and

$$\nabla^2 \mathbf{A} = \nabla (\nabla \cdot \mathbf{A}) - \nabla \times (\nabla \times \mathbf{A}), \tag{1.29b}$$

where

$$\nabla^2 \mathbf{A} \equiv (\nabla^2 A_x) \mathbf{e}_x + (\nabla^2 A_y) \mathbf{e}_y + (\nabla^2 A_z) \mathbf{e}_z. \tag{1.29c}$$

Use (1.24) through (1.27) to prove these three rules.

If $\mathbf{A}$ lies along one of the Cartesian axes, then $\nabla \times \mathbf{A}$ is a vector that lies in the plane perpendicular to the given axis. For example, if $\mathbf{A} = A\mathbf{e}_x$, then $\nabla \times \mathbf{A} = (\partial A/\partial z)\mathbf{e}_y - (\partial A/\partial y)\mathbf{e}_z$. Also, if $\mathbf{A} = A(x, y, z)\mathbf{e}$ where $\mathbf{e}$ is a constant unit vector in some arbitrary direction (so that $\mathbf{A}$ points in the same direction for all points $\mathbf{x}$), then $\nabla \times \mathbf{A}$ is perpendicular to $\mathbf{A}$. In general, though, $\nabla \times \mathbf{A}$ is not perpendicular to $\mathbf{A}$.

A common scalar function used in potential theory and wave theory is

$$\psi = \frac{1}{r} = \frac{1}{\sqrt{x^2 + y^2 + z^2}}. \tag{1.30}$$

It can be shown, by working out the derivatives using (1.30), that

$$\nabla \left(\frac{1}{r} \right) = -\frac{1}{r^2} \mathbf{e}_r \qquad \text{and} \qquad \nabla^2 \left(\frac{1}{r} \right) = 0, \tag{1.31}$$

where $\mathbf{e}_r$ is the unit vector in the direction of the position vector $\mathbf{r}$. The equation $\nabla^2 \psi = 0$ is called Laplace's equation. It arises in potential theory.

Maxwell's four equations for the electromagnetic field can be written in a succinct form through the use of the divergence and curl as

$$\nabla \cdot \mathbf{E} = \frac{\rho}{\epsilon_0}, \quad \nabla \cdot \mathbf{B} = 0, \quad \nabla \times \mathbf{E} = -\frac{\partial \mathbf{B}}{\partial t}, \quad c^2 \nabla \times \mathbf{B} = \frac{\mathbf{J}}{\epsilon_0} + \frac{\partial \mathbf{E}}{\partial t}, \tag{1.32}$$

where $\mathbf{E}$ is the electric field, $\mathbf{B}$ is the magnetic induction, ρ is the total charge density in the matter, c is the speed of light in vacuum, ϵ_0 is the permittivity of free space, and **J** is the total electric current per unit area in the matter. One also has $\epsilon_0 \mu_0 c^2 = 1$, where μ_0 is the permeability of free space.

These can be used to show that **E** and **B** satisfy standard classical wave equations. In the case of free space, $\rho = 0$ and $\mathbf{J} = \mathbf{0}$. Hence, using the fact that the order of partial differentiation does not matter, and using (1.32) and (1.29b), one has

$$\nabla \times \nabla \times \mathbf{E} = -\frac{\partial}{\partial t} \nabla \times \mathbf{B} = -\frac{\partial}{\partial t} \left(\frac{1}{c^2} \frac{\partial \mathbf{E}}{\partial t} \right) = -\frac{1}{c^2} \frac{\partial^2 \mathbf{E}}{\partial t^2}, \qquad \text{and}$$

$$\nabla(\nabla \cdot \mathbf{E}) - \nabla^2 \mathbf{E} = \nabla \times (\nabla \times \mathbf{E}) = -\frac{1}{c^2} \frac{\partial^2 \mathbf{E}}{\partial t^2} \qquad \Longrightarrow$$

$$\nabla^2 \mathbf{E} = \frac{1}{c^2} \frac{\partial^2 \mathbf{E}}{\partial t^2} \quad \text{(the wave equation for } \mathbf{E}\text{)}$$

because $\nabla \cdot \mathbf{E} = \rho/\epsilon_0 = 0$. A similar calculation can be done to show that **B** also satisfies a wave equation. Hence, electromagnetic waves in free space travel at the speed of light. These results were also used historically to suggest that light is an electromagnetic wave.

1.2 Directional Derivatives

Equation (1.20) can also be written as

$$d\psi = \nabla \psi \cdot d\mathbf{r}. \tag{1.33}$$

ψ changes by an amount $d\psi$ when we move from point **r** to point $\mathbf{r} + d\mathbf{r}$. Equation (1.33) shows that $d\psi$ is a maximum if $d\mathbf{r}$ is parallel to $\nabla \psi$. Hence, $\nabla \psi$ lies in the direction of the maximum space rate of change of ψ, and $|\nabla \psi|$ is the magnitude of this maximum rate of change.

Let $\psi = \psi(x, y, z) = constant$ be a surface in space (for example, the surface of a sphere of unit radius, $x^2 + y^2 + z^2 = 1$). Let **r** and $\mathbf{r} + d\mathbf{r}$ be the position vectors of two points on the surface. Hence, ψ has the same value at these two points, i.e., $d\psi = 0$. Also $d\mathbf{r}$ is tangential to the surface at **r**. Therefore, (1.33) shows that $\nabla \psi$ is perpendicular to the surface at **r**. To summarize, $\nabla \psi$ is everywhere perpendicular to the surface $\psi = const$. For example, on a topography map, if $E(x, y) = E_0$ is a contour line along which the elevation has the constant value E_0, then ∇E, which is everywhere perpendicular to the contour, gives the horizontal direction in which the elevation changes maximally, and $|\nabla E|$ gives the magnitude of this maximum rate of change. See Figure 1.5.

If a line or curve in three-dimensional (3D)-space is specified parametrically in terms of the arc length parameter s, i.e., $x = x(s)$, $y = y(s)$, and $z = z(s)$, then as we move along the line, ψ changes with s at the rate

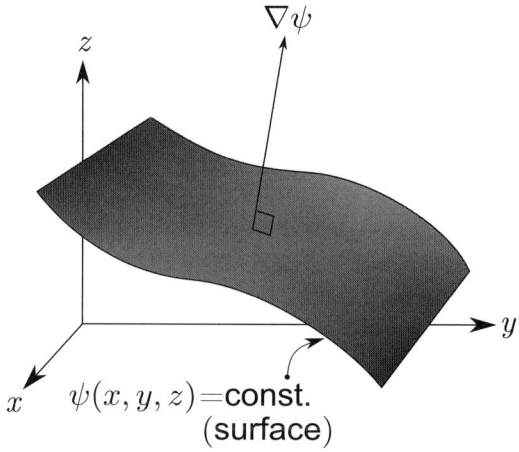

Figure 1.5 $\nabla\psi$ is normal to $\psi = const.$

$$\frac{d\psi}{ds} = \nabla\psi \cdot \frac{d\mathbf{r}}{ds}.$$ (1.34)

$d\psi/ds$ is a **directional derivative** because it gives the rate of change of ψ in the direction specified by the line. The magnitude of the maximum value of the directional derivative is $|\nabla\psi|$, since $|d\mathbf{r}/ds| = 1$. In the same way, $\nabla\psi \cdot \mathbf{e}$, where $\mathbf{e}$ is a unit vector, is a directional derivative because it gives the spatial rate of change of ψ in the direction of $\mathbf{e}$.

1.3 Curl and Rotation

We know, from basic physics, that a finite rotation cannot be represented by a single vector (unlike a finite linear displacement). For example, if we rotate an object by 90 degrees about the z axis of a Cartesian coordinate system and then rotate it by 90 degrees about the x axis, the orientation of the object after the two rotations is not the same as the orientation it has if we rotate first about the x axis by 90 degrees and then about the z axis by 90 degrees. Since vector addition is commutative, these rotations cannot be described by vectors. However, an infinitesimal rotation *can* be represented by a vector. By convention, the direction of the rotation vector is determined by the right-hand rule: if the thumb of the right hand lies along the rotation axis and the fingers of the right hand point in the direction of rotation, then the thumb points in the direction of the rotation vector.

Let $\boldsymbol{\Theta}$ be the rotation vector for a rotation by the small (infinitesimal) angle Θ, and suppose a particle suffers a small displacement $\mathbf{u}$ due to $\boldsymbol{\Theta}$ (see Figure 1.6). Then

$$|\mathbf{u}| \approx \ell|\boldsymbol{\Theta}| = |\boldsymbol{\Theta}|\,|\mathbf{r}|\,\sin\gamma = |\boldsymbol{\Theta} \times \mathbf{r}|,$$

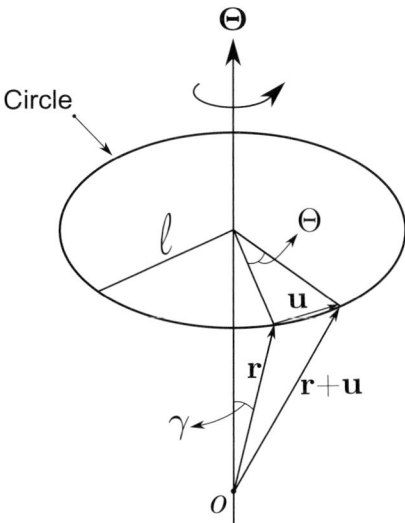

Figure 1.6 Particle displacement due to rotation.

which implies (as we know the directions of $\mathbf{u}$, $\boldsymbol{\Theta}$, and $\mathbf{r}$)

$$\mathbf{u} = \boldsymbol{\Theta} \times \mathbf{r} \qquad (1.35)$$

for an infinitesimal rotation. Taking the curl of $\mathbf{u}$ gives

$$\boldsymbol{\Theta} = \tfrac{1}{2}\nabla \times \mathbf{u}. \qquad (1.36)$$

This shows that the curl of $\mathbf{u}$ has something to do with the rotational properties of the displacement field $\mathbf{u}$. If the particle suffers only a translation, but no rotation, then $\nabla \times \mathbf{u} = \mathbf{0}$, in which case $\mathbf{u}$ is called an **irrotational vector field**.

1.4 Divergence and Flux

Consider a volume V bounded by a surface S, and let $\mathbf{n}$ be a unit vector that is everywhere normal (perpendicular) to the surface S. Suppose a vector field $\mathbf{A} = \mathbf{A}(x, y, z)$ exists in the region. The **flux** of $\mathbf{A}$ through the surface S is defined by

$$\text{Flux of } \mathbf{A} = \int_S \mathbf{A} \cdot \mathbf{n}\, dS. \qquad (1.37)$$

The flux is the integral over the surface of the component of $\mathbf{A}$, which is normal to the surface and is a measure of the total "flow" of $\mathbf{A}$ through the surface. The flux can be related to the divergence through **Gauss' divergence theorem**, which states that

$$\int_V \nabla \cdot \mathbf{A}\, dV = \int_S \mathbf{A} \cdot \mathbf{n}\, dS. \qquad (1.38)$$

This theorem can be used to obtain a physical interpretation of $\nabla \cdot \mathbf{A}$. Consider V to be small enough so that $\nabla \cdot \mathbf{A}$ does not change much spatially throughout V. Then

$$\nabla \cdot \mathbf{A} \approx \frac{1}{V} \int_S \mathbf{A} \cdot \mathbf{n} \, dS, \qquad (1.39)$$

i.e., at a given point (or small region) in space, the divergence of $\mathbf{A}$ is the flux of $\mathbf{A}$ per unit volume.

If the flux is a positive number, it represents a net outward flow, and the closed surface S is said to contain a "source." If the flux is negative, the net flow is inward and S contains a "sink." If the flux is zero, there is no net flow outward or inward (the outward flow equals the inward flow) and S contains neither sources nor sinks – also, $\nabla \cdot \mathbf{A} = 0$ in this case. Fields for which $\nabla \cdot \mathbf{A} = 0$ are called **solenoidal** fields. The magnetic induction $\mathbf{B}$ is such a field (see (1.32)), since $\nabla \cdot \mathbf{B} = 0$. In effect, this states that there are no isolated magnetic poles.

An example of an application of the divergence theorem is the calculation of the electric field $\mathbf{E}$ of a point charge Q at a distance r from the charge. Let S be the surface of a sphere of radius r centered on the point charge, and let V be the enclosed volume. The symmetry of the situation implies that $\mathbf{E}$ is normal to S (everywhere on S), and that E is constant on S. Hence, the flux of $\mathbf{E}$ through S is

$$\int_S \mathbf{E} \cdot \mathbf{n} \, dS = \int_S (E)(1)(\cos 0) \, dS = E \int_S dS = E(4\pi r^2).$$

Also, from the first of Maxwell's equations (1.32), we have

$$\int_V \nabla \cdot \mathbf{E} \, dV = \int_V \left(\frac{\rho}{\epsilon_0} \right) dV = \frac{1}{\epsilon_0} \int_V \rho \, dV = \frac{Q}{\epsilon_0}.$$

Hence, upon equating the two, the divergence theorem implies

$$E = \frac{Q}{4\pi \epsilon_0 r^2}. \qquad (1.40)$$

If $\mathbf{e}_r$ is the unit vector in the radial direction, then $\mathbf{E} = E\mathbf{e}_r$. The first equation of (1.31) implies that

$$\psi = \frac{Q}{4\pi \epsilon_0 r} \qquad \text{and} \qquad \mathbf{E} = -\nabla \psi, \qquad (1.41)$$

where ψ is the electric potential, and that ψ satisfies Laplace's equation.

1.5 Line Integrals

Consider a curve C defined parametrically as $x = x(s)$, $y = y(s)$, and $z = z(s)$, where s is the arc length along C. Referring to Figure 1.7, the component of $\mathbf{A}$ that is tangential to the curve C at the point $\mathbf{r}$, is $A_{\text{tang}} = A \cos \theta$. In the infinitesimal

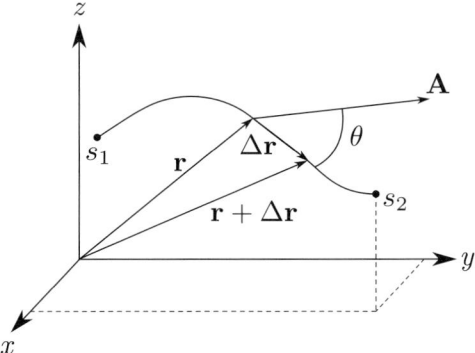

Figure 1.7 Curve for line integral of **A**.

limit, $\Delta\mathbf{r}$ becomes $d\mathbf{r}$, which is tangential to C at $\mathbf{r}$. The **line integral** of **A** along C is defined as

$$\int_C A_{\text{tang}}\, dr = \int_C A\cos\theta\, dr = \int_C \mathbf{A}\cdot d\mathbf{r}\,. \tag{1.42}$$

In a Cartesian coordinate system, it can also be written as

$$\int_C \mathbf{A}\cdot d\mathbf{r} = \int_C (A_x dx + A_y dy + A_z dz) = \int_{s_1}^{s_2} f(s)\, ds, \tag{1.43}$$

where $f(s)$ is obtained by substituting $x = x(s)$, etc., into $A_x(x, y, z)$, etc., and using $dx = (dx/ds)ds$, etc. Line integrals satisfy the rule

$$\int_{s_1}^{s_2} \mathbf{A}\cdot d\mathbf{r} = -\int_{s_2}^{s_1} \mathbf{A}\cdot d\mathbf{r}\,. \tag{1.44}$$

A familiar example of a line integral is the work done by a force **F** in moving a particle along a curve C in space. The work is given by (1.42) with **A** replaced by the force **F** acting on the particle.

If C is a closed curve, then the line integral of **A** along C is called the **circulation** of **A** around C and is given a special notation:

$$\text{Circulation of } \mathbf{A} \text{ around } C = \oint_C \mathbf{A}\cdot d\mathbf{r}\,. \tag{1.45}$$

Consider the case in which **A** can be written as the gradient of a scalar function $\psi(x, y, z)$, i.e., $\mathbf{A} = \nabla\psi$. In this case, **A** is called a **conservative** vector field and ψ is called the **scalar potential** of **A**. Equation (1.29a) implies that $\nabla \times \mathbf{A} = \mathbf{0}$, i.e., a conservative field is an irrotational field. Also, for $\mathbf{A} = \nabla\psi$, (1.33) implies

$$\int_{s_1}^{s_2} \mathbf{A}\cdot d\mathbf{r} = \int_{s_1}^{s_2} d\psi = \psi(s_2) - \psi(s_1)\,, \tag{1.46}$$

i.e., for a conservative vector field, the value of the line integral along a curve is equal to the difference between the potential values at the endpoints of the curve. In other words, the value of the line integral is independent of the path. If two points s_1 and s_2 are connected by a curve C_1 and also by a different curve C_2, then if $\mathbf{A}$ is a conservative vector field, the line integral of $\mathbf{A}$ along C_1 is equal to the line integral of $\mathbf{A}$ along C_2. Furthermore, if C is a closed curve (e.g., $C = C_1 + C_2$, say), then (1.44) implies that the line integral of $\mathbf{A}$ along C is zero.

Equations (1.41) show that the electric field due to a point charge is a conservative field. Other examples of conservative vector fields are the Newtonian gravitational field and the particle displacement field $\mathbf{u}$ produced by seismic P waves generated by a spherical explosion.

Let S be a surface in 3D space bounded by the curve C, and let $\mathbf{n}$ be the unit vector that is everywhere normal to S. The line integral of any vector field $\mathbf{A}$ can be related to the curl of $\mathbf{A}$ through **Stokes' theorem**, which states that

$$\oint_C \mathbf{A} \cdot d\mathbf{r} = \int_S (\nabla \times \mathbf{A}) \cdot \mathbf{n} \, dS, \tag{1.47}$$

where C is traversed in the positive (counterclockwise) direction. $(\nabla \times \mathbf{A}) \cdot \mathbf{n}$ is the component of $\nabla \times \mathbf{A}$ that is normal to the surface S, i.e., it is the projection of $\nabla \times \mathbf{A}$ along $\mathbf{n}$. Stokes' theorem implies that if S is small enough so that $\nabla \times \mathbf{A}$ does not vary appreciably over S, then,

$$(\nabla \times \mathbf{A}) \cdot \mathbf{n} \approx \frac{1}{S} \oint_C \mathbf{A} \cdot d\mathbf{r}. \tag{1.48}$$

An example of an application of Stokes' theorem is the deduction of Faraday's law in electromagnetic theory. Assume a magnetic field $\mathbf{B}$ is passing through a closed loop C, and that $\mathbf{B}$ is changing with time. This will induce an electric current in the loop C. The current will generate another magnetic field whose direction will be the opposite of $\mathbf{B}$, i.e., it will oppose the change in $\mathbf{B}$ (Lenz' law). Using Stokes' theorem together with the third Maxwell equation (1.32) gives

$$\mathcal{E} \equiv \oint_C \mathbf{E} \cdot d\mathbf{r} = \int_S (\nabla \times \mathbf{E}) \cdot \mathbf{n} \, dS = -\int_S \frac{\partial \mathbf{B}}{\partial t} \cdot \mathbf{n} \, dS = -\frac{d}{dt} \int_S \mathbf{B} \cdot \mathbf{n} \, dS \equiv -\frac{d\Phi}{dt},$$
$$\tag{1.49}$$

where $\mathcal{E}$ is the induced electromotive force (EMF), which produces the current in C, and Φ is the magnetic flux through S. The equation $\mathcal{E} = -(d\Phi/dt)$ is a statement of Faraday's law.

1.6 Curvilinear Coordinates

Sometimes it is convenient to use curvilinear coordinates instead of Cartesian coordinates. For example, when analyzing the seismic waves produced by a spherically symmetric explosion, it is usually easier to use the spherical polar coordinates r, θ, and ϕ than to use x, y, and z. Let q_1, q_2, and q_3 represent general curvilinear coordinates. The coordinates of a point in one coordinate system can be obtained from the coordinates of the point in another system through transformation equations, e.g., $x = x(q_1, q_2, q_3)$, or $q_1 = q_1(x, y, z)$, with similar equations holding for the other coordinates. A familiar set of transformation equations is the set giving the Cartesian coordinates x and y in terms of the two-dimensional (2D) polar coordinates r and ϕ : $x = r \cos \phi$ and $y = r \sin \phi$.

The equations $x = x_0$, $y = y_0$, and $z = z_0$, where x_0, y_0, and z_0 are constants, describe planes that are perpendicular to the x, y, and z axes, respectively. They are called **coordinate planes**. Similarly, the equations $q_1 = a$, $q_2 = b$, and $q_3 = c$, where a, b, and c are constants, describe the **coordinate surfaces** in the curvilinear system. Two given coordinate surfaces intersect along a curve called a **coordinate curve**. For example, a q_1 coordinate curve is produced by the intersection of $q_2 = b$ and $q_3 = c$. In the Cartesian system, $y = y_0$ and $z = z_0$ intersect to produce an x coordinate curve (which is parallel to the x axis). A collection of coordinate curves can be used to form a coordinate "grid." For example, in a 2D plane, the x coordinate curves are straight lines parallel to the x axis, and the y coordinate curves are straight lines parallel to the y axis, and they form a rectangular grid. One could also use the 2D polar coordinates: the ϕ coordinate curves are concentric circles about the origin, and the r coordinate curves are straight lines radiating in different directions from the origin, and together they form a polar grid.

From the equation $\mathbf{r} = \mathbf{r}(q_1, q_2, q_3)$ for the position vector, we obtain

$$dr = \frac{\partial \mathbf{r}}{\partial q_1} dq_1 + \frac{\partial \mathbf{r}}{\partial q_2} dq_2 + \frac{\partial \mathbf{r}}{\partial q_3} dq_3. \tag{1.50}$$

The vector $\partial \mathbf{r}/\partial q_j$, $j = 1, 2, 3$, is tangent to the q_j coordinate curves. For example, in the Cartesian system, $\mathbf{r} = x\mathbf{e}_x + y\mathbf{e}_y + z\mathbf{e}_z$ implies $\partial \mathbf{r}/\partial x = \mathbf{e}_x$, which is tangent to the x coordinate curves, i.e., to the x axis. In fact, it is the unit vector. In a general curvilinear coordinate system, $\partial \mathbf{r}/\partial q_j$ is a vector in the q_j direction (at a given point in space), and can be written as

$$\frac{\partial \mathbf{r}}{\partial q_j} = h_j \mathbf{e}_j , \qquad h_j \equiv \left| \frac{\partial \mathbf{r}}{\partial q_j} \right| , \qquad j = 1, 2, 3 \tag{1.51}$$

and where $\mathbf{e}_j$ is a unit vector in the q_j direction. h_j is called a **scale factor**. In general, both $\mathbf{e}_j$ and h_j are functions of q_1, q_2, and q_3. If the coordinate surfaces intersect at right angles, then the curvilinear system is an **orthogonal** one and the unit vectors are perpendicular to each other.

Equation (1.50) can now be written as

$$d\mathbf{r} = h_1 dq_1 \mathbf{e}_1 + h_2 dq_2 \mathbf{e}_2 + h_3 dq_3 \mathbf{e}_3. \tag{1.52}$$

If $\mathbf{r}$ and $\mathbf{r} + d\mathbf{r}$ are the position vectors of two points on some curve in space, then $ds = |d\mathbf{r}|$ is the arc length between the two points. For an orthogonal coordinate system,

$$ds^2 = d\mathbf{r} \cdot d\mathbf{r} = h_1^2 dq_1^2 + h_2^2 dq_2^2 + h_3^2 dq_3^2. \tag{1.53}$$

For a Cartesian coordinate system, the scale factors are all equal to one.

In a Cartesian coordinate system, the volume element is $dV = dx\,dy\,dz$. In a curvilinear coordinate system, it is

$$dV = \left| (h_1 dq_1 \mathbf{e}_1) \cdot (h_2 dq_2 \mathbf{e}_2) \times (h_3 dq_3 \mathbf{e}_3) \right|, \tag{1.54}$$

which is the volume of an infinitesimally small parallelopiped whose sides are the vectors in the parentheses in the preceding equation. If the curvilinear system is orthogonal, then

$$dV = h_1 h_2 h_3 \, dq_1 dq_2 dq_3 \tag{1.55}$$

since in that case, $\mathbf{e}_1 \cdot \mathbf{e}_2 \times \mathbf{e}_3 = \mathbf{e}_1 \cdot \mathbf{e}_1 = 1$.

Note that the determination of dV from (1.54) requires a knowledge of the scale factors h_1, h_2, and h_3. One can determine dV more directly, without working out the scale factors first, as follows. Substitute (1.51) into (1.54), and use the vector rule

$$\mathbf{A} \cdot (\mathbf{B} \times \mathbf{C}) = \begin{vmatrix} A_x & A_y & A_z \\ B_x & B_y & B_z \\ C_x & C_y & C_z \end{vmatrix} \tag{1.56}$$

to obtain

$$dV = \left| \frac{\partial(x, y, z)}{\partial(q_1, q_2, q_3)} \right| dq_1 \, dq_2 \, dq_3, \tag{1.57}$$

where

$$\frac{\partial(x, y, z)}{\partial(q_1, q_2, q_3)} \equiv \begin{vmatrix} \dfrac{\partial x}{\partial q_1} & \dfrac{\partial x}{\partial q_2} & \dfrac{\partial x}{\partial q_3} \\[2mm] \dfrac{\partial y}{\partial q_1} & \dfrac{\partial y}{\partial q_2} & \dfrac{\partial y}{\partial q_3} \\[2mm] \dfrac{\partial z}{\partial q_1} & \dfrac{\partial z}{\partial q_2} & \dfrac{\partial z}{\partial q_3} \end{vmatrix} \tag{1.58}$$

is known as the **Jacobian determinant** of the coordinate transformation. The fact that the determinant of a matrix is equal to the determinant of its transpose has also been used to obtain (1.58). The Jacobian can be used to transform triple integrals from rectangular to curvilinear coordinates:

$$\iiint_V F(x, y, z) \, dx \, dy \, dz = \iiint_V F_q(q_1, q_2, q_3) \left| \frac{\partial(x, y, z)}{\partial(q_1, q_2, q_3)} \right| dq_1 \, dq_2 \, dq_3, \tag{1.59}$$

where $F_q(q_1, q_2, q_3)$ is the function obtained when one substitutes the coordinate transformation equations ($x = x(q_1, q_2, q_3)$, etc.) into $F(x, y, z)$. A similar equation can be used to transform double integrals on a plane (e.g., for the xy plane, drop z, dz, q_3, and dq_3 from (1.59)).

For double integrals over a curved surface (a coordinate surface), one can use the elements of area of the curvilinear system. The area element dS_1 on the q_1 coordinate surface ($q_1 = const$) is

$$dS_1 = \left| (h_2 dq_2 \mathbf{e}_2) \times (h_3 dq_3 \mathbf{e}_3) \right| . \tag{1.60}$$

This can be seen by considering the parallelopiped mentioned in connection with (1.54). The volume of the parallelopiped is the volume element dV, and the areas of the three different sides, which are parallelograms, are the area elements. Since the area of a parallelogram with sides $\mathbf{A}$ and $\mathbf{B}$ is $|\mathbf{A} \times \mathbf{B}|$, (1.60) follows. If the curvilinear system is orthogonal, then

$$dS_1 = h_2 h_3 \, dq_2 dq_3. \tag{1.61}$$

Equations similar to (1.60) and (1.61) hold for dS_2 and dS_3. For rectangular (Cartesian) coordinates, one has $dS_x = dy\,dz$, etc.

Let $\psi = \psi(q_1, q_2, q_3)$ be a scalar field and $\mathbf{A} = \mathbf{A}(q_1, q_2, q_3) = A_1 \mathbf{e}_1 + A_2 \mathbf{e}_2 + A_3 \mathbf{e}_3$ be a vector field. The gradient, divergence, curl, and Laplacian can be written in terms of curvilinear coordinates as follows:

$$\nabla \psi = \frac{1}{h_1} \frac{\partial \psi}{\partial q_1} \mathbf{e}_1 + \frac{1}{h_2} \frac{\partial \psi}{\partial q_2} \mathbf{e}_2 + \frac{1}{h_3} \frac{\partial \psi}{\partial q_3} \mathbf{e}_3 \tag{1.62}$$

$$\nabla \cdot \mathbf{A} = \frac{1}{h_1 h_2 h_3} \left[\frac{\partial}{\partial q_1} (h_2 h_3 A_1) + \frac{\partial}{\partial q_2} (h_3 h_1 A_2) + \frac{\partial}{\partial q_3} (h_1 h_2 A_3) \right] \tag{1.63}$$

$$\nabla \times \mathbf{A} = \frac{1}{h_1 h_2 h_3} \begin{vmatrix} h_1 \mathbf{e}_1 & h_2 \mathbf{e}_2 & h_3 \mathbf{e}_3 \\ \dfrac{\partial}{\partial q_1} & \dfrac{\partial}{\partial q_2} & \dfrac{\partial}{\partial q_3} \\ h_1 A_1 & h_2 A_2 & h_3 A_3 \end{vmatrix} \tag{1.64}$$

$$\nabla^2 \psi = \frac{1}{h_1 h_2 h_3} \left[\frac{\partial}{\partial q_1} \left(\frac{h_2 h_3}{h_1} \frac{\partial \psi}{\partial q_1} \right) + \frac{\partial}{\partial q_2} \left(\frac{h_3 h_1}{h_2} \frac{\partial \psi}{\partial q_2} \right) + \frac{\partial}{\partial q_3} \left(\frac{h_1 h_2}{h_3} \frac{\partial \psi}{\partial q_3} \right) \right]. \tag{1.65a}$$

In curvilinear coordinates, $\nabla^2 \mathbf{A}$ is calculated using (1.29b):

$$\nabla^2 \mathbf{A} = \nabla(\nabla \cdot \mathbf{A}) - \nabla \times (\nabla \times \mathbf{A}), \tag{1.65b}$$

where the right-hand side is calculated from (1.62) through (1.64) (see (1.83)).

Two orthogonal curvilinear coordinate systems, the **cylindrical** coordinate system and the **spherical polar** coordinate system, are often used in seismic wave propagation theory, and are outlined in the following two sections. For orthogonal systems, the scale factors h_1, h_2, and h_3 can be calculated by using the transformation equations to obtain (1.53) from $ds^2 = dx^2 + dy^2 + dz^2$ (by substituting $dx = \sum_j (\partial x / \partial q_j) dq_j$, etc., into it – see the examples in the following two sections).

1.7 Cylindrical Coordinates

These are used to study problems in which there is cylindrical symmetry, such as the seismic response of a medium to an infinitely long line source. Such problems are essentially two-dimensional, which simplifies the mathematics. The curvilinear coordinates q_1, q_2, and q_3 are ρ, ϕ, and z respectively (see Figure 1.8a), and the transformation equations are

$$
\begin{aligned}
x &= \rho \cos \phi & & & \rho \geq 0 \\
y &= \rho \sin \phi & \text{where} & & 0 \leq \phi < 2\pi \\
z &= z & & & -\infty < z < \infty.
\end{aligned}
\tag{1.66}
$$

These are essentially 2D polar coordinates in the xy plane extended uniformly in the z direction. $\rho = const$ is the equation of the surface of a right circular cylinder whose axis is the z axis.

From the transformation equations, one obtains

$$
dx = \frac{\partial x}{\partial \rho} d\rho + \frac{\partial x}{\partial \phi} d\phi + \frac{\partial x}{\partial z} dz = (\cos \phi) d\rho - (\rho \sin \phi) d\phi,
\tag{1.67a}
$$

$$
dy = \frac{\partial y}{\partial \rho} d\rho + \frac{\partial y}{\partial \phi} d\phi + \frac{\partial y}{\partial z} dz = (\sin \phi) d\rho + (\rho \cos \phi) d\phi.
\tag{1.67b}
$$

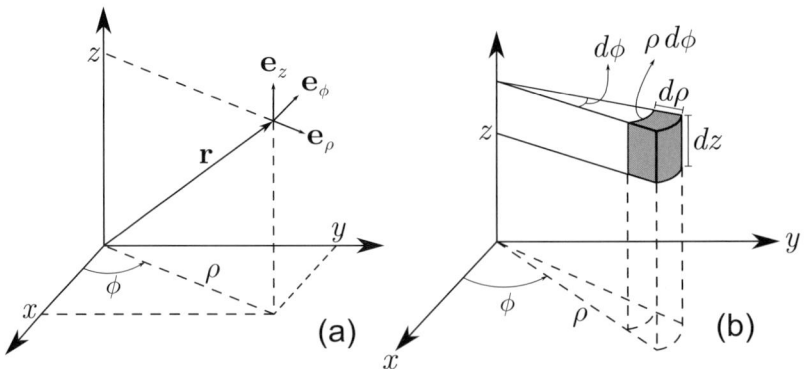

Figure 1.8 Cylindrical coordinates: (a) definition; (b) volume and surface elements.

Substitution of these into $ds^2 = dx^2 + dy^2 + dz^2$ yields, after some algebra,

$$ds^2 = d\rho^2 + \rho^2 d\phi^2 + dz^2, \tag{1.68}$$

from which the scale factors can be read off:

$$h_\rho = 1, \qquad h_\phi = \rho, \qquad h_z = 1. \tag{1.69}$$

We then have, for example, from (1.52), (1.55), and (1.61),

$$d\mathbf{r} = d\rho\,\mathbf{e}_\rho + \rho\,d\phi\,\mathbf{e}_\phi + dz\,\mathbf{e}_z, \tag{1.70}$$
$$dV = \rho\,d\rho\,d\phi\,dz, \qquad \text{and} \qquad dS_\rho = \rho\,d\phi\,dz. \tag{1.71}$$

The position vector $\mathbf{r}$ can be obtained from inspection of Figure 1.8a:

$$\mathbf{r} = \rho\mathbf{e}_\rho + z\mathbf{e}_z. \tag{1.72}$$

It can also be obtained in a more systematic way by using $\mathbf{r} = x\mathbf{e}_x + y\mathbf{e}_y + z\mathbf{e}_z$ and substituting the transformation equations along with equations expressing the Cartesian unit vectors in terms of the cylindrical unit vectors.

Note that the unit vectors $\mathbf{e}_\rho$ and $\mathbf{e}_\phi$ vary with ϕ, and that $\mathbf{r} \neq \sum_n h_n q_n \mathbf{e}_n$ in general. Taking the differential $d\mathbf{r}$ of (1.72) implies, from (1.70), that $d\mathbf{e}_\rho = d\phi\mathbf{e}_\phi$.

The formulas for dV, dS_ρ, and the other area elements can be derived more directly from inspection of Figure 1.8b — dV is the volume of the elementary box shown, and dS_ρ is the area of the face of the box that lies on the ρ coordinate surface ($\rho = const$). Similarly, $dS_z = \rho\,d\rho\,d\phi$.

Equations (1.62) through (1.65) can be used to obtain expressions for the gradient, divergence, curl, and Laplacian in cylindrical coordinates.

A simple example of the use of cylindrical coordinates is the computation of the area of a circle of radius R in the xy plane. This area is

$$\iint dS_z = \int_0^{2\pi} \int_0^R \rho\,d\rho\,d\phi = \pi R^2 . \tag{1.73}$$

One could also use the 2D Jacobian

$$\frac{\partial(x, y)}{\partial(\rho, \phi)} = \begin{vmatrix} \cos\phi & -\rho\sin\phi \\ \sin\phi & \rho\cos\phi \end{vmatrix} = \rho, \tag{1.74}$$

in the 2D version of (1.59):

$$\iint dx\,dy = \iint \left| \frac{\partial(x, y)}{\partial(\rho, \phi)} \right| d\rho\,d\phi = \int_0^{2\pi} \int_0^R \rho\,d\rho\,d\phi = \pi R^2 . \tag{1.75}$$

1.8 Spherical Coordinates

These are used to study problems in which there is spherical symmetry, such as the seismic response of a medium to a spherically symmetric explosive source. The curvilinear coordinates q_1, q_2, and q_3 are r, θ, and ϕ respectively (see Figure 1.9a), and the transformation equations are

$$
\begin{aligned}
x &= r\sin\theta\cos\phi & & r \geq 0 \\
y &= r\sin\theta\sin\phi & \text{where} \quad & 0 \leq \phi < 2\pi \\
z &= r\cos\theta & & 0 \leq \theta \leq \pi.
\end{aligned}
\tag{1.76}
$$

The scale factors can be obtained as they were for cylindrical coordinates, and are

$$
h_r = 1, \qquad h_\theta = r, \qquad h_\phi = r\sin\theta \,.
\tag{1.77}
$$

Equations (1.52), (1.55), and (1.61) give

$$
d\mathbf{r} = dr\,\mathbf{e}_r + r\,d\theta\,\mathbf{e}_\theta + r\sin\theta\,d\phi\,\mathbf{e}_\phi \,,
\tag{1.78}
$$

$$
dV = r^2\sin\theta\,dr\,d\theta\,d\phi \qquad \text{and} \qquad dS_r = r^2\sin\theta\,d\theta\,d\phi \,.
\tag{1.79}
$$

The other area elements, dS_θ and dS_ϕ, can be obtained from similar general formulas. The position vector $\mathbf{r}$ can be obtained from inspection of Figure 1.9a:

$$
\mathbf{r} = r\mathbf{e}_r \,.
\tag{1.80}
$$

It can also be obtained in a more systematic way by using $\mathbf{r} = x\mathbf{e}_x + y\mathbf{e}_y + z\mathbf{e}_z$ and substituting the transformation equations along with equations expressing the Cartesian unit vectors in terms of the spherical unit vectors.

Note that the unit vectors $\mathbf{e}_r$, $\mathbf{e}_\theta$, and $\mathbf{e}_\phi$ vary with θ and ϕ. Note also that $d\mathbf{r} = d(r\mathbf{e}_r)$ implies, from (1.78), that $d\mathbf{e}_r = d\theta\,\mathbf{e}_\theta + \sin\theta\,d\phi\,\mathbf{e}_\phi$.

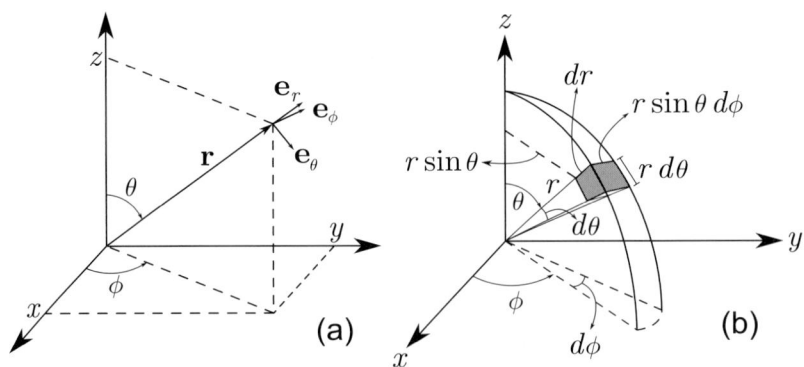

Figure 1.9 Spherical coordinates: (a) definition; (b) volume and surface elements.

The elements dV, dS_r, dS_θ, and dS_ϕ can be derived more directly from inspection of Figure 1.9b. dV is the volume of the elementary box shown, and dS_r is the area of the face of the elementary box, which lies on the r coordinate surface ($r = const$), etc. Integration of dS_r over all values of θ and ϕ would give the surface area of a sphere, $4\pi r^2$.

Equations (1.62) through (1.65) can be used to obtain expressions for the gradient, divergence, curl, and Laplacian in spherical coordinates (the subscripts 1, 2, and 3 on all quantities would be replaced by r, θ, and ϕ respectively).

A simple example of the use of spherical coordinates is the computation of the surface area of a right circular cone. The length of the side of the cone (from the tip to any point on the perimeter of the base) is L, and the side makes an angle θ with the vertical. Since $\theta = const$ is the equation of the surface of the cone in spherical coordinates, integration of dS_θ gives the result. From Figure 1.9b, $dS_\theta = r \sin\theta \, dr \, d\phi$ (this could also be obtained from $dS_\theta = h_r h_\phi \, dr \, d\phi$), so the surface area of the cone is

$$\iint dS_\theta = \int_0^{2\pi} \int_0^L r \sin\theta \, dr \, d\phi = \pi L^2 \sin\theta = \pi L R, \tag{1.81}$$

where $R = L \sin\theta$ is the radius of the base of the cone.

Another example is the computation of the mass of a sphere of radius R with a given density distribution $\rho = \rho(r, \theta, \phi)$. For instance, if ρ decreases linearly with r from a value of ρ_c at the center of the sphere to a value of zero at the surface, i.e., $\rho = \rho_c(1 - r/R)$, then the mass M is

$$M = \iiint \rho \, dV = \int_0^{2\pi} \int_0^\pi \int_0^R \rho_c \left(1 - \frac{r}{R}\right) r^2 \sin\theta \, dr \, d\theta \, d\phi = \frac{1}{3}\pi R^3 \rho_c, \tag{1.82}$$

which is a quarter of the value it would have if the density had the constant value ρ_c throughout the sphere.

1.9 Laplacian of a Vector

Unlike for Cartesian coordinates, for curvilinear coordinates, $\nabla^2 \mathbf{A}$ must be computed from the formulas for $\nabla \cdot \mathbf{A}$ and $\nabla \times \mathbf{A}$ in (1.63) and (1.64), as is indicated by (1.65b), repeated here:

$$\nabla^2 \mathbf{A} = \nabla(\nabla \cdot \mathbf{A}) - \nabla \times (\nabla \times \mathbf{A}). \tag{1.83}$$

More specifically, in Cartesian coordinates, the x component of $\nabla^2 \mathbf{A}$ is $\nabla^2 A_x$. The same goes for the y and z components. But in spherical coordinates, for example, the r component of $\nabla^2 \mathbf{A}$ is *not* $\nabla^2 A_r$, but rather the r component of $\nabla(\nabla \cdot \mathbf{A}) - \nabla \times (\nabla \times \mathbf{A})$. The same goes for the θ and ϕ components. If we derive a vector equation that contains $\nabla^2 \mathbf{A}$, which happens in seismic wave propagation theory (the equation of motion in an isotropic medium), and if we want to express the equation in curvilinear coordinates, then we must remember to evaluate $\nabla^2 \mathbf{\Lambda}$ via (1.65b), i.e., (1.83). Or, equivalently, we could replace $\nabla^2 \mathbf{A}$ with $\nabla(\nabla \cdot \mathbf{A}) - \nabla \times (\nabla \times \mathbf{A})$ in the equation. In Cartesian coordinates, it is not necessary to do this.

1.10 Rotation of Coordinates

Consider a set of seismic data recorded by three-component geophones or seismometers. Let u_x, u_y, and u_z be the three Cartesian components of the particle displacement vector $\mathbf{u}$ (standard geophones actually measure the particle velocity $d\mathbf{u}/dt$, but the displacement can be obtained by applying a $90°$ phase shift to the particle velocity data). It is sometimes desirable to know what the data would look like if the seismometers had been tilted so that $\mathbf{u}$ was parallel to one of the Cartesian axes (two of the components of $\mathbf{u}$ would then be zero). This can be done by mathematically rotating the coordinate system of the data so that the components of $\mathbf{u}$ in the rotated system can be computed from a knowledge of the components of $\mathbf{u}$ in the original system (the data). This technique is sometimes used in the analysis of three-component vertical seismic profiling data and other data analysis methods.

At this point, we change the notation that we have been using for Cartesian coordinates from x, y, and z to x_1, x_2, and x_3, respectively. Similarly, A_x, A_y, and A_z become A_1, A_2, and A_3, and the unit vectors $\mathbf{e}_x$, $\mathbf{e}_y$, and $\mathbf{e}_z$ become $\mathbf{e}_1$, $\mathbf{e}_2$, and $\mathbf{e}_3$. In this notation, the position vector $\mathbf{r}$ is also sometimes written as $\mathbf{x}$. This notation allows us to write expressions involving vectors in a more compact form that makes use of the summation symbol. For example, the scalar product of two vectors $\mathbf{A}$ and $\mathbf{B}$ can be written as

$$\mathbf{A} \cdot \mathbf{B} = A_1 B_1 + A_2 B_2 + A_3 B_3 = \sum_{j=1}^{3} A_j B_j \,. \tag{1.84}$$

Imagine a rotation of the coordinate system from one whose coordinates are x_j, $j = 1, 2, 3$ to one whose coordinates are x'_j, $j = 1, 2, 3$, as shown in Figure 1.10a (more correctly, we should put the prime on the j, i.e., $x_{j'}$, but for our purposes, x'_j will suffice).

A vector $\mathbf{A}$ whose components are $A_j = \mathbf{A} \cdot \mathbf{e}_j$ in the unrotated system has the components $A'_j = \mathbf{A} \cdot \mathbf{e}'_j$ in the rotated (primed) system, and the primed components will be different from the unprimed ones. Hence, any vector $\mathbf{A}$ can be written as either

$$\mathbf{A} = \sum_{j=1}^{3} A_j \mathbf{e}_j = \sum_{j=1}^{3} (\mathbf{A} \cdot \mathbf{e}_j) \mathbf{e}_j \quad \text{or} \quad \mathbf{A} = \sum_{j=1}^{3} A'_j \mathbf{e}'_j = \sum_{j=1}^{3} (\mathbf{A} \cdot \mathbf{e}'_j) \mathbf{e}'_j \,. \tag{1.85}$$

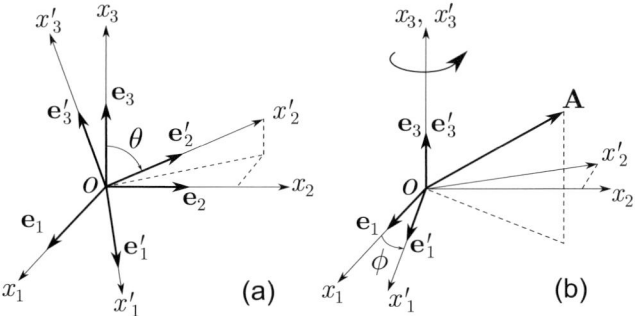

Figure 1.10 Rotation of coordinates (a) in general, and (b) about the z axis.

Applying the first equation of (1.85) to the primed unit vector $\mathbf{e}'_i$, i.e., letting $\mathbf{A}$ be $\mathbf{e}'_i$, gives

$$\mathbf{e}'_i = \sum_{j=1}^{3} R_{ij}\mathbf{e}_j, \qquad \text{where} \qquad R_{ij} \equiv \mathbf{e}'_i \cdot \mathbf{e}_j, \qquad i, j = 1, 2, 3. \tag{1.86}$$

R_{ij} is the **direction cosine** associated with the x'_i and x_j axes (see Equation 1.4), i.e., it is the cosine of the angle between the x'_i axis and the x_j axis. For example, in Figure 1.10a, $R_{23} = \cos\theta$. Similarly, applying the second equation of (1.85) to $\mathbf{e}_i$, one obtains

$$\mathbf{e}_i = \sum_{j=1}^{3} R_{ji}\mathbf{e}'_j, \qquad i = 1, 2, 3. \tag{1.87}$$

Remember that equations such as (1.87) are really three equations (for $i = 1$, 2, and 3). If one writes them out in full, one appreciates the convenience of the summation notation.

The coordinates of the position vector $\mathbf{r}$ are x_1, x_2, and x_3 in the original system, and x'_1, x'_2, and x'_3 in the rotated system. Given the values of the coordinates in the original system, what are the values of the coordinates in the rotated system? Since these two sets of coordinates refer to the same vector $\mathbf{r}$, even though they are different, one can write

$$\mathbf{r} = \sum_{i=1}^{3} x_i\mathbf{e}_i = \sum_{i=1}^{3} x'_i\mathbf{e}'_i. \tag{1.88}$$

Substituting (1.87) into the first sum in (1.88) gives

$$\mathbf{r} = \sum_{i=1}^{3} x_i \left(\sum_{j=1}^{3} R_{ji}\mathbf{e}'_j \right) = \sum_{j=1}^{3} \sum_{i=1}^{3} R_{ji}x_i\mathbf{e}'_j = \sum_{i=1}^{3} \left(\sum_{j=1}^{3} R_{ij}x_j \right) \mathbf{e}'_i. \tag{1.89}$$

In the second step in (1.89), the order of summation was reversed (since the order doesn't matter in this case) and the x_i is brought inside the sum over j since it does not change when summing over j, and in the third step, the indices i and j were switched (which can be done since they are just dummy summation indices). Comparing (1.88) and (1.89) gives

$$x_i' = \sum_{j=1}^{3} R_{ij} x_j, \qquad i = 1, 2, 3. \tag{1.90}$$

These equations allow one to compute the coordinates x_i' of the position vector in the primed system, given the coordinates x_i of the position vector in the unprimed system.

The profusion of summation symbols make the derivation in (1.89) appear cumbersome. To make such derivations clearer and easier to follow, it is customary in seismic wave theory literature to use the **summation convention**, which deletes the summation symbols and states that if a subscript appears exactly twice in a term, then a sum is performed over that subscript. For example, with the summation convention, the scalar product $\mathbf{A} \cdot \mathbf{B}$ in (1.84) can simply be written as $A_j B_j$ instead of $\sum_{j=1}^{3} A_j B_j$. As the "j" appears twice, a sum from $j = 1$ to $j = 3$ is performed.

The summation convention is particularly useful for double sums, triple sums, or higher-order sums. For example, the sum

$$\sum_{i=1}^{3} \sum_{j=1}^{3} a_{ij} b_{ij} = a_{11} b_{11} + a_{12} b_{12} + a_{13} b_{13} + a_{21} b_{21} + a_{22} b_{22} + \cdots + a_{33} b_{33}$$

can simply be written as $a_{ij} b_{ij}$, rather than as $\sum_{i=1}^{3} \sum_{j=1}^{3} a_{ij} b_{ij}$.

The summation convention also implicitly includes the fact that in double sums, triple sums, etc., the summations can be performed in any order. For instance, in the preceding sum, the sum over i can be performed first followed by the sum over j, or j first followed by i. The expression $a_{ij} b_{ij}$ does not imply or indicate an order of summation.

When using the summation convention, the squared length of a vector must be written as $A_i A_i$ rather than A_i^2 to conform to the convention. Also, sometimes one may wish to use the symbol A_{ii} to refer to *either* A_{11}, A_{22}, *or* A_{33}, rather than $A_{11} + A_{22} + A_{33}$. In that case, one must specify explicitly that no sum is implied.

Using the summation convention, the derivation in (1.89) could have been written more simply as

$$\mathbf{r} = x_i R_{ji} \mathbf{e}_j' = R_{ji} x_i \mathbf{e}_j' = R_{ij} x_j \mathbf{e}_i' \, .$$

When compared with (1.88), this leads to $x_i' = R_{ij} x_j$, i.e., (1.90).

We will use the summation convention occasionally in this chapter and the next, for calculations and expressions involving a relatively large number of summation symbols.

The corresponding equations for a general vector $\mathbf{A}$ can be obtained in the same way – just replace $\mathbf{r}$ with $\mathbf{A}$ and x with A in the derivation for (1.89). One obtains

$$\mathbf{A} = \sum_{i=1}^{3} A_i \mathbf{e}_i = \sum_{i=1}^{3} A_i' \mathbf{e}_i', \qquad \text{where} \tag{1.91}$$

$$A_i' = \sum_{j=1}^{3} R_{ij} A_j, \qquad i = 1, 2, 3. \tag{1.92}$$

By a similar analysis, i.e., substituting $\mathbf{e}_i' = \sum_{j=1}^{3} R_{ij} \mathbf{e}_j$ into $\mathbf{r} = \sum_{i=1}^{3} x_i' \mathbf{e}_i'$, one obtains the unprimed components in terms of the primed components as

$$x_i = \sum_{j=1}^{3} R_{ji} x_j', \qquad A_i = \sum_{j=1}^{3} R_{ji} A_j', \qquad i = 1, 2, 3. \tag{1.93}$$

The above equations can also be written in matrix form. Let

$$\mathbf{r} = \begin{pmatrix} x_1 \\ x_2 \\ x_3 \end{pmatrix}, \qquad \mathbf{r}' = \begin{pmatrix} x_1' \\ x_2' \\ x_3' \end{pmatrix}, \qquad \mathbf{A} = \begin{pmatrix} A_1 \\ A_2 \\ A_3 \end{pmatrix}, \qquad \mathbf{A}' = \begin{pmatrix} A_1' \\ A_2' \\ A_3' \end{pmatrix} \tag{1.94}$$

$$\text{and} \qquad \mathbf{R} = \begin{pmatrix} R_{11} & R_{12} & R_{13} \\ R_{21} & R_{22} & R_{23} \\ R_{31} & R_{32} & R_{33} \end{pmatrix}, \tag{1.95}$$

where $\mathbf{R}$ is the **rotation matrix**. Then we have

$$\mathbf{r}' = \mathbf{R}\mathbf{r}, \qquad \mathbf{A}' = \mathbf{R}\mathbf{A}, \qquad \mathbf{r} = \mathbf{R}^T \mathbf{r}', \qquad \text{and} \quad \mathbf{A} = \mathbf{R}^T \mathbf{A}', \tag{1.96}$$

where $\mathbf{R}^T$ is the **transpose** of $\mathbf{R}$, i.e., $R_{ij}^T = R_{ji}$.

Note that there is a slight ambiguity in the use of the prime (′) notation: $\mathbf{A}$ and $\mathbf{A}'$ are the same vector, whereas $\mathbf{e}_j$ and $\mathbf{e}_j'$ are not. We could fix this by introducing a more complicated notation, but it is easier to not do this and just keep the ambiguity in mind.

As an example, consider a rotation of the coordinate system about the x_3 axis (the z axis) by an angle ϕ (see Figure 1.10b). The rotation matrix $\mathbf{R}$ is

$$\mathbf{R} = \begin{pmatrix} \cos\phi & \sin\phi & 0 \\ -\sin\phi & \cos\phi & 0 \\ 0 & 0 & 1 \end{pmatrix}. \tag{1.97a}$$

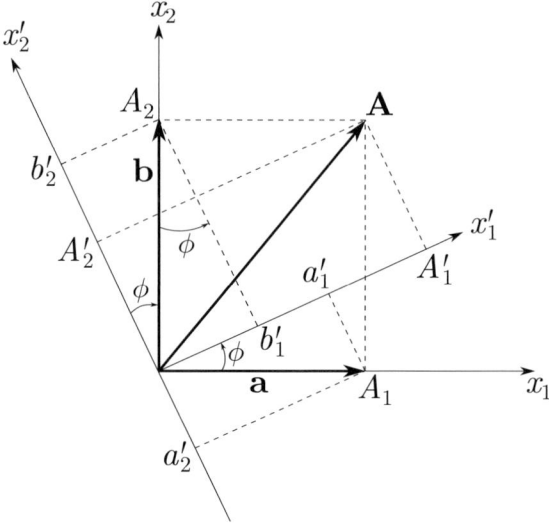

Figure 1.11 A rotation about the x_3 axis, and a vector **A** lying in the x_1–x_2 plane.

Its elements can be computed from $R_{ij} \equiv \mathbf{e}'_i \cdot \mathbf{e}_j$ in (1.86). For example, $R_{11} = \mathbf{e}'_1 \cdot \mathbf{e}_1 = \cos\phi$, $R_{12} = \mathbf{e}'_1 \cdot \mathbf{e}_2 = \cos(90° - \phi) = \sin\phi$, etc. The transformation equations $\mathbf{A}' = \mathbf{R}\mathbf{A}$ are

$$A'_1 = A_1 \cos\phi + A_2 \sin\phi, \qquad A'_2 = -A_1 \sin\phi + A_2 \cos\phi, \qquad A'_3 = A_3. \quad (1.97\text{b})$$

Because these equations are for the relatively simple case of a rotation about one of the coordinate axes, they can also be obtained in a simpler way, as follows. In Figure 1.11, we see that $\mathbf{A} = \mathbf{a} + \mathbf{b}$. Writing this equation in component form, with respect to the rotated (primed) frame of reference in Figure 1.11, we have

$$A'_1 = a'_1 + b'_1 = a \cos\phi + b \sin\phi = A_1 \cos\phi + A_2 \sin\phi, \qquad (1.97\text{c})$$
$$A'_2 = a'_2 + b'_2 = -a \sin\phi + b \cos\phi = -A_1 \sin\phi + A_2 \cos\phi. \qquad (1.97\text{d})$$

As an example, suppose $\mathbf{A} = (2, 0, 1)$, and that $\phi = 20°$ in Figure 1.10b). Then, from (1.97b), the components of **A** in the rotated system are $\mathbf{A}' = (1.88, -0.684, 1)$.

For a general rotation of the coordinate system about some arbitrary axis, the elements of the rotation matrix **R** can be computed using **Euler angles**. See Appendix 1A.

The elements of the rotation matrix are not independent, i.e., they are related to each other. For example, in the case of the rotation about the z axis, we see from (1.97a) that $R_{11}^2 + R_{21}^2 + R_{31}^2 = \cos^2\phi + \sin^2\phi = 1$. This is just one member of a set of equations relating the elements of **R**. This set, given by (1.99) or (1.100) or (1.102), can be obtained as follows: from (1.96) we have

$$\mathbf{A} = \mathbf{R}^T \mathbf{A}' = \mathbf{R}^T \mathbf{R}\mathbf{A}, \qquad (1.98)$$

which implies

$$\mathbf{R}^T\mathbf{R} = \mathbf{I} = \begin{pmatrix} 1 & 0 & 0 \\ 0 & 1 & 0 \\ 0 & 0 & 1 \end{pmatrix}, \tag{1.99}$$

where $\mathbf{I}$ is the identity matrix. Equation (1.99) is known as the **orthogonality condition**. Using the matrix multiplication rule

$$\mathbf{C} = \mathbf{AB} \ \Rightarrow \ C_{jk} = \sum_i A_{ji}B_{ik}$$

and the fact that $A_{ml}^T = A_{lm}$, one may write (1.99) in indicial notation as

$$\sum_{i=1}^{3} R_{ij}R_{ik} = I_{jk} = \delta_{jk} \,, \quad j,k = 1,2,3, \tag{1.100}$$

where δ_{jk} is known as the **Kronecker delta** and is defined by

$$\delta_{jk} = \begin{cases} 1, & \text{if } j = k; \\ 0, & \text{if } j \neq k. \end{cases} \tag{1.101}$$

A linear transformation $x_i' = \sum_j R_{ij}x_j$ is called an **orthogonal transformation** if the quantities R_{ij} satisfy the orthogonality condition.

It is sometimes useful to express (1.100) as $\sum_i \sum_m R_{mj}R_{ik}\delta_{mi} = \delta_{jk}$. One can see that summing first over m in this equation results in (1.100) – the term $R_{mj}\delta_{mi}$ in the sum over m is zero except for $m = i$. In general, the rule is

$$\sum_m a_m \delta_{mn} = a_n.$$

In words, considering the values of the Kronecker delta given in (1.101), in performing the sum over m, each term is zero except for the one with $m = n$ (for which the Kronecker delta equals one), leading to the result a_n.

An equivalent version of the orthogonality condition can be obtained as follows:

$$\mathbf{A}' = \mathbf{RA} = \mathbf{RR}^T\mathbf{A}' \ \Rightarrow \ \mathbf{RR}^T = \mathbf{I} \ \text{ i.e., } \ \sum_{i=1}^{3} R_{ji}R_{ki} = \delta_{jk}. \tag{1.102}$$

Note that since $\mathbf{R}^{-1}\mathbf{R} = \mathbf{RR}^{-1} = \mathbf{I}$, where $\mathbf{R}^{-1}$ is the **inverse** matrix of $\mathbf{R}$, we have

$$\mathbf{R}^{-1} = \mathbf{R}^T, \tag{1.103}$$

i.e., the inverse of $\mathbf{R}$ is given by its transpose. In general, though, the inverse and transpose of an arbitrary matrix $\mathbf{C}$ are not the same, i.e., $\mathbf{C}^{-1} \neq \mathbf{C}^T$.

Another way to view the orthogonality condition $\mathbf{R}\mathbf{R}^T = \mathbf{I}$ is that it is the mathematical expression of the fact that the basis vectors $\mathbf{e}'_i$ are orthogonal and have unit length. For example, from (1.86), i.e., $\mathbf{e}'_i = \sum_{j=1}^{3} R_{ij}\mathbf{e}_j$, $i = 1, 2, 3$, we may write

$$\mathbf{e}'_1 = R_{11}\mathbf{e}_1 + R_{12}\mathbf{e}_2 + R_{13}\mathbf{e}_3.$$

As $\mathbf{e}'_1$ is a unit vector, we have

$$R_{11}{}^2 + R_{12}{}^2 + R_{13}{}^2 = 1.$$

But this is just the relation we obtain from (1.102) for $j = k = 1$. Similarly, taking the dot product between $\mathbf{e}'_1$ and

$$\mathbf{e}'_2 = R_{21}\mathbf{e}_1 + R_{22}\mathbf{e}_2 + R_{23}\mathbf{e}_3$$

gives

$$R_{11}R_{21} + R_{12}R_{22} + R_{13}R_{23} = 0$$

because they are orthogonal or perpendicular to each other. But this is just the relation we obtain from (1.102) for $j = 1$ and $k = 2$.

1.11 Invariance

When we discussed the definitions of various vector-related quantities, such as the dot product and cross product, etc., we assumed that those definitions held in any Cartesian reference frame, in particular, that they were **invariant** under a rotation of coordinates. For example, for the dot product $\mathbf{A} \cdot \mathbf{B}$, we should have

$$A_1B_1 + A_2B_2 + A_3B_3 = A'_1B'_1 + A'_2B'_2 + A'_3B'_3 . \tag{1.104}$$

In words, even though $A_1 \neq A'_1$, $B_1 \neq B'_1$, etc., the preceding two sums are the same. That this is true can be demonstrated as follows. Using the summation convention (see the explanation following Equation 1.90), we have, from (1.92) and (1.100),

$$A'_iB'_i = R_{ij}A_jR_{ik}B_k = R_{ij}R_{ik}A_jB_k = \delta_{jk}A_jB_k = A_kB_k. \tag{1.105}$$

Note that in the second term, the summation index (k) used for B'_i must be different from the one (j) used for A'_i. This calculation shows that the expression for the scalar product has the same form in the primed and unprimed frames, i.e., the scalar product is invariant under a rotation of coordinates. If it wasn't, it would not be a useful quantity. If we put $B = A$ in (1.104) and take the square root of both sides, we see that the length of a vector $\mathbf{A}$ is invariant, which is intuitively obvious.

The invariance of the dot product can also be demonstrated using matrices as follows. First, note that the dot product $\mathbf{A} \cdot \mathbf{B}$ can be expressed in matrix form as $\mathbf{A}^T\mathbf{B}$. Hence, using the matrix rules

$$(\mathbf{AB})^T = \mathbf{B}^T\mathbf{A}^T \quad \text{and} \quad (\mathbf{A}^T)^T = \mathbf{A},$$

we have

$$\mathbf{A}^T\mathbf{B} = (\mathbf{R}^T\mathbf{A}')^T(\mathbf{R}^T\mathbf{B}') = \mathbf{A}'^T\mathbf{R}\mathbf{R}^T\mathbf{B}' = \mathbf{A}'^T\mathbf{I}\,\mathbf{B}' = \mathbf{A}'^T\mathbf{B}'$$

showing that $\mathbf{A}^T\mathbf{B}$ is invariant.

Also, the quantity $ds^2 = dx^2 + dy^2 + dz^2 = \sum_i dx_i dx_i$ is invariant. In addition, it can be shown that the cross product $\mathbf{A} \times \mathbf{B}$ is invariant.

As another example, consider the gradient $\nabla \psi$ of the scalar field ψ. We should have

$$\sum_{i=1}^{3} \frac{\partial \psi'}{\partial x_i'} \mathbf{e}_i' = \sum_{i=1}^{3} \frac{\partial \psi}{\partial x_i} \mathbf{e}_i \,. \tag{1.106}$$

To demonstrate that this is so, let $\psi'(x_1', x_2', x_3')$, which is the quantity obtained by substituting the first equation of (1.93), i.e., $x_i = \sum_{j=1}^{3} R_{ji}x_j'$, into $\psi(x_1, x_2, x_3)$, be the same field as $\psi(x_1, x_2, x_3)$ but in a rotated frame. For instance, they could represent the temperature at some point. The prime on ψ' indicates that ψ' is not necessarily the same function of x_1', x_2', and x_3' as ψ is of x_1, x_2, and x_3. Since the primed and unprimed coordinates refer to the same point in space, we must have $\psi'(x_1', x_2', x_3') = \psi(x_1, x_2, x_3)$. Differentiating with respect to x_i' on both sides of this, using the chain rule of partial differentiation, and letting $\mathbf{G} \equiv \nabla \psi$ then gives

$$G_i' = \frac{\partial \psi'}{\partial x_i'} = \sum_{j=1}^{3} \frac{\partial \psi}{\partial x_j} \frac{\partial x_j}{\partial x_i'} = \sum_{j=1}^{3} R_{ij} \frac{\partial \psi}{\partial x_j} \equiv \sum_{j=1}^{3} R_{ij} G_j, \tag{1.107}$$

where the following rule, obtainable from (1.90) and the first equation of (1.93), was used:

$$R_{ij} = \frac{\partial x_i'}{\partial x_j} = \frac{\partial x_j}{\partial x_i'} \,. \tag{1.108}$$

Since (1.107) has the form of (1.92) with "$\mathbf{A}$" being the gradient "$\mathbf{G}$," (1.106) follows from (1.91). This also shows that it is only necessary to prove that the transformation rule $A_i' = \sum_j R_{ij}A_j$ is satisfied.

Equation (1.107), when compared with (1.92), shows that the primed and unprimed components of the gradient are related in the same way that the primed and unprimed components of an arbitrary vector $\mathbf{A}$ are related. This proves that the gradient $\nabla \psi$ is indeed a vector. In general, it is necessary to demonstrate that a given quantity $\mathbf{A}$ is actually a vector, since not every three-component quantity we may choose to define is a vector. For example, we can use the procedure shown in (1.107) to show that the quantity

$$\left(\frac{\partial \psi}{\partial x_1}\right)^2 \mathbf{e}_1 + \left(\frac{\partial \psi}{\partial x_2}\right)^2 \mathbf{e}_2 + \left(\frac{\partial \psi}{\partial x_3}\right)^2 \mathbf{e}_3, \tag{1.109}$$

is *not* a vector, since the components do not satisfy (1.92). In the same way, one can show that the cross product and curl are invariant vectors.

It is generally desirable to express physical laws in vector form. However, before doing so, one must first make sure that the quantities involved are actually vectors. With this in mind, (1.92) is usually used in the definition of a true vector. Consider three quantities that take the values A_1, A_2, and A_3 in one Cartesian frame and the values A_1', A_2', and A_3' in another, which is rotated with respect to the first. Then these three quantities are said to be the components of a **vector** if they are related by (1.92), i.e., by $A_i' = \sum_j R_{ij} A_j$.

1.12 Rotation of Vectors

In the equation $\mathbf{r}' = \mathbf{R}\mathbf{r}$, ((1.96)), $\mathbf{R}$ can be thought of as an operator that describes the rotation of the coordinate frame. However, one may also think of $\mathbf{R}$ as an operator that rotates the vector $\mathbf{r}$ into a *new* vector $\mathbf{r}' = \sum_j r_j' \mathbf{e}_j$, rather than one that rotates the coordinate frame. In other words, we think of two different vectors, $\mathbf{r}$ and $\mathbf{r}'$, in the same coordinate frame. Similarly, $\mathbf{R}^{-1}$ or $\mathbf{R}^T$ would rotate $\mathbf{r}'$ back to $\mathbf{r}$ through the operation $\mathbf{r} = \mathbf{R}^T \mathbf{r}' = \mathbf{R}^{-1} \mathbf{r}'$.

Vector rotations can be combined or cascaded. For instance, consider a double rotation from $\mathbf{r}$ to $\mathbf{r}'$ to $\mathbf{r}''$ via the rotation operators $\mathbf{R}$ and $\mathbf{S}$ respectively. Then

$$\mathbf{r}'' = \mathbf{S}\mathbf{r}' = \mathbf{S}\mathbf{R}\mathbf{r} . \tag{1.110}$$

$\mathbf{S}\mathbf{R}$ is then the rotation operator that rotates $\mathbf{r}$ to $\mathbf{r}''$. Quantitatively, it is represented by the product of the matrices $\mathbf{S}$ and $\mathbf{R}$. Since matrix multiplication is not commutative, i.e., $\mathbf{S}\mathbf{R} \neq \mathbf{R}\mathbf{S}$, changing the order of the rotation of $\mathbf{r}$ (rotating with $\mathbf{S}$ first, then with $\mathbf{R}$) yields a final position vector that is *different* from $\mathbf{r}''$. As we have seen before, in discussing the curl, combining finite rotations is not a commutative process.

1.13 Translation of Coordinates

Figure 1.10a shows a general rotation of coordinates. If a **translation** of the origin from point O to point O' occurs as well, along with a rotation, then the transformation equations become

$$x_i' = \sum_{j=1}^{3} R_{ij}x_j + c_i, \qquad i = 1, 2, 3, \tag{1.111}$$

where the c_i are constants that quantify the translation. However, we can still use (1.90) and (1.92) to analyze vectors, rather than the more complicated Equation (1.111). To explain this, let x_i and y_i be the coordinates of two points P and Q, respectively, in space. The three numbers $z_i \equiv y_i - x_i$ are the components of the **displacement** vector $\mathbf{u}_{PQ}$ going from P to Q. Let x_i' and y_i' be the new coordinates of points P and Q in a new frame that is rotated and translated with respect to the first. The new coordinates are given in terms of the old by (1.111). Hence, the new components z_i' of the displacement vector $\mathbf{u}_{PQ}$ are given by

$$z_i' = y_i' - x_i' = \sum_{j=1}^{3} R_{ij}y_j - \sum_{j=1}^{3} R_{ij}x_j = \sum_{j=1}^{3} R_{ij}z_j, \tag{1.112}$$

i.e., they satisfy the transformation rule for vectors. Hence we can use (1.90), or (1.92), for a reference frame that is translated as well as rotated with respect to another if we think of x_j, or A_j, more generally as the components of a displacement vector connecting two points in space, rather than the components of the vector connecting the origin with some point in space. The previous comments about invariance, and vector definitions and equations having the same form in any Cartesian reference frame, apply in general to a rotation and/or translation of coordinates.

1.14 Cartesian Tensors

As we have seen, if the quantities A_i, $i = 1, 2, 3$, are the components of a vector **A**, then $A_i' = \sum_j R_{ij}A_j$ (see (1.92)), are the components of **A** in another Cartesian reference frame. In general, this holds for an N-dimensional space, i.e., i (and j in the sum of (1.92)) can range from 1 to N. For our purposes though, $N = 3$.

We can extend the theory to quantities that have more than one index. We have already seen an example of such quantities: the Kronecker delta. For example, consider the nine quantities that take the values C_{ij}, $i, j = 1, 2, 3$, in one Cartesian reference frame, and the values C_{ij}' in another frame. If these quantities are related by the equations

$$C_{ij}' = \sum_{k=1}^{3} \sum_{l=1}^{3} R_{ik}R_{jl}C_{kl} = \sum_{k=1}^{3} \sum_{l=1}^{3} \frac{\partial x_i'}{\partial x_k}\frac{\partial x_j'}{\partial x_l}C_{kl}, \tag{1.113a}$$

where (1.108) has been used, then they are said to be the components of a **tensor** of rank 2. Similarly, this definition can be extended to tensors of higher rank, e.g., $C_{ijk...}$ A vector is now seen to be a tensor of rank 1, and a scalar is a tensor of rank 0. However, the transformation equation for a scalar is simply $\psi'(x_1', x_2', x_3') = \psi(x_1, x_2, x_3)$ (see the comments preceding (1.107)).

Equation (1.113a) is often written more simply using the summation convention as

$$C_{ij}' = R_{ik}R_{jl}C_{kl} = \frac{\partial x_i'}{\partial x_k}\frac{\partial x_j'}{\partial x_l}C_{kl} \tag{1.113b}$$

Equation (1.113a) can also be written in matrix form as follows. First, rewrite (1.113a) as

$$C_{ij}' = \sum_{k=1}^{3}\sum_{l=1}^{3} R_{ik}C_{kl}R_{jl} = \sum_{k=1}^{3}\sum_{l=1}^{3} R_{ik}C_{kl}R_{lj}^{T} \, .$$

Inspection of the placement of the indices in the last expression shows that it is the indicial formula for the multiplication of the three matrices $\mathbf{R}$, $\mathbf{C}$ and $\mathbf{R}^T$. Hence,

$$\mathbf{C}' = \mathbf{RCR}^T \, . \tag{1.113c}$$

This formula can be used to compute, via matrix multiplication, the components of the tensor $\mathbf{C}$ in a rotated frame from those in the unrotated frame.

To simplify discussions, it is also conventional to say "the tensor $C_{ijk...}$" rather than "the components $C_{ijk...}$ of the tensor." We will use this convention.

If a physical law can be expressed as a tensor equation, such as $A_{ijk} = B_{ijk}$, then this law has the same form, i.e., $A_{ijk}' = B_{ijk}'$, in all Cartesian reference frames. The tensor equation is said to be **covariant** under a rotation and/or translation of axes. For example, consider Newton's second law $\mathbf{F} = m\mathbf{a}$. As both $\mathbf{F}$ and $\mathbf{a}$ are vectors, the transformation rule applies to them. Consequently, one may write

$$F_i' - ma_i' = \sum_j R_{ij}F_j - m\sum_j R_{ij}a_j = \sum_j R_{ij}(F_j - ma_j) = \sum_j R_{ij}(0) = 0,$$

showing that $\mathbf{F}' = m\mathbf{a}'$ if $\mathbf{F} = m\mathbf{a}$.

We can show that a quantity is or is not a tensor by demonstrating that the transformation equations are or are not satisfied. For instance, consider the quantity C_{ij} defined by $C_{ij} = A_iB_j$, where A_i and B_i are vectors. Is C_{ij} a tensor of rank 2? If it is, it should transform like (1.113a) or (1.113b). The following calculation proves that it does. Since A_i and B_j are tensors of rank one, i.e., vectors, we have, using the summation convention,

$$A_i' = R_{ik}A_k, \quad B_j' = R_{jl}B_l,$$

and multiplying them together gives

$$C_{ij}' \equiv A_i'B_j' = R_{ik}A_kR_{jl}B_l = R_{ik}R_{jl}A_kB_l = R_{ik}R_{jl}C_{kl}, \tag{1.114}$$

showing that the product C_{ij} satisfies the tensor transformation rule (1.113b), and is therefore also a tensor. Note that the components C'_{ij} are obtained from the components of the two vectors in the same way that the components C_{ij} are. The equation $C_{ij} = A_i B_j$ is covariant, i.e., it has the same form in all Cartesian reference frames.

In general, any product of tensors gives another tensor. For example, if A_{ijk}, B_i, and C_{ij} are tensors of rank 3, 1, and 2, respectively, then $D_{ijklmn} \equiv A_{ijk} B_l C_{mn}$ is a tensor of rank 6. Such a product is often called an **outer product**.

The addition of two tensors of the same rank gives another tensor of that rank, e.g., $C_{ij} \equiv A_{ij} + B_{ij}$ is a tensor if A_{ij} and B_{ij} are tensors. The same can be said for the subtraction of two tensors.

Consider the Kronecker delta defined in (1.101). Is it a tensor? The calculation

$$\delta'_{ij} = \sum_k \sum_l R_{ik} R_{jl} \delta_{kl} = \sum_k R_{ik} R_{jk} = \delta_{ij}, \tag{1.115}$$

where (1.102) has been used in the rightmost equation, shows that it is indeed a tensor, and also that its components have the same values in all frames.

Consider the partial derivative of the tensor A_{ij} with respect to x_k, i.e., $\partial A_{ij} / \partial x_k$. Is this a tensor? To answer this, we differentiate the tensor A_{ij} in another reference frame. Using the summation convention, we obtain

$$\frac{\partial A'_{ij}}{\partial x'_k} = \frac{\partial}{\partial x'_k} R_{il} R_{jm} A_{lm} = R_{il} R_{jm} \frac{\partial A_{lm}}{\partial x_n} \frac{\partial x_n}{\partial x'_k} = R_{il} R_{jm} R_{kn} \frac{\partial A_{lm}}{\partial x_n}, \tag{1.116}$$

where the chain rule of partial differentiation and (1.108) were used. Note that the last term is a triple sum over l, m, and n. Hence, $\partial A_{ij} / \partial x_k$ is a tensor of rank 3 since it transforms according to the transformation rules of a tensor of rank 3. In general, the nth-order derivative of a tensor of rank m is a tensor of rank $m + n$. When doing calculations involving high-order derivatives of tensors, it is convenient to use simpler notations for the partial derivatives, to avoid having to write many partial derivative symbols. Hence, $\partial A_{ij} / \partial x_k$ is often written as "$\partial_k A_{ij}$" or as "$A_{ij,k}$." The latter notation is obviously the simplest. For higher-order derivatives, only one comma is usually used. For example,

$$\frac{\partial^2 A_{ij}}{\partial x_k \partial x_l} \equiv \partial_k \partial_l A_{ij} \equiv A_{ij,kl}. \tag{1.117}$$

This notation is quite common in the advanced literature on seismic wave theory and other subjects.

Let A_{ij} be a tensor. If $A_{ij} = A_{ji}$ for all values of i and j, then the tensor A_{ij} is said to be **symmetric** with respect to the indices i and j. Similarly, if $A_{ij} = -A_{ji}$, then it is said to be **antisymmetric** with respect to i and j. We will see in the next chapter that the stress tensor for a homogeneous isotropic elastic medium is symmetric.

Symmetry can also exist for tensors of higher rank, and can exist with respect to any pair of indices. For example, if $A_{ijk} = A_{ikj}$, then A_{ijk} is symmetric with respect to j and k. The transformation rules can be used to show that symmetry and antisymmetry are preserved upon transformation, i.e., if $A_{ijk} = A_{ikj}$, then $A'_{ijk} = A'_{ikj}$.

If a tensor property, such as symmetry, is not preserved upon transformation, e.g., if it is true in only one reference frame, then it is basically useless, because it is desireable to express physical quantities and laws as tensors and tensor equations that are independent of the reference frame used, i.e., that hold in all reference frames.

A useful fact is that any second-rank tensor C_{ij} can be written as a sum of a symmetric tensor S_{ij} and an antisymmetric tensor A_{ij}, i.e.,

$$C_{ij} = S_{ij} + A_{ij}, \quad \text{where} \quad S_{ij} \equiv \tfrac{1}{2}(C_{ij} + C_{ji}) \quad \text{and} \quad A_{ij} \equiv \tfrac{1}{2}(C_{ij} - C_{ji}).$$

Consider a procedure in which one index of a tensor is made equal to another index and a sum is performed over that index. This procedure is called a **contraction** of the tensor. For example,

$$C_{ij} = A_i B_j \quad \Rightarrow \quad \sum_i C_{ii} = \sum_i A_i B_i = \mathbf{A} \cdot \mathbf{B}. \tag{1.118}$$

Note that contraction has reduced a tensor of rank 2 to a scalar. In general, the contraction of a tensor produces a new tensor whose rank is lower by two. For instance, let A_{ijk} be a tensor of rank 3. Using the summation convention, we have

$$A'_{ijk} = R_{il} R_{jm} R_{kn} A_{lmn}, \tag{1.119}$$

and contraction with respect to j and k yields

$$A'_{ijj} = R_{il} R_{jm} R_{jn} A_{lmn} = R_{il} \delta_{mn} A_{lmn} = R_{il} A_{lmm}, \tag{1.120}$$

where the orthogonality condition was used. The contraction of A_{ijk} produced a new tensor $B_i \equiv \sum_j A_{ijj}$ whose rank is 1. B_i is a tensor since $B'_i = \sum_l R_{il} B_l$. If we repeat the calculation in Equation (1.120) for a second-rank tensor A_{ij}, we find that $\sum_i A'_{ii} = \sum_i A_{ii}$, i.e., we obtain an invariant scalar (the trace of the matrix whose elements are A_{ij}). For another example, consider the vector A_i. We know, from earlier discussions, that $\partial A_i / \partial x_j$ is a second-rank tensor. Contraction then yields the divergence of A_i:

$$\sum_i \frac{\partial A_i}{\partial x_i} = \nabla \cdot \mathbf{A}. \tag{1.121}$$

Forming the outer product of two tensors, followed by a contraction, gives an **inner product** of the two tensors. For example, from the outer product $A_{ij} B_{klm}$, which is a tensor of rank 5, one may set $j = k$ and sum over k to obtain the inner product $\sum_k A_{ik}B_{klm}$, which is a tensor of rank 3. The dot product in (1.118) is also an inner product.

Another quantity that is sometimes useful is the **permutation symbol** ε_{ijk}, also known as the **Levi–Civita symbol**, defined by

$$\varepsilon_{ijk} : \begin{cases} \varepsilon_{123} = \varepsilon_{231} = \varepsilon_{312} = +1, \\ \varepsilon_{213} = \varepsilon_{132} = \varepsilon_{321} = -1, \\ \varepsilon_{ijk} = 0 \quad \text{if two or more indices are equal.} \end{cases} \tag{1.122}$$

Then the ith component of the cross product $\mathbf{C} = \mathbf{A} \times \mathbf{B}$ can be written as

$$C_i = \sum_j \sum_k \varepsilon_{ijk} A_j B_k. \tag{1.123}$$

It can be shown that ε_{ijk} is actually a tensor if the only coordinate frame changes that one considers are rotations of coordinates. But more generally, if one also considers reflections and inversions of coordinates, then ε_{ijk} turns out to be a **pseudotensor** or a **tensor density**, and the cross product in (1.123) is a **pseudovector**. More on these topics can be found in texts containing substantial sections on tensor analysis.

Another useful rule is the **quotient rule**, which works as follows. Suppose it is not known whether or not a quantity A is a tensor. If an inner product of A with a tensor is itself a tensor, then A is a tensor. For example, suppose that B_n is a tensor, but it is not known if A_{ijk} is a tensor. Form the outer product $A_{ijk}B_n = C_{ijkn}$, then contract with respect to k and n to obtain the inner product $\sum_k A_{ijk}B_k = C_{ij}$. If C_{ij} is a tensor, then so is A_{ijk}.

1.15 General Tensors

Tensors can be analyzed with reference to curvilinear coordinate systems and **non-Euclidean** spaces as well (a plane surface is an example of a 2D Euclidean space, whereas the surface of a sphere is an example of a 2D non-Euclidean space).

In our discussion of curvilinear coordinates, we saw that a vector $\mathbf{A}$ could be written as $\mathbf{A} = \sum_j A_j \mathbf{e}_j$, where $\mathbf{e}_j = h_j^{-1}(\partial\mathbf{r}/\partial q_j)$, since $(\partial\mathbf{r}/\partial q_j)$ is a vector tangent to the q_j coordinate curve. Instead of forming unit vectors $\mathbf{e}_j$, though, we could have simply used the $(\partial\mathbf{r}/\partial q_j)$, $j = 1, 2, 3$, as basis vectors, so that we could write $\mathbf{A} = \sum_j C_j(\partial\mathbf{r}/\partial q_j)$. Furthermore, since the vector ∇q_j is also tangent to the q_j coordinate curve (it is perpendicular to $q_j = const$), we could also write $\mathbf{A} = \sum_j c_j \nabla q_j$. In other words, in a general coordinate system, a vector $\mathbf{A}$ really has two different sets of components: the C_js are called the **contravariant** components of $\mathbf{A}$ and the c_js are called the **covariant** components of $\mathbf{A}$. It can be shown that

$C_i' = \sum_j (\partial q_i'/\partial q_j)C_j$ and $c_i' = \sum_j (\partial q_j/\partial q_i')c_j$. These are the transformation equations giving the contravariant and covariant components of $\mathbf{A}$ in another coordinate system (q_1', q_2', q_3').

A different notation is conventionally used: since "i" appears in the numerator of the partial derivative in the contravariant transformation equation and in the denominator in the covariant equation, superscripts and subscripts are used for the contravariant and covariant components, respectively, of a vector. In other words, "A^i" denotes the ith contravariant component of $\mathbf{A}$, and "A_i" denotes the ith covariant component of $\mathbf{A}$. Note that this "A_i" is different from the "A_i" in the equation $\mathbf{A} = \sum_i A_i \mathbf{e}_i$ (an ambiguity exists here). Since the coordinate transformation equations $q_i' = q_i'(q_1, q_2, q_3)$, $i = 1, 2, 3$, imply $dq_i' = \sum_j (\partial q_i'/\partial q_j)dq_j$, which is a contravariant transformation equation, the coordinates themselves should have superscripts. So, in general, we let $(x^1, \ldots, x^N)$ denote a general set of coordinates (either Cartesian or curvilinear). We then have

$$A'^i = \sum_j \frac{\partial x'^i}{\partial x^j}A^j \quad \text{and} \quad A_i' = \sum_j \frac{\partial x^j}{\partial x'^i}A_j, \tag{1.124}$$

which give the contravariant and covariant components of the vector $\mathbf{A}$ in another frame.

Let $x'^i = x'^i(x^1, \ldots, x^N)$, $i = 1, \ldots, N$, be a general set of coordinate transformation equations. Then dx'^i, given by

$$dx'^i = \sum_j \frac{\partial x'^i}{\partial x^j}dx^j, \tag{1.125}$$

is an example of a contravariant tensor of rank 1. Let $\psi'(x'^1, \ldots, x'^N) = \psi(x^1, \ldots, x^N)$ be a scalar field. Then the gradient $\partial \psi'/\partial x'^i$, given by (using the chain rule of differentiation

$$\frac{\partial \psi'}{\partial x'^i} = \sum_j \frac{\partial x^j}{\partial x'^i}\frac{\partial \psi}{\partial x^j}, \tag{1.126}$$

is an example of a covariant tensor of rank 1. For Cartesian tensors, (1.108) shows that

$$\frac{\partial x'^i}{\partial x^j} = \frac{\partial x^j}{\partial x'^i}. \tag{1.127}$$

Hence, there is no distinction between contravariant and covariant Cartesian tensors, i.e., superscripted indices are not necessary. However, this is not true in general.

General tensors of higher rank occur in three types: **contravariant**, **covariant**, and **mixed**. For example, for rank 2, using the summation convention, they transform respectively as

$$A'^{ij} = \frac{\partial x'^i}{\partial x^k}\frac{\partial x'^j}{\partial x^l}A^{kl}, \quad A'_{ij} = \frac{\partial x^k}{\partial x'^i}\frac{\partial x^l}{\partial x'^j}A_{kl}, \quad A'^i_{\ j} = \frac{\partial x'^i}{\partial x^k}\frac{\partial x^l}{\partial x'^j}A^k_{\ l}. \tag{1.128}$$

General tensor analysis has been developed much further than what is contained in this brief introduction. Its best-known application is the mathematical description of the gravitational effects of curved spacetime in Einstein's general theory of relativity. In theoretical seismology, it is needed to perform calculations involving stress and strain tensors in curvilinear coordinate frames (e.g., spherical polar coordinates are often used to describe stress, strain, and wave motion in a spherical Earth).

1.16 Fourier Transforms

Suppose the function $g(t)$ is defined on the interval $-T/2 \leq t \leq T/2$, and suppose that $g(t)$ is periodic with period T, i.e., $g(t \pm jT) = g(t), j = 1, 2, \ldots$ Then $g(t)$ can be expressed as a **Fourier series**, i.e.,

$$g(t) = \frac{1}{2}a_0 + \sum_{n=1}^{\infty}\left[a_n\cos(\omega_n t) + b_n\sin(\omega_n t)\right], \tag{1.129}$$

where

$$\omega_n = 2\pi f_n = \frac{2\pi n}{T}, \quad a_n = \frac{2}{T}\int_{-T/2}^{T/2} g(t)\cos(\omega_n t)\,dt, \quad \text{and}$$

$$b_n = \frac{2}{T}\int_{-T/2}^{T/2} g(t)\sin(\omega_n t)\,dt. \tag{1.130}$$

Using the rules

$$2\cos\theta = e^{i\theta} + e^{-i\theta} \quad \text{and} \quad 2i\sin\theta = e^{i\theta} - e^{-i\theta}, \quad \text{where} \quad e^{i\theta} = \cos\theta + i\sin\theta,$$

we can write the Fourier series in *complex* form as

$$g(t) = \sum_{n=-\infty}^{\infty}\bar{g}_n e^{-i\omega_n t}, \quad \text{where} \quad \bar{g}_n = \frac{1}{T}\int_{-T/2}^{T/2} g(t)e^{i\omega_n t}\,dt, \tag{1.131}$$

where $\bar{g}_n$, a_n, and b_n are related by

$$2\bar{g}_n = a_n + ib_n, \quad 2\bar{g}_{-n} = a_n - ib_n, \quad 2\bar{g}_0 = a_0, \quad n = 1, 2, \ldots \tag{1.132}$$

Although the interval $(-T/2, T/2)$ was assumed in the preceding equations, any interval of length T could be used. In particular, if $g(t)$ is defined for the range $0 \leq t \leq T$, then the limits on the integrals would be 0 and T.

Since $\bar{g}_n$ is a complex number, it can be written as $\bar{g}_n = |\bar{g}_n| \exp(i\psi_n)$, where $|\bar{g}_n|$ is the magnitude (or amplitude) of $\bar{g}_n$ and ψ_n is the phase angle of $\bar{g}_n$ in the complex plane. Equation (1.131) expresses $g(t)$ as a sum of sinusoids (represented by complex exponentials). The nth sinusoid is $\bar{g}_n \exp(-i\omega_n t) = |\bar{g}_n| \exp[-i(\omega_n t - \psi_n)]$. Its amplitude is $|\bar{g}_n|$, its frequency is $f_n = \omega_n/2\pi = n/T$, and its phase is ψ_n. The sequence of numbers $\{|\bar{g}_n|\}$, all n, forms the discrete amplitude spectrum of $g(t)$, and the sequence of numbers $\{\psi_n\}$, all n, forms the discrete phase spectrum of $g(t)$.

Consider the case in which $T \to \infty$. In this case, $g(t)$ is no longer explicitly periodic – it is defined for the range $-\infty < t < \infty$. Also, the interval between discrete frequency values, $1/T$, approaches zero, meaning that the range of frequency values becomes continuous (and the amplitude and phase spectra become continuous). In this case, $g(t)$ and its frequency spectrum are related not by the Fourier series formulas (Equation 1.131), but by the **Fourier transform**, i.e.,

$$\bar{g}(\omega) = \int_{-\infty}^{\infty} g(t) e^{i\omega t}\, dt \quad \text{and} \quad g(t) = \frac{1}{2\pi} \int_{-\infty}^{\infty} \bar{g}(\omega) e^{-i\omega t}\, d\omega, \quad (1.133\text{a,b})$$

where ω is the angular frequency $\omega = 2\pi f$. The integral over t is the **forward** Fourier transform, and the integral over ω is the **inverse** Fourier transform.

Another way to write the Fourier transform formulas is to use the actual frequency f:

$$\bar{g}(f) = \int_{-\infty}^{\infty} g(t) e^{2\pi i f t}\, dt \quad \text{and} \quad g(t) = \int_{-\infty}^{\infty} \bar{g}(f) e^{-2\pi i f t}\, df, \quad (1.133\text{c,d})$$

with the advantage being that they appear more "symmetric" in that the "$1/2\pi$" is not present. It should also be noted that $\bar{g}(f)$ is not quite the same function of f that $\bar{g}(\omega)$ is of ω. In fact, $\bar{g}(f)$ can be obtained from the formula for $\bar{g}(\omega)$ by replacing ω with $2\pi f$ (not just with f).

$\bar{g}(\omega)$ is the **frequency spectrum** of $g(t)$. Since it is in general a complex number, it can be written as

$$\bar{g}(\omega) = \text{Re}(\bar{g}) + i\text{Im}(\bar{g}) = |\bar{g}(\omega)| e^{i\psi(\omega)}, \quad (1.134\text{a})$$

where $|\bar{g}(\omega)|$ is the **amplitude spectrum** of $g(t)$ and $\psi(\omega)$ is the **phase spectrum** of $g(t)$, and where

$$|\bar{g}(\omega)| = \sqrt{\left[\text{Re}(\bar{g})\right]^2 + \left[\text{Im}(\bar{g})\right]^2}, \quad \psi(\omega) = \tan^{-1}\left[\text{Im}(\bar{g})/\text{Re}(\bar{g})\right], \quad (1.134\text{b})$$

where the principal branch of the arctangent is used if $\text{Re}(\bar{g})$ is positive and the upper or lower branch if $\text{Re}(\bar{g})$ is negative (the principal branch lies between $-90°$ and $+90°$, but the phase angle ψ of $\bar{g}$ can range between $-180°$ and $+180°$ on the complex plane). Putting it another way, if $\text{Re}(\bar{g})$ is negative, then add

$\pm 180°$ to the principal branch value of $\psi(\omega)$. For example, the phase of $3 - 2i$ is $\psi = \tan^{-1}(-2/3) = -33.7°$, but the phase of $-3 + 2i$ is $\psi = \tan^{-1}(-2/3) = -33.7° + 180° = 146.3°$.

Example: *Fourier Transform of a Delta Function*. The Dirac delta function $\delta(t)$ is defined by the following three properties:

$$\delta(t) = 0, \quad t \neq 0; \qquad \int_{-\infty}^{\infty} \delta(t)\, dt = 1; \quad \text{and} \qquad (1.135\text{a,b})$$

$$\int_a^b y(t)\delta(t - t_0)\, dt = \begin{cases} y(t_0), & \text{if } a \leq t_0 \leq b \\ 0, & \text{otherwise} \end{cases} \qquad (1.135\text{c})$$

where $y(t)$ is an arbitrary function of t. $\delta(t)$ is a "spike" of infinite height and zero width at $t = 0$, such that the area under the curve $\delta(t)$ is one.

Consider the time-shifted delta function

$$g(t) = \delta(t - t_0). \qquad (1.136\text{a})$$

This is a "spike" located at $t = t_0$. Then, using (1.135c),

$$\overline{g}(\omega) = \int_{-\infty}^{\infty} \delta(t - t_0)e^{i\omega t}\, dt = e^{i\omega t_0} \qquad \Longrightarrow \qquad |\overline{g}(\omega)| = 1 \quad \text{and} \quad \psi(\omega) = \omega t_0, \qquad (1.136\text{b})$$

i.e., the amplitude at all frequencies is one (the amplitude spectrum is "flat") and the phase spectrum is linear in the frequency ω.

Equation (1.133b) implies

$$\delta(t - t_0) = \frac{1}{2\pi} \int_{-\infty}^{\infty} e^{-i\omega(t - t_0)}\, d\omega. \qquad (1.137\text{a})$$

Since $\delta(t - t_0)$ is a real number and is therefore equal to its complex conjugate, we also have

$$\delta(t - t_0) = \frac{1}{2\pi} \int_{-\infty}^{\infty} e^{i\omega(t - t_0)}\, d\omega. \qquad (1.137\text{b})$$

Example: *Fourier Transform of $g(t - t_0)$*. If $\overline{g}(\omega)$ is the Fourier transform of $g(t)$, then $e^{i\omega t_0}\overline{g}(\omega)$ is the Fourier transform of the time-shifted function $g(t - t_0)$. This can be easily shown using the substitution $u = t - t_0$ in Equation (1.133a). This means that $g(t)$ and $g(t - t_0)$ have the same amplitude spectrum, but that if $\psi(\omega)$ is the phase spectrum of $g(t)$, then $\psi(\omega) + \omega t_0$ is the phase spectrum of $g(t - t_0)$.

Example: *Fourier Transform of a Cosine Function*. Let

$$g(t) = \cos(\omega_0 t), \qquad (1.138\text{a})$$

i.e., a cosine of angular frequency ω_0 and period $T_0 = 2\pi/\omega_0$. Then, using $2\cos\theta = e^{i\theta} + e^{-i\theta}$, we have

$$\bar{g}(\omega) = \int_{-\infty}^{\infty} \cos(\omega_0 t) e^{i\omega t} \, dt = \frac{1}{2} \int_{-\infty}^{\infty} \left[e^{i\omega_0 t} + e^{-i\omega_0 t} \right] e^{i\omega t} \, dt$$

$$= \frac{1}{2} \int_{-\infty}^{\infty} \left[e^{i(\omega+\omega_0)t} + e^{i(\omega-\omega_0)t} \right] dt$$

$$= \pi \delta(\omega + \omega_0) + \pi \delta(\omega - \omega_0), \tag{1.138b}$$

where (1.137b) has been used with t and ω switched. The frequency spectrum of a cosine of frequency ω_0 is a pair of "spikes" in the frequency domain located at $\omega = \pm\omega_0$.

Example: *Fourier Transform of a Decaying Exponential Function.* Let

$$g(t) = e^{-at} H(t) = \begin{cases} 0, & \text{if } t < 0 \\ e^{-at}, & \text{if } t \geq 0, \end{cases} \tag{1.139a}$$

where a is a positive real constant, and where $H(t)$ is the **Heaviside step function**, which is 0 if $t < 0$ and 1 if $t \geq 0$. Then

$$\bar{g}(\omega) = \int_0^{\infty} e^{-at} e^{i\omega t} \, dt = \int_0^{\infty} e^{-(a-i\omega)t} \, dt = \left[\frac{e^{-(a-i\omega)t}}{-(a-i\omega)} \right]_{t=0}^{t=\infty} = \frac{1}{a - i\omega}, \tag{1.139b}$$

where the fact that $e^{-a\infty} = 0$ was used. To obtain the real and imaginary parts of G, we first write

$$\bar{g}(\omega) = \frac{1}{a - i\omega} \times \frac{a + i\omega}{a + i\omega} = \frac{a + i\omega}{a^2 + \omega^2}, \tag{1.140}$$

which implies

$$\text{Re}(\bar{g}) = \frac{a}{a^2 + \omega^2}, \quad \text{Im}(\bar{g}) = \frac{\omega}{a^2 + \omega^2}. \tag{1.141}$$

The amplitude spectrum is given by

$$\left| \bar{g}(\omega) \right| = \left| \frac{1}{a - i\omega} \right| = \frac{1}{|a - i\omega|} = \frac{1}{\sqrt{a^2 + \omega^2}} \tag{1.142}$$

and the phase spectrum is given by

$$\psi(\omega) = \tan^{-1}\left[\frac{\text{Im}(\bar{g})}{\text{Re}(\bar{g})} \right] = \tan^{-1}\left[\frac{\omega}{a} \right]. \tag{1.143}$$

Substituting (1.134a) into (1.133b) allows us to write (1.133b) as

$$g(t) = \frac{1}{2\pi} \int_{-\infty}^{\infty} |\bar{g}(\omega)| e^{-i(\omega t - \psi)} \, d\omega = \frac{1}{2\pi} \int_{-\infty}^{\infty} |\bar{g}(\omega)| e^{-i\omega(t - \psi/\omega)} \, d\omega,$$

which expresses $g(t)$ as a superposition of sinusoids of different frequencies, with the sinusoid of frequency ω having the amplitude $|\bar{g}(\omega)|$ and being delayed in time by an amount $\psi(\omega)/\omega$ or equivalently, having a phase delay of $\psi(\omega)$.

Equation (1.141) shows that $\mathrm{Re}(\bar{g}) > 0$ for both positive and negative frequencies, i.e., the phase angle ψ of $\bar{g}$ is always between $-90°$ and $+90°$, i.e., on the principal branch of the arctangent. Hence, there is no need to add or subtract $180°$ from any of the principal values of the arctangent.

Figure 1.12 shows a graph of the decaying exponential and its spectra.

The **convolution** $c(t)$ of two continuous signals $x(t)$ and $g(t)$ is defined as

$$c(t) = x(t) * g(t) \equiv \int_{-\infty}^{\infty} x(\tau)g(t-\tau)\,d\tau \; . \tag{1.144}$$

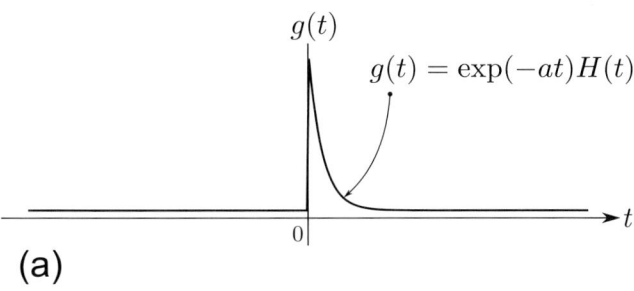

(a)

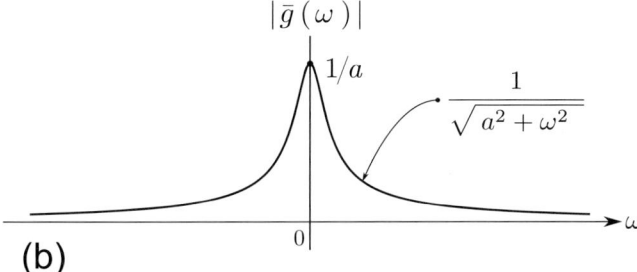

(b)

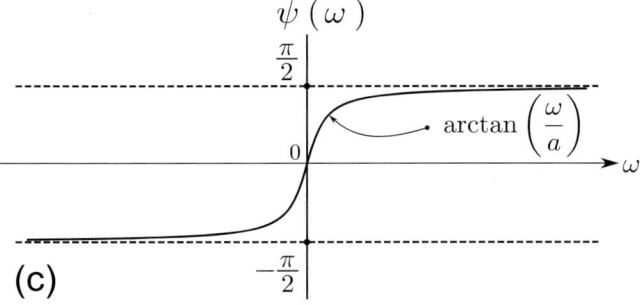

(c)

Figure 1.12 (a) The function $e^{-at}H(t)$, (b) its amplitude spectrum, and (c) its phase spectrum.

Convolution is used to model seismic traces. The Fourier transform of $c(t)$ is calculated as follows:

$$\bar{c}(\omega) = \int_{-\infty}^{\infty} x(t) * g(t)\, e^{i\omega t}\, dt = \int_{-\infty}^{\infty} \int_{-\infty}^{\infty} x(\tau) g(t - \tau) e^{i\omega t}\, d\tau\, dt$$

$$= \int_{-\infty}^{\infty} x(\tau) \left[\int_{-\infty}^{\infty} g(t - \tau) e^{i\omega t}\, dt \right] d\tau = \int_{-\infty}^{\infty} x(\tau) \left[\int_{-\infty}^{\infty} g(u) e^{i\omega u}\, du \right] e^{i\omega \tau}\, d\tau$$

$$\Rightarrow \quad \bar{c}(\omega) = \int_{-\infty}^{\infty} x(\tau) e^{i\omega \tau}\, d\tau \ \times \ \bar{g}(\omega) = \bar{x}(\omega)\, \bar{g}(\omega), \tag{1.145}$$

where the substitution $u = t - \tau$ was used in the fourth step. $\bar{c}(\omega) = \bar{x}(\omega)\bar{g}(\omega)$ states that the spectrum of the convolution of $x(t)$ and $g(t)$ is just the product of the spectra of $x(t)$ and $g(t)$.

Convolution of continuous time functions, one or both of which could be finite or semi-infinite in duration, is sometimes used in seismic wave modeling (see, e.g., Equation 2.95). Appendix 1B shows an example of how to calculate such convolutions.

From (1.133a), it can be seen that if $g(t)$ is a real-valued function of the time t, then

$$\bar{g}(-\omega) = \left[\bar{g}(\omega) \right]^*, \tag{1.146}$$

where the asterisk denotes the complex conjugate.

What does this say about the amplitude and phase spectra? Write $\bar{g}(\omega)$ as $A(\omega) \exp\left[i\psi(\omega) \right]$, where $A(\omega) = |\bar{g}(\omega)|$ is the amplitude spectrum and $\psi(\omega)$ is the phase spectrum. Equation (1.146) then gives, for real $g(t)$,

$$A(-\omega) e^{i\psi(-\omega)} = A(\omega) e^{-i\psi(\omega)}, \tag{1.147}$$

which implies

$$A(-\omega) = A(\omega), \quad \psi(-\omega) = -\psi(\omega), \tag{1.148}$$

i.e., if $g(t)$ is real, then its amplitude spectrum is an even function of frequency and its phase spectrum is an odd function of frequency. The decaying exponential previously discussed (Figure 1.12) is an example of this rule.

The Fourier space (x) transform pair is defined as follows:

$$\tilde{h}(\kappa) = \int_{-\infty}^{\infty} h(x) e^{-i\kappa x}\, dx \quad \text{and} \quad h(x) = \frac{1}{2\pi} \int_{-\infty}^{\infty} \tilde{h}(\kappa) e^{i\kappa x}\, d\kappa, \tag{1.149}$$

where κ is the angular wavenumber, i.e., the "spatial angular frequency." Note that the opposite sign convention for the exponential is used compared to (1.133). This is not necessary, but is convenient for the 2D analysis of seismic data (see Chapter 6).

Lastly, a useful rule is that concerning the Fourier transform of a derivative. Let $g_{,t} = dg/dt$ and $h_{,x} = dh/dx$, where the Fourier transforms of $g(t)$ and $h(x)$ are given in (1.133a) and (1.149), respectively. Then

$$\overline{g_{,t}}(\omega) = -i\omega \overline{g}(\omega) \qquad \text{and} \qquad \widetilde{h_{,x}}(\kappa) = i\kappa \widetilde{h}(\kappa). \tag{1.150}$$

The proof of the first equation goes as follows, and the proof of the second is similar. Using (1.133b),

$$g(t) = \frac{1}{2\pi} \int_{-\infty}^{\infty} \overline{g}(\omega) e^{-i\omega t} \, d\omega \quad \Rightarrow \quad \frac{dg(t)}{dt} = \frac{1}{2\pi} \int_{-\infty}^{\infty} [-i\omega \overline{g}(\omega)] \, e^{-i\omega t} \, d\omega, \tag{1.151}$$

which shows by definition that $[-i\omega \overline{g}(\omega)]$ is the Fourier time transform of dg/dt.

In words, the Fourier time transform of dg/dt is just $-i\omega$ times the Fourier time transform of $g(t)$, and a similar statement can be made for the Fourier space transform of dh/dx. These rules make it easy to calculate the Fourier transforms of derivatives of functions if one knows the Fourier transforms of the functions themselves.

1.17 Appendix 1A: General Rotation with Euler Angles

A rotation about an arbitrary axis (i.e., other than the x_1, x_2, or x_3 axes) can be described with **Euler angles**.

Let us say that a "positive" rotation means that if the thumb of the right hand points in the direction of the rotation axis (with the base of the thumb at the origin), then the fingers point in the direction in which the rotation angle is increasing. For example, for a positive rotation about the x axis, the y axis moves toward the z axis.

Consider a positive rotation about the x_3 axis, i.e., the z axis, by an angle ϕ. The rotation matrix for this rotation is

$$\mathbf{R}_1 = \begin{pmatrix} \cos\phi & \sin\phi & 0 \\ -\sin\phi & \cos\phi & 0 \\ 0 & 0 & 1 \end{pmatrix}. \tag{1.152}$$

Then, follow this with a positive rotation about the x' axis by an angle θ. The rotation matrix for this rotation is

$$\mathbf{R}_2 = \begin{pmatrix} 1 & 0 & 0 \\ 0 & \cos\theta & \sin\theta \\ 0 & -\sin\theta & \cos\theta \end{pmatrix}. \tag{1.153}$$

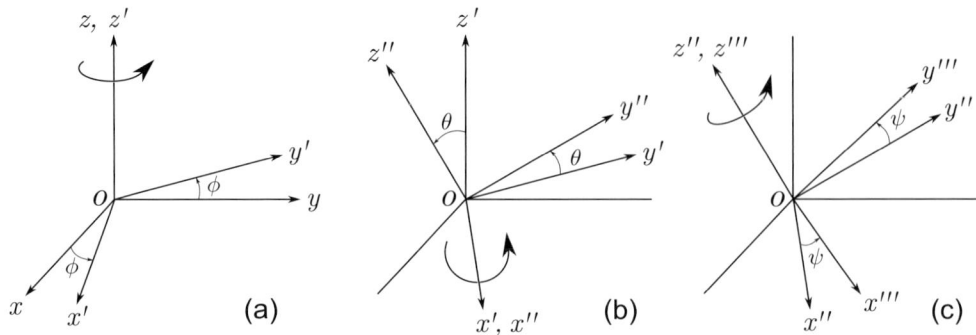

Figure 1.13 The Euler angles ϕ, θ, and ψ, showing rotations about the (a) z axis, (b) x' axis, and (c) z'' axis.

Lastly, follow this with a positive rotation about the z'' axis by an angle ψ. The rotation matrix for this rotation is

$$\mathbf{R}_3 = \begin{pmatrix} \cos\psi & \sin\psi & 0 \\ -\sin\psi & \cos\psi & 0 \\ 0 & 0 & 1 \end{pmatrix}. \tag{1.154}$$

The components of a vector $\mathbf{A}$ can then be expressed in the three rotated frames as follows:

$$\mathbf{A}' = \mathbf{R}_1\mathbf{A}. \tag{1.155}$$

$$\mathbf{A}'' = \mathbf{R}_2\mathbf{A}' = \mathbf{R}_2\mathbf{R}_1\mathbf{A}. \tag{1.156}$$

$$\mathbf{A}''' = \mathbf{R}_3\mathbf{A}'' = \mathbf{R}_3\mathbf{R}_2\mathbf{R}_1\mathbf{A} \equiv \mathbf{R}\mathbf{A} \tag{1.157}$$

$$\text{with} \quad \mathbf{R} \equiv \mathbf{R}_3\mathbf{R}_2\mathbf{R}_1 . \tag{1.158}$$

The rotation matrix $\mathbf{R}$ takes us directly from the unrotated frame to the final rotated $x'''y'''z'''$ frame.

The three angles ϕ, θ, and ψ (see Figure 1.13) are called the Euler angles. Any arbitrary rotation can be described with these three angles. Variations exist, i.e., instead of rotating about the x' axis in the second rotation, the y' axis could have been used, etc.

1.18 Appendix 1B: The Convolution of Continuous Functions

To convolve two time signals, one or both of which could be nonzero for only part of the time domain, we first review how to plot signals that are time shifted and/or reversed.

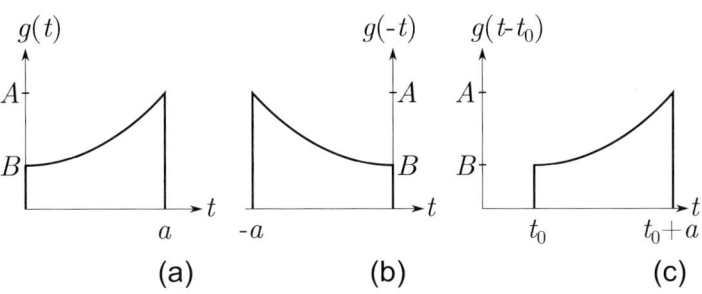

Figure 1.14 Plots of (a) $g(t)$, (b) $g(-t)$, and (c) $g(t - t_0)$.

1.18.1 *Reversing and Time Shifting of Signals*

In the following discussion, we will need to consider reversed and time-shifted signals. Suppose we are given a plot of the continuous signal $g(t)$ versus t. What do the plots of $g(-t)$, $g(t - t_0)$, $g(t_0 - t)$, etc., look like ? To answer this, consider the signal $g(t)$ in Figure 1.14a. The values $g(0) = B$ and $g(a) = A$ occur where t, the argument of $g(t)$, is zero and a respectively, i.e., at $t = 0$ and $t = a$. Hence, to plot $g(-t)$ versus t, for instance, we note that the same values B and A occur where $-t$ is zero and a respectively, i.e., at $t = 0$ and $t = -a$ (see Figure 1.14b). Similarly, to plot $g(t - t_0)$ versus t, we note that the values B and A occur where $t - t_0$ is zero and a respectively, i.e., at $t = t_0$ and $t = t_0 + a$ (see Figure 1.14c).

The same method may be used to plot $g(t_0 - t)$, $g(t + t_0)$, etc.

1.18.2 *Convolution of Continuous Signals*

To compute the convolution of two continuous signals $x(t)$ and $g(t)$, we must evaluate the following integral:

$$y(t) = x(t) * g(t) \equiv \int_{-\infty}^{\infty} x(\tau)g(t - \tau)\, d\tau. \tag{1.159}$$

Note that the integration variable is τ, i.e., the integrand is a function of τ, and that t is a "constant" parameter in the integrand. We must evaluate the integral for all possible values of the "constant" t, i.e., typically $-\infty < t < +\infty$. The following example will demonstrate how this is done. Consider the signals

$$x(t) = \begin{cases} e^{-t}, & \text{if } 0 \leq t \leq 1; \\ 0, & \text{otherwise.} \end{cases} \quad \text{and} \quad g(t) = \begin{cases} 2e^{-2t}, & \text{if } 0 \leq t < \infty; \\ 0, & \text{otherwise.} \end{cases} \tag{1.160}$$

The signals $x(t)$ and $g(t)$, and the factors $x(\tau)$ and $g(t - \tau)$ of the convolution integrand, are shown in Figure 1.15.

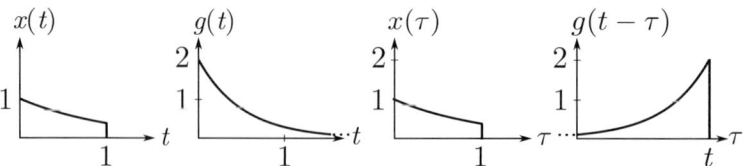

Figure 1.15 $x(t)$ and $g(t)$ versus t (see Equation 1.160) and $x(\tau)$ and $g(t - \tau)$ versus τ. $g(t - \tau)$ is shown for some positive value of t.

As t takes on all values from $-\infty$ to $+\infty$, Figure 1.15 shows that $g(t - \tau)$ slides along the τ axis from the negative end to the positive end. For some values of t, the nonzero parts of the integrand factors $x(\tau)$ and $g(t - \tau)$ will overlap. For $t \leq 0$, Figure 1.16a shows that there is no such overlap between $x(\tau)$ and $g(t - \tau)$, hence the integrand of Equation (1.159) is zero, which makes $y(t) = 0$. As t becomes positive, the nonzero parts of $x(\tau)$ and $g(t - \tau)$ begin to overlap. Figure 1.16b shows that the overlap zone is $0 \leq \tau \leq t$, and that this zone exists for $0 \leq t \leq 1$. The integrand $x(\tau)g(t - \tau)$ is nonzero for values of τ inside this zone, and zero for values of τ outside of it. Hence,

$$y(t) = \int_{-\infty}^{\infty} x(\tau)g(t - \tau)\,d\tau = \int_{0}^{t} e^{-\tau}\big[2e^{-2(t-\tau)}\big]\,d\tau = 2e^{-2t}\int_{0}^{t} e^{\tau}\,d\tau \quad \Longrightarrow$$
$$y(t) = 2e^{-2t}\big(e^{t} - 1\big) = 2\big(e^{-t} - e^{-2t}\big). \tag{1.161}$$

As we slide t past 1 on the τ axis, Figure 1.16c shows that the overlap zone changes to $0 \leq \tau \leq 1$, and that this zone exists for the remaining values of t, i.e., $t \geq 1$. Hence

$$y(t) = \int_{-\infty}^{\infty} x(\tau)g(t - \tau)\,d\tau = \int_{0}^{1} e^{-\tau}\big[2e^{-2(t-\tau)}\big]\,d\tau = 2e^{-2t}\int_{0}^{1} e^{\tau}\,d\tau \quad \Longrightarrow$$
$$y(t) = 2(e - 1)e^{-2t}. \tag{1.162}$$

To summarize, the result of the convolution is

$$y(t) = x(t) * g(t) = \begin{cases} 0, & \text{if } t \leq 0; \\ 2\big(e^{-t} - e^{-2t}\big), & \text{if } 0 \leq t \leq 1; \\ 2(e - 1)e^{-2t}, & \text{if } t \geq 1. \end{cases} \tag{1.163}$$

$y(t)$ is shown in Figure 1.16d.

In the preceding example, three distinct ranges of t had to be treated individually. If the infinitely long tail of $g(t)$ would be cut off at some point, then we would find that five distinct ranges of t would have to be treated.

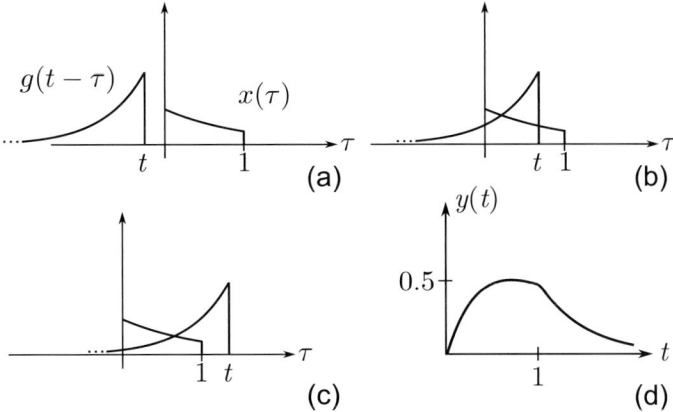

Figure 1.16 The factors of the integrand of Equation (1.159) (see Figure 1.14) for (a) $t \leq 0$, (b) $0 \leq t \leq 1$, and (c) $t \geq 1$. (d) $y(t)$ is the result of the convolution of $x(t)$ and $g(t)$.

Exercises

1. Given $\mathbf{A} = \mathbf{e}_x + 2\mathbf{e}_y + 3\mathbf{e}_z$ and $\mathbf{B} = 2\mathbf{e}_x - 5\mathbf{e}_y + 4\mathbf{e}_z$, find $\mathbf{A} \cdot \mathbf{B}$ and $\mathbf{A} \times \mathbf{B}$. Show that $\mathbf{A} \times \mathbf{B}$ is perpendicular to both $\mathbf{A}$ and $\mathbf{B}$. Find the angle between $\mathbf{A}$ and $\mathbf{B}$.

2. Given $\psi = xy^2 z - x^2$ and $\mathbf{A} = 3xyz^2\mathbf{e}_x + 4xy^3\mathbf{e}_y - x^2yz\mathbf{e}_z$, find $\nabla\psi$, $\nabla \cdot \mathbf{A}$, and $\nabla \times \mathbf{A}$. Verify the rules $\nabla \times \nabla\psi = 0$ and $\nabla \cdot (\nabla \times \mathbf{A}) = 0$ and $\nabla^2\mathbf{A} = \nabla(\nabla \cdot \mathbf{A}) - \nabla \times (\nabla \times \mathbf{A})$ (where $\nabla^2\mathbf{A} \equiv (\nabla^2 A_x)\mathbf{e}_x + (\nabla^2 A_y)\mathbf{e}_y + (\nabla^2 A_z)\mathbf{e}_z$).

3. Consider the following statement: "Equation (1.22), which states that $\mathbf{A} \cdot d\mathbf{A}/dt = A\, dA/dt$, implies that $\mathbf{A}$ and $d\mathbf{A}/dt$ are parallel (see the sentence that precedes Equation 1.6)". The statement cannot, however, be true. For instance, if $\mathbf{A}$ is the position vector $\mathbf{r}$ of a moving particle, then

$$\frac{d\mathbf{A}}{dt} = \frac{d\mathbf{r}}{dt} = \mathbf{v} = \text{the velocity of the particle}$$

As Figure 1.17 shows, $\mathbf{r}$ and $d\mathbf{r}/dt$ need not be parallel. Find the flaw in the statement.

4. Derive Equation (1.36).

5. Verify Gauss' divergence theorem for $\mathbf{A} = y^2 z\mathbf{e}_x - 2x^3 y\mathbf{e}_y + xyz^2\mathbf{e}_z$ and the volume V being the unit cube shown in Figure 1.18.

6. Verify Stokes' theorem for $\mathbf{A} = (y + 2z)\mathbf{e}_x + (yz - 2x^2 y)\mathbf{e}_y + (xyz + x^2)\mathbf{e}_z$ and the closed curve C being the square in Figure 1.19.

7. Use Equations (1.62) through (1.65) to write down the formulas for $\nabla\psi$, $\nabla \cdot \mathbf{A}$, $\nabla \times \mathbf{A}$, and $\nabla^2\psi$ in cylindrical coordinates.

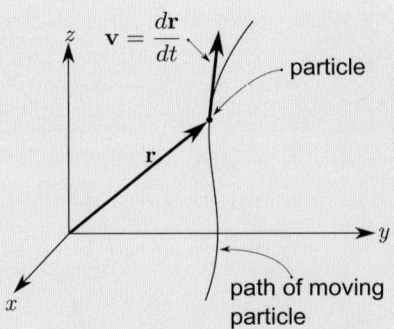

Figure 1.17 See exercise 3.

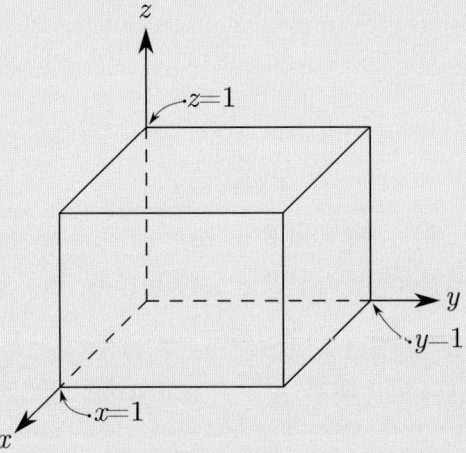

Figure 1.18 The volume V for exercise 5.

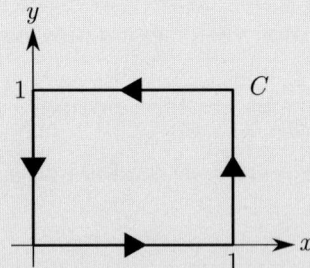

Figure 1.19 The curve C for exercise 6.

8. Use Equations (1.62) through (1.65) to write down the formulas for $\nabla\psi$, $\nabla \cdot \mathbf{A}$, $\nabla \times \mathbf{A}$, and $\nabla^2 \psi$ in spherical coordinates.

9. Assume the Earth is a perfect sphere whose radius is 6,378 km. The province of Saskatchewan in Canada is approximately bordered by the 102° and

110° lines of longitude and the 49° and 60° lines of latitude. What is the approximate surface area of Saskatchewan in square kilometers? (Use spherical coordinates.)

10. Spherical coordinates:

    (a) Obtain the formulas for $\mathbf{e}_r$, $\mathbf{e}_\theta$, and $\mathbf{e}_\phi$ in terms of $\mathbf{e}_x$, $\mathbf{e}_y$, and $\mathbf{e}_z$, and vice versa.
    (b) If $\mathbf{A} = z\mathbf{e}_x - 3y\mathbf{e}_y + 2x\mathbf{e}_z = A_r\mathbf{e}_r + A_\theta\mathbf{e}_\theta + A_\phi\mathbf{e}_\phi$, work out the formulas for A_r, A_θ, and A_ϕ (in terms of r, θ, and ϕ).

11. The mass per unit surface area of a hollow hemispherical dome is $\rho = \rho_0 + a\theta$, where ρ_0 and a are constants, and θ is the second spherical coordinate. Derive a formula for the mass of the dome. The radius of the dome is R.

12. An oblate spheroid is a sphere that is flattened at the poles, i.e., its polar radius R_p is less than its equatorial radius R_e. It is described by the equation

$$\left(\frac{x}{R_e}\right)^2 + \left(\frac{y}{R_e}\right)^2 + \left(\frac{z}{R_p}\right)^2 = 1$$

if the poles are on the z axis. Due to centrifugal force (from rotation), the Earth is approximately an oblate spheroid. The transformation equations for *oblate spheroidal coordinates* are

$$
\begin{aligned}
x &= a \cosh u \cos v \cos\phi \\
y &= a \cosh u \cos v \sin\phi \\
z &= a \sinh u \sin v
\end{aligned}
\qquad
\begin{aligned}
&u \geq 0 \\
&-\tfrac{\pi}{2} \leq v \leq \tfrac{\pi}{2} \\
&0 \leq \phi \leq 2\pi \\
&a = \text{constant}
\end{aligned}
$$

u, v, and ϕ form an orthogonal coordinate system.

    (a) Show that the coordinate surface $u = constant$ is an oblate spheroid. What are R_e and R_p in terms of a and u?
    (b) Show that the scale factors h_j are given by

$$h_1 = h_u = a\sqrt{\sinh^2 u + \sin^2 v} = h_2 = h_v\,, \quad h_3 = h_\phi = a\cosh u \cos v$$

    You may need the rule $\cosh^2 x - \sinh^2 x = 1$.
    (c) Obtain formulas for the volume and surface area of an oblate spheroid in terms of R_e and R_p (use oblate spheroidal coordinates to do the volume and surface integrals). For the surface integral, you will need the rule:

$$\int \sqrt{1 + p^2 \sin^2 x} \, \cos x \, dx = \frac{1}{2} \sqrt{1 + p^2 \sin^2 x} \, \sin x$$
$$+ \frac{1}{2p} \ln \left(p \sin x + \sqrt{1 + p^2 \sin^2 x} \right),$$

where p is a constant.

(d) Apply your formulas to the Earth. Assume $R_e = 6{,}378$ km and $R_p = 6{,}357$ km. Compare your answers to what you would get if you assumed the Earth was a sphere with radius R_e. Make sure all your numerical calculations are good to four significant digits (don't round off any numbers during intermediate steps in the calculations) so that your comparisons are accurate.

13. (a) Write Equations (1.86) and (1.87) out in full for the special case of a rotation of coordinates through an angle ϕ about the z axis.

   (b) What are they for $\phi = 90°$?

14. Show that the rotation matrix for rotation about the z axis satisfies the orthogonality condition.

15. $\mathbf{A} \times \mathbf{B}$ is invariant under a rotation of coordinates, i.e.,

$$(A_2' B_3' - A_3' B_2')\mathbf{e}_1' + (A_3' B_1' - A_1' B_3')\mathbf{e}_2' + (A_1' B_2' - A_2' B_1')\mathbf{e}_3'$$

$$= (A_2 B_3 - A_3 B_2)\mathbf{e}_1 + (A_3 B_1 - A_1 B_3)\mathbf{e}_2 + (A_1 B_2 - A_2 B_1)\mathbf{e}_3.$$

Verify this equation for the special case of a 90° rotation of coordinates about the z axis (a general proof of the invariance of $\mathbf{A} \times \mathbf{B}$ can be produced using the permutation symbol ε_{ijk} — see exercise 21).

16. Show that $\nabla^2 \psi$ is invariant under a rotation of coordinates.

17. A new vector product is defined as $\mathbf{A} \tilde{\times} \mathbf{B} \equiv A_1 B_3 \mathbf{e}_1 + A_2 B_2 \mathbf{e}_2 + A_3 B_1 \mathbf{e}_3$. Show that $\mathbf{A} \tilde{\times} \mathbf{B}$ is not really a vector. (Hint: show that $\mathbf{A} \tilde{\times} \mathbf{B}$ is not invariant under a 90° rotation of coordinates about the z axis).

18. Say that someone proposes the following definition for vector "division":

$$\mathbf{A} \div \mathbf{B} \equiv \frac{A_1}{B_1}\mathbf{e}_1 + \frac{A_2}{B_2}\mathbf{e}_2 + \frac{A_3}{B_3}\mathbf{e}_3.$$

Show that $\mathbf{A} \div \mathbf{B}$ is not really a vector. (Hint: show that $\mathbf{A} \div \mathbf{B}$ is not invariant under a 90° rotation of coordinates about the z axis).

19. Sometimes, for special purposes, the following much more restrictive definition of scalars and vectors is used:

$$\psi(x_1', x_2', x_3') = \psi(x_1, x_2, x_3) \quad \text{and}$$

$$\sum_{j=1}^{3} A_j(x_1', x_2', x_3')\mathbf{e}_j' = \sum_{j=1}^{3} A_j(x_1, x_2, x_3)\mathbf{e}_j,$$

i.e., under a rotation of coordinates, ψ' is the same function of x_i' as ψ is of x_i, and A_j' is the same function of x_i' as A_j is of x_i. For example

$$\psi = x_1^2 + x_2^2 + x_3^2 = x_1'^2 + x_2'^2 + x_3'^2 = \psi'.$$

For a rotation about the z axis, determine whether or not $-x_2\mathbf{e}_1 + x_1\mathbf{e}_2$ and $x_1\mathbf{e}_1 - x_2\mathbf{e}_2$ are vectors under this restrictive definition, i.e., does $-x_2'\mathbf{e}_1' + x_1'\mathbf{e}_2' = -x_2\mathbf{e}_1 + x_1\mathbf{e}_2$, etc.? (Do exercise 13 first.)

20. A new scalar product is defined as $\mathbf{A} \circ \mathbf{B} \equiv A_1B_2 + A_2B_3 + A_3B_1$. Show that $\mathbf{A} \circ \mathbf{B}$ is not really a scalar. (Hint: show that $\mathbf{A} \circ \mathbf{B}$ is not invariant under a $90°$ rotation of coordinates about the z axis.)

21. Prove that the cross product $\mathbf{A} \times \mathbf{B}$ is an invariant quantity, and hence a true vector. (Hint: use the permutation symbol ε_{ijk}.)

22. The displacement of a geophone at a given time, due to a P wave striking it, is $\mathbf{u} = \mathbf{e}_x - 2\mathbf{e}_z$ in arbitrary units (the z axis points down and the x axis points to the right). The geophone records both the x and z components of displacement. In other words, two traces are produced, one for u_x and one for u_z. The P wave pulse appears on both of them – on the trace for u_x, it has an amplitude of 1, and on the trace for u_z, it has an amplitude of 2 (with negative polarity). It is desired to "rotate" the data so that the P wave appears only on the trace for u_z. What is the rotation matrix needed to accomplish this? What is the amplitude of the pulse on the trace for u_z after rotation?

23. Calculate the Fourier transforms of the functions $g(t)$ that follow, defined for $-\infty < t < \infty$, as well as their amplitude and phase spectra. In the following formulas, ω_0 and a are positive real constants, and "sgn(t)" is the sign of t (i.e., ± 1). In each case, sketch or plot graphs of $g(t)$ and its amplitude and phase spectra.

   (a) $g(t) = \sin(\omega_0 t)$.
   (b) $g(t) = e^{-a|t|}$.
   (c) $g(t) = \text{sgn}(t)e^{-a|t|}$.

24. Prove that $\nabla\psi$ points in the direction in which ψ is *increasing* most rapidly.

25. Verify Gauss' divergence theorem for $\mathbf{A} = (2r\cos\theta)\mathbf{e}_r + (5r\sin\theta)\mathbf{e}_\theta$, where r and θ are spherical coordinates (see Section 1.8), and where the volume V is a sphere of radius R.

26. Verify Gauss' divergence theorem for $\mathbf{A} = (2r)\mathbf{e}_r + (5r\sin\theta)\mathbf{e}_\theta$, where r and θ are spherical coordinates (see Section 1.8), and where the volume V is a sphere of radius R.

27. Calculate the Fourier transforms of the functions $g(t)$ that follow, defined for $-\infty < t < \infty$, as well as their amplitude and phase spectra. In the following formulas, ω_0 and a are positive real constants. In each case, sketch or plot graphs of $g(t)$ and its amplitude and phase spectra.

    (a) $g(t) = \{\sin(\omega_0 t),\ 0 \leq t \leq \pi/\omega_0;\ 0,\ \text{otherwise.}\}$
    (b) $g(t) = e^{-a|t|}\sin(\omega_0 t)$.
    (c) $g(t) = \{e^{-at}\sin(\omega_0 t),\ t \geq 0;\ 0,\ t < 0.\}$

28. Prove that if $u(t)$ is a real function (e.g., a seismic trace), then

$$u(t) = \frac{1}{2\pi}\int_{-\infty}^{\infty}\bar{u}(\omega)e^{-i\omega t}\,d\omega = \frac{1}{\pi}\text{Re}\int_{0}^{\infty}\bar{u}(\omega)e^{-i\omega t}\,d\omega.$$

    The rightmost expression for $u(t)$ eliminates the need to deal with negative frequencies.

29. Suppose that the frequency spectrum of a certain waveform $g(t)$ is given by

$$\bar{g}(\omega) = \frac{1 + \exp(i\pi\omega/\omega_0)}{\omega^2 - \omega_0^2},$$

    where ω_0 is a constant.

    (a) What are the values of the amplitude and phase spectra at $\omega = \omega_0/2$?
    (b) What are the values of the amplitude and phase spectra at $\omega = \omega_0$?

30. The residue theorem from complex variables is sometimes used in seismic wave theory to calculate integrals, such as those arising in Fourier transforms. Review the residue theorem, then use it to calculate the Fourier transform $\bar{g}(\omega)$ (i.e., the frequency spectrum) of $g(t) = t/(t^2 + c^2)$, where c is a positive real constant. Obtain $\bar{g}(\omega)$ for all frequencies (positive, negative, and zero). Derive the formulas for the amplitude and phase spectra, and sketch or plot graphs of them versus ω.

31. Convolve the signals $x(t)$ and $g(t)$ given in the following, and sketch or plot a graph of your result.

$$x(t) = \begin{cases} \sin t, & \text{if } 0 \leq t \leq \pi \\ 0, & \text{otherwise.} \end{cases}\ ,\qquad g(t) = \begin{cases} 1, & \text{if } 0 \leq t \leq \pi \\ 0, & \text{otherwise.} \end{cases}.$$

2

Stress, Strain, and Seismic Waves

In this chapter, we discuss the concepts of stress and strain in the theory of elasticity and the relation between them. We also discuss seismic waves and the conservation of energy. More careful, rigorous, or detailed discussions of these topics can be found in Achenbach (1973); Aki and Richards (1980, 2002); Båth and Berkhout (1984); Ben-Menahem and Singh (2000); Brekhovskikh (1980); Bullen and Bolt (1985); Carcione (2007); Červený (2001); Chapman (2004); Dahlen and Tromp (1998); Elmore and Heald (1969); Ewing, Jardetzky, and Press (1957); Fung (1965); Hudson (1980); Kennett (2001, 2002); Kennett and Bunge (2018); Lay and Wallace (1995); Pujol (2003); Slawinski (2015); Shearer (2009); Stein and Wysession (2003); Udías and Buforn (2018); among others, as well as in books on continuum mechanics.

2.1 Stress

Consider a volume V inside an infinite medium – see Figure 2.1a. Let S represent the surface of the volume and dS an element of surface area. Let $\mathbf{n}$ be the unit vector that is normal to dS. Consider an element of force $d\mathbf{F}$ acting on dS in an arbitrary direction, due to a deformation of the medium. The deformation could be due to gravitational compaction or compression, tectonic forces, elastic wave motion, etc., or a combination of such forces. Then we define the **traction T** as the force per unit surface area: $\mathbf{T} = d\mathbf{F}/dS$.

The traction basically represents a contact force between the material particles on the two sides of the surface. More precisely, it is the force per unit area exerted by the material on the positive side of dS (the side to which $\mathbf{n}$ is pointing) on the material on the negative side. Sometimes $\mathbf{T}$ is called the **stress vector**.

In general, $\mathbf{T}$ varies with position on the surface S, i.e., $\mathbf{T} = \mathbf{T}(\mathbf{n}) = \mathbf{T}(n_x, n_y, n_z)$ (since $\mathbf{n}$ also varies across the surface). But precisely how does $\mathbf{T}$ vary with n_x, n_y, and n_z? The simplest relationship between the components of $\mathbf{T}$ and the components

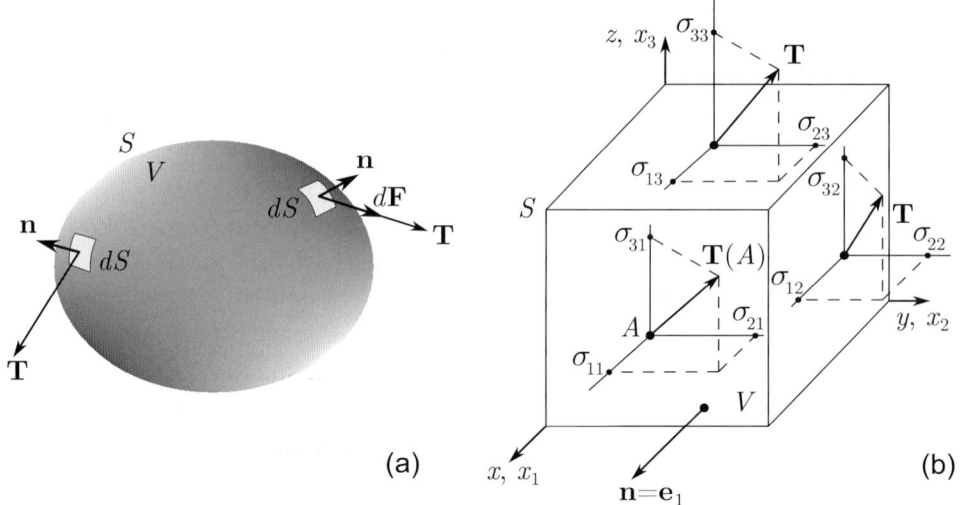

Figure 2.1 The traction (a) and its components (b).

of $\mathbf{n}$ would be a linear one. In fact, it can be shown (see texts on continuum mechanics) that the components of $\mathbf{T}$ can each be written as a linear combination of the components of $\mathbf{n}$, e.g., $T_x = an_x + bn_y + cn_z$, where a, b, and c are parameters whose values vary with location in space. As we have three components of traction, we will need nine such parameters altogether. If we denote them by σ_{ij}, $i,j = 1,2,3$, then we have, using numerical subscripts,

$$T_i = \sum_{j=1}^{3} \sigma_{ij}n_j, \qquad i = 1,2,3. \tag{2.1}$$

By the quotient rule, explained in Chapter 1, the nine parameters σ_{ij} form the components of a Cartesian tensor of rank 2 known as the **stress tensor**.

Consider Figure 2.1a again. By Newton's third law, the particles on the negative side of dS exert an equal and opposite force on the particles on the positive side, i.e., $\mathbf{T}(-\mathbf{n}) = -\mathbf{T}(\mathbf{n})$. This can also be shown by applying the equation of motion (2.10) to a thin disk whose volume is allowed to shrink to zero (see, e.g., Aki and Richards, 2002, and Hudson, 1980).

If the medium is a fluid undergoing deformation, then $\mathbf{T}$ is due to a pressure p, normal to S, produced by the deformation. By convention, a positive value of p is associated with an inward pressure on S, i.e., $\mathbf{T} = -p\mathbf{n}$. If the medium is a solid, the **tension** is $\mathbf{T} \cdot \mathbf{n}$ and the pressure is $-\mathbf{T} \cdot \mathbf{n}$.

Consider an infinitesimally small box in a medium under stress (see Figure 2.1b). Apply Equation (2.1) to point A on the face normal to the x axis. On this face, $\mathbf{n} = \mathbf{e}_1 = (1,0,0)$, so we obtain, for the three components of traction at A,

$$T_1 = \sigma_{11}, \quad T_2 = \sigma_{21}, \quad T_3 = \sigma_{31}.$$

The traction components for the other faces can be obtained in the same way. We see that in the quantity "σ_{ij}", the first index "i" denotes the direction of the traction component, and the second index "j" denotes the direction of the normal to the face on which the traction is acting. In general, we may write $\sigma_{ij} = T_i(\mathbf{e}_j)$. *A note of caution*: many texts use the opposite convention, i.e., T_i is defined as $\sum_{j=1}^{3} \sigma_{ji} n_j$, meaning that the first component refers to the face and the second to the direction. Practically, however, it makes no difference which convention is used, because the stress tensor is symmetric, i.e., $\sigma_{ij} = \sigma_{ji}$, as we will see later.

The components σ_{11}, σ_{22}, and σ_{33} are called **normal** stresses (because they are normal to the faces) and the components σ_{12}, σ_{13}, σ_{21}, σ_{23}, σ_{31}, and σ_{32} are called **shear** stresses (because they are tangential to the faces). There are nine components σ_{ij}, but only six are independent. This is because the stress tensor is symmetric, i.e.,

$$\sigma_{12} = \sigma_{21}, \quad \sigma_{13} = \sigma_{31}, \quad \text{and} \quad \sigma_{23} = \sigma_{32}. \tag{2.2}$$

If this were not so, the box in Figure 2.1b would be experiencing angular accelerations. For example, if $\sigma_{12} \neq \sigma_{21}$, the box would rotate faster and faster about the z axis.

For a fluid, $\mathbf{T} = -p\mathbf{n}$ implies

$$T_i = \sum_j \sigma_{ij} n_j = -p n_i = -\sum_j p \delta_{ij} n_j \quad \Longrightarrow \quad \sum_j (\sigma_{ij} + p\delta_{ij}) n_j = 0,$$

where δ_{ij} is the Kronecker delta, i.e., $\delta_{ij} = 1$ for $i = j$ and 0 for $i \neq j$. This must be true for all $\mathbf{n}$ (which is nonzero), hence, $\sigma_{ij} = -p\delta_{ij}$ for a fluid (this is similar to stating that if $\mathbf{a} \cdot \mathbf{b} = \sum_j a_j b_j = 0$ for *all* vectors $\mathbf{b}$, then $\mathbf{a}$ must be zero). In words, the normal stresses in a fluid are all equal to $-p$, and the shear stresses are all zero.

Equation (2.1), $T_i = \sum_j \sigma_{ij} n_j$, is also recognizable as the formula for the multiplication of a matrix $\boldsymbol{\sigma}$, whose elements are σ_{ij}, with a column vector $\mathbf{n}$, whose elements are n_j, i.e., $\mathbf{T} = \boldsymbol{\sigma}\mathbf{n}$, or

$$\begin{bmatrix} T_1 \\ T_2 \\ T_3 \end{bmatrix} = \begin{bmatrix} \sigma_{11} & \sigma_{12} & \sigma_{13} \\ \sigma_{21} & \sigma_{22} & \sigma_{23} \\ \sigma_{31} & \sigma_{32} & \sigma_{33} \end{bmatrix} \begin{bmatrix} n_1 \\ n_2 \\ n_3 \end{bmatrix}.$$

In any solid, one can find a set of orthogonal planes across which the shear stresses are zero. Hence, the traction across any one of these orthogonal planes is parallel to $\mathbf{n}$ for the plane, i.e., $\mathbf{T} = \boldsymbol{\sigma}\mathbf{n} = \tau\mathbf{n}$. This is an eigenvalue equation. The three eigenvalues τ_1, τ_2, and τ_3 are the **principal stresses** and the three corresponding orthogonal eigenvectors $\mathbf{n}_1$, $\mathbf{n}_2$, and $\mathbf{n}_3$ are parallel to the **principal axes**. The planes normal to these eigenvectors are the **principal planes** across which there are no shear stresses. In the Cartesian coordinate system with $\mathbf{n}_1$, $\mathbf{n}_2$, and $\mathbf{n}_3$ as the basis unit vectors, one has $\sigma_{11} = \tau_1$, $\sigma_{22} = \tau_2$, $\sigma_{33} = \tau_3$, and $\sigma_{ij} = 0$ for $i \neq j$, i.e., σ is a diagonal matrix in the principal axis frame of reference.

From (1.113c), if a rotation of coordinates is performed, specified by the rotation matrix $\mathbf{R}$, then the components σ' of the stress tensor in the rotated frame can be obtained from $\sigma' = \mathbf{R}\sigma\mathbf{R}^T$.

2.2 Strain

Consider two infinitesimally close points, P and Q, in a medium (see Figure 2.2). Point P is located at $\mathbf{x} = (x_1, x_2, x_3)$ and point Q at $\mathbf{x} + d\mathbf{x}$. Under stress, P undergoes a displacement $\mathbf{u}$ to P', and Q undergoes a displacement $\mathbf{u} + d\mathbf{u}$ to Q'. Mathematically, the displacement of point P is $\mathbf{u} = \mathbf{u}(x_1, x_2, x_3) \equiv \mathbf{u}(\mathbf{x})$, and the displacement of point Q is $\mathbf{u}(\mathbf{x} + d\mathbf{x}) = \mathbf{u}(\mathbf{x}) + d\mathbf{u}(\mathbf{x})$, where

$$du_i(\mathbf{x}) = \nabla u_i \cdot d\mathbf{x} = \sum_{j=1}^{3} \frac{\partial u_i}{\partial x_j} dx_j, \quad i = 1, 2, 3. \tag{2.3}$$

du_i can be obtained from the 3D Taylor series formula for $\mathbf{u}(\mathbf{x}+d\mathbf{x})$ (with terms of second order and higher dropped), or from the definition of a differential. In general, $\mathbf{u}$ is a function of the time t also (e.g., when strain is produced by wave motion), but for our purposes here, we can consider, without loss of generality, a static strain. We can rewrite du_i as

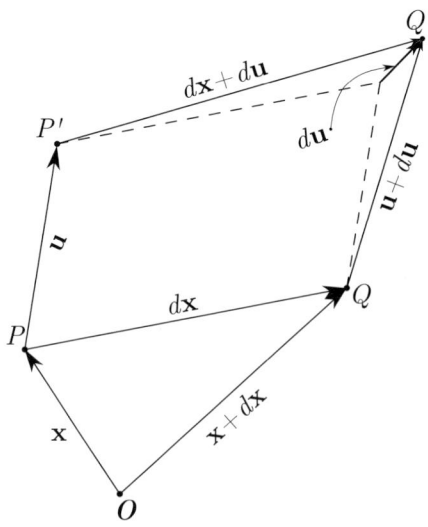

Figure 2.2 Strain in 3D.

$$du_i - \sum_j (e_{ij} - \eta_{ij})\, dx_j, \quad \text{where} \quad e_{ij} = \frac{1}{2}\left(\frac{\partial u_i}{\partial x_j} + \frac{\partial u_j}{\partial x_i}\right) \qquad (2.4\text{a,b})$$

$$\text{and} \quad \eta_{ij} = \frac{1}{2}\left(\frac{\partial u_j}{\partial x_i} - \frac{\partial u_i}{\partial x_j}\right). \qquad (2.4\text{c})$$

Note that e_{ij} is symmetric, i.e., $e_{ij} = e_{ji}$, whereas η_{ij} is antisymmetric, i.e., $\eta_{ij} = -\eta_{ji} \cdot e_{ij}$ is a Cartesian tensor of rank 2 and is called the **strain** tensor. e_{11}, e_{22}, and e_{33} are called the **normal** strains, and e_{12}, e_{13}, e_{21}, e_{23}, e_{31}, and e_{32} are called the **shear** strains. Like the stress tensor, the strain tensor has only six independent components, since $e_{12} = e_{21}$, $e_{13} = e_{31}$, and $e_{23} = e_{32}$.

The normal strains, sometimes called **longitudinal** strains, have simpler formulas than the shear strains. For example, from the preceding formulas, e_{11} and e_{12} are

$$e_{11} = \frac{\partial u_1}{\partial x_1} \quad \text{and} \quad e_{12} = \frac{1}{2}\left(\frac{\partial u_1}{\partial x_2} + \frac{\partial u_2}{\partial x_1}\right).$$

What are the physical meanings of the normal and shear strains?

The normal strain e_{11} is the relative change, due to strain, in the x_1 component of the vector distance between two infinitesimally close points that both lie on the x_1 axis prior to being strained. To see this, imagine that points P and Q in Figure 2.2 both lie on the x_1 axis. Then $dx_2 = dx_3 = 0$, and the change in the x_1 component of the vector distance between them is

$$(dx_1 + du_1) - (dx_1) = du_1 = \frac{\partial u_1}{\partial x_1} dx_1, \qquad (2.5)$$

where Equation (2.3) has been used for du_1. Dividing by their original separation, dx_1, to obtain the relative change, gives the normal strain e_{11}. Similarly, the normal strain e_{22} is the relative change in the x_2 component of the vector distance between two infinitesimally close points that both lie on the x_2 axis prior to being strained. A similar statement can be made for e_{33}.

The shear strain e_{12} can be understood by referring to the strained box in Figure 2.3. This is a box whose lower-left edge is fixed in place. The strain is only in the $x_1 x_2$ plane (there is no displacement in the x_3 direction, and all variables are independent of x_3) – this is known as the case of **plane strain**. Since P is fixed, its displacement $\mathbf{u}$ is zero, meaning that the displacement of Q_1 is $\mathbf{u} + d\mathbf{u} = d\mathbf{u}$ evaluated for Q_1, and similarly the displacement of Q_2 is $d\mathbf{u}$ evaluated for Q_2. The shear strain e_{12} is then simply the average of the (small) angles γ_1 and γ_2 (a natural measure of the shear strain in this case). To see this, we first remember that $\tan \gamma \approx \gamma$ for small angles γ. Referring to Figure 2.3, we then obtain

$$\frac{1}{2}(\gamma_2 + \gamma_1) \approx \frac{1}{2}(\tan \gamma_2 + \tan \gamma_1) \approx \frac{1}{2}\left(\frac{du_1}{dx_2} + \frac{du_2}{dx_1}\right)$$

$$= \frac{1}{2}\left(\frac{\partial u_1}{\partial x_2} + \frac{\partial u_2}{\partial x_1}\right) = e_{12}, \qquad (2.6)$$

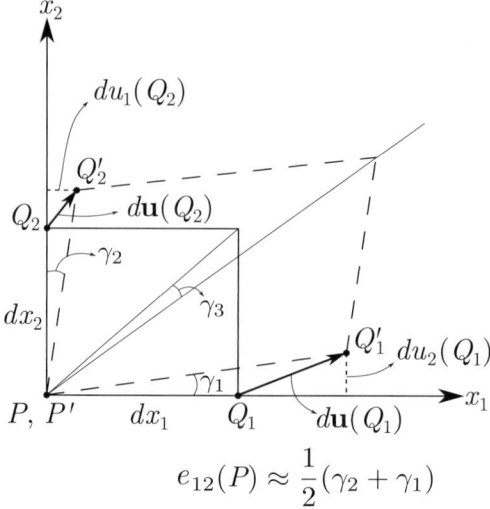

$$e_{12}(P) \approx \frac{1}{2}(\gamma_2 + \gamma_1)$$

Figure 2.3 Strain in 2D (the *xy* plane). The edge at *P* is fixed, so $\mathbf{u} = \mathbf{0}$.

where Equation (2.3) was used to obtain $du_1 = (\partial u_1/\partial x_2)\, dx_2$ (since $dx_1 = dx_3 = 0$ for Q_2) and $du_2 = (\partial u_2/\partial x_1)\, dx_1$ (since $dx_2 = dx_3 = 0$ for Q_1). It is straightforward to extend the analysis to the more general case of plane strain in which the lower-left edge is not fixed in place. The same results are obtained (Equation 2.6).

A strain involves a change of shape, i.e., a change of volume.

As for η_{ij}, it is not difficult to show that

$$\sum_j \eta_{ij}dx_j = i\text{th component of } \frac{1}{2}\, d\mathbf{x} \times (\nabla \times \mathbf{u}). \tag{2.7}$$

From Equations (1.35) and (1.36), we know that this describes a rotation of $d\mathbf{x}$ with the rotation vector being given by $\frac{1}{2}\nabla \times \mathbf{u}$ (note that the line between P and Q in Figure 2.2 has not only been stretched, but also rotated). Hence, η_{ij} is a **rotation** tensor. Equation (2.4c) implies that $\eta_{11} = \eta_{22} = \eta_{33} = 0$, and that η_{ij} is antisymmetric. Note also that η_{23}, η_{31}, and η_{12} (or $-\eta_{32}$, $-\eta_{13}$, and $-\eta_{21}$) are the x, y, and z components of $\frac{1}{2}\nabla \times \mathbf{u}$, respectively. The angle γ_3 in Figure 2.3 is a natural measure of the rotation:

$$\gamma_3 \approx \frac{1}{2}(\gamma_2 - \gamma_1) \approx \frac{1}{2}\left(\frac{\partial u_1}{\partial x_2} - \frac{\partial u_2}{\partial x_1}\right) = \eta_{21}. \tag{2.8}$$

The rotation tensor does not describe a change in volume, i.e., it does not describe a strain, but merely a rotation.

A strain is necessary for wave motion to exist. If $e_{ij} = 0$, then in Figure 2.2, the line between P and Q is merely translated and rotated, but not stretched, and hence there is no strain.

2.3 Dilatation

Consider an infinitesimally small box of volume $\Delta V = \Delta x\, \Delta y\, \Delta z$. Suppose that the box is then strained, so that it attains a new volume $\Delta V'$. Since the normal strain components represent relative changes in length, the new volume is given by

$$\Delta V' = (1 + e_{11})\Delta x\, (1 + e_{22})\Delta y\, (1 + e_{33})\Delta z.$$

Therefore, the relative change in volume is

$$\frac{\Delta V' - \Delta V}{\Delta V} \approx e_{11} + e_{22} + e_{33} = \nabla \cdot \mathbf{u} \equiv \mathcal{D}, \tag{2.9}$$

where second- and third-order small terms ($e_{11}e_{22}$, etc.) have been dropped in the numerator. $\mathcal{D}$, i.e., $\nabla \cdot \mathbf{u}$, is known as the **dilatation**.

Equivalently, the preceding formula can be derived as follows. Consider a box with sides x, y, and z. The volume of the box is $V = xyz$. Assume the box experiences a small (infinitesimal) strain, so that the displacement $\mathbf{u}$ does not change much over the box, meaning the volume does not change much. Then the relative change in the volume is given by

$$\frac{dV}{V} = \frac{1}{V}\left(\frac{\partial V}{\partial x}dx + \frac{\partial V}{\partial y}dy + \frac{\partial V}{\partial z}dz\right) = \frac{yz\,dx + xz\,dy + xy\,dz}{xyz}$$

$$= \frac{dx}{x} + \frac{dy}{y} + \frac{dz}{z} = e_{11} + e_{22} + e_{33} = \nabla \cdot \mathbf{u} \equiv \mathcal{D}$$

since dx/x is just the relative change in the length of the x-edge of the box, i.e., it is e_{11}, etc.

If $\mathcal{D}$ is positive (negative), then it represents a volume increase (decrease). In general, if a mathematical formula is available for the volume V, then $\mathcal{D} = dV/V$.

2.4 Equation of Motion

Consider a volume V with surface S in an infinite medium. We want to obtain an equation that describes the motion of V when it is subjected to stresses and strains that vary in space and time (such as those due to the passage of a seismic wave). To do this, we apply Newton's second law of motion, *force = mass × acceleration*. The forces acting on V are surface forces (due to the traction $\mathbf{T}$) and **body forces** (e.g., gravity).

Consider an element of volume dV whose position is $\mathbf{x} = (x_1, x_2, x_3)$ at a reference time t_0 (see Figure 2.4).

The mass of dV is $\rho\, dV$, where $\rho = \rho(x_1, x_2, x_3)$ is the density. Let $\mathbf{u}(x_1, x_2, x_3, t)$ be the displacement of dV at time t, i.e., the vector distance from $\mathbf{x}$ to the position of dV at time t. In other words, $\mathbf{u}(x_1, x_2, x_3, t)$ is the displacement at time t of a material

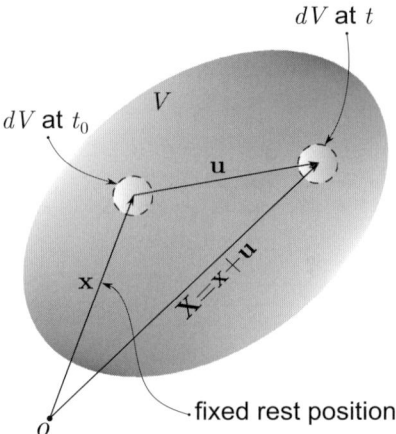

Figure 2.4 Displacement of dV.

particle, whose fixed rest position is $\mathbf{x} = (x_1, x_2, x_3)$. Then, since $\mathbf{x}$ is independent of t, meaning $d\mathbf{x}/dt = 0$, the velocity of dV is given by

$$\dot{\mathbf{u}} \equiv \frac{d\mathbf{u}}{dt} = \sum_{j=1}^{3} \frac{\partial \mathbf{u}}{\partial x_j} \frac{dx_j}{dt} + \frac{\partial \mathbf{u}}{\partial t} = \frac{\partial \mathbf{u}}{\partial t}$$

and hence the acceleration of dV is $\ddot{\mathbf{u}} = \partial^2 \mathbf{u}/\partial t^2$. Therefore, the force $d\mathbf{F}$ acting on dV is $d\mathbf{F} = \rho \ddot{\mathbf{u}} \, dV$, and the total force acting on V is $\int_V d\mathbf{F}$. This total force is produced by the sum of all the contact force elements over the surface plus the sum of all the body force elements. Hence, applying Newton's second law gives

$$\int_S \mathbf{T} \, dS + \int_V \mathbf{f} \, dV = \int_V \rho \ddot{\mathbf{u}} \, dV, \qquad (2.10)$$

where $\mathbf{f} = \mathbf{f}(\mathbf{x}, t)$ is the **body force density**. Writing this vector equation in terms of components, and using Equation (2.1), we have, for $i = 1, 2, 3$,

$$\int_S \sum_{j=1}^{3} \sigma_{ij} n_j \, dS + \int_V f_i \, dV = \int_V \rho \, \ddot{u}_i \, dV. \qquad (2.11)$$

Next, note that Gauss' divergence theorem (1.38) can be written in indicial form as

$$\int_V \sum_{j=1}^{3} \frac{\partial A_j}{\partial x_j} \, dV = \int_S \sum_{j=1}^{3} A_j n_j \, dS.$$

We can apply this to (2.11) because the right-hand side of this equation has the form of the surface integral in (2.11) (except for the extra index i, which does not invalidate the application). Consequently, we obtain

$$\int_S \sum_{j=1}^{3} \sigma_{ij} n_j \, dS = \int_V \sum_{j=1}^{3} \frac{\partial \sigma_{ij}}{\partial x_j} \, dV. \tag{2.12}$$

Equation (2.11) can now be written as

$$\int_V \left(\sum_{j=1}^{3} \frac{\partial \sigma_{ij}}{\partial x_j} + f_i - \rho \, \ddot{u}_i \right) dV = 0. \tag{2.13}$$

The only way this equation can hold for any arbitrary volume V is if the integrand itself is zero. Hence we have

$$\sum_{j=1}^{3} \frac{\partial \sigma_{ij}}{\partial x_j} + f_i = \rho \frac{\partial^2 u_i}{\partial t^2}, \quad i = 1, 2, 3. \tag{2.14a}$$

This is the **equation of motion**. If the summation convention and the simplified notation for partial derivatives is used (see Section 1.10 and Equation 1.117), as it commonly is in the literature, then (2.14a) is written as

$$\sigma_{ij,j} + f_i = \rho \ddot{u}_i, \quad i = 1, 2, 3. \tag{2.14b}$$

A closer look at Equation (2.12) reveals that, in applying the divergence theorem, we should have written $\partial \sigma_{ij} / \partial X_j$, where $\mathbf{X} = \mathbf{x} + \mathbf{u}$, instead of $\partial \sigma_{ij} / \partial x_j$. This is because at time t the particles constituting S have been displaced from their original positions – a given particle originally at $\mathbf{x}$ is now at $\mathbf{X}$. Taking this strictly into account, i.e., using $\partial / \partial X_j$, is known as the **Euler** approach, whereas the approach used in the preceding equations, involving $\partial / \partial x_j$, is known as the **Lagrange** approach. In seismology, however, there is no *practical* difference between the two approaches because stresses, accelerations, etc., do not vary much in going from $\mathbf{x}$ to $\mathbf{X}$, and so $\partial / \partial X_j \approx \partial / \partial x_j$. In other fields, such as fluid mechanics, the Euler approach is more useful.

The equation of motion was derived by applying Newton's second law of motion, i.e., *force = rate of change of linear momentum*, to the volume V. If we apply the *angular* version of this law, i.e., *torque = rate of change of angular momentum*, to the volume V, we would obtain the result $\sigma_{ij} = \sigma_{ji}$, which we have already obtained in a more intuitive way.

A note on body forces: As mentioned previously, body forces are forces acting on the body of a medium. An example is gravity, which normally has a negligible effect on seismic wave propagation and can be ignored. However, physical sources that produce seismic waves, such as explosions or fault slips, can also be represented in the equation of motion as body forces, by using appropriate mathematical formulas for the body force density $\mathbf{f}$. For example, a force exerted at the point $\mathbf{x}_0$ and in the direction of unit vector $\mathbf{e}$, and which varies with time as $g(t)$, can be written using the 3D delta function as $\mathbf{f} = g(t)\delta(\mathbf{x} - \mathbf{x}_0)\mathbf{e}$.

2.5 Conservation of Energy

We will use the summation convention (explained after Equation 1.90) for the calculations in this section.

In basic physics, it is standard procedure to derive a conservation of energy equation for a particle moving in a force field, once we know the equation of motion. For example, consider a projectile of mass m following a parabolic trajectory above the Earth's surface. Assume the z axis points upward from $z = 0$ at the Earth's surface. The equation of motion ($\mathbf{F} = m\mathbf{a}$) can be written as

$$m\frac{d\mathbf{v}}{dt} - m\mathbf{g} = \mathbf{0},$$

where $\mathbf{v}$ is the velocity of the projectile and $\mathbf{g}$ is the vector acceleration of gravity. One way to derive a conservation of energy equation is to take the dot (scalar) product of the equation of motion with the velocity:

$$m\frac{d\mathbf{v}}{dt}\cdot\mathbf{v} - m\mathbf{g}\cdot\mathbf{v} = 0 \qquad \Longrightarrow \qquad m\frac{dv_i}{dt}v_i - mg_iv_i = 0.$$

Since $g_1 = g_2 = 0$ and $g_3 = -g$, where g is the acceleration of gravity, and since $v_3 = dz/dt$ where z is the height of the projectile, this becomes

$$m\frac{dv_i}{dt}v_i + mg\frac{dz}{dt} = 0 \qquad \Longrightarrow \qquad \frac{d}{dt}\left(\frac{1}{2}mv_iv_i + mgz\right) = 0,$$

which states that the total energy (kinetic plus potential) is conserved.

As a reminder, $v_iv_i = v_1{}^2 + v_2{}^2 + v_3{}^2 = v^2$ in the preceding equation, because we are using the summation convention in this section.

In the same way, we can derive a conservation of energy equation for our elastic equation of motion (2.14) by taking the scalar product of the equation of motion with the velocity $\dot{\mathbf{u}}$. We obtain

$$\frac{\partial \sigma_{ij}}{\partial x_j}\,\dot{u}_i + f_i\dot{u}_i = \rho\,\ddot{u}_i\dot{u}_i = \frac{\partial}{\partial t}\left(\tfrac{1}{2}\rho\dot{u}_i\dot{u}_i\right) = \frac{\partial K}{\partial t}, \qquad (2.15a)$$

where

$$K = \frac{1}{2}\rho\left(\frac{\partial u_i}{\partial t}\right)\left(\frac{\partial u_i}{\partial t}\right). \qquad (2.15b)$$

We note again that $(dK/dt) = (\partial K/\partial t)$, as $\mathbf{x}$ is a fixed reference position that does not vary with t. As $\partial u_i/\partial t$ is the particle velocity (or material velocity), K is recognizable as the kinetic energy per unit volume, i.e., the **kinetic energy density**. Using the product rule of differentiation, the first term on the left of (2.15a) can be written as

$$\frac{\partial \sigma_{ij}}{\partial x_j}\,\dot{u}_i = \frac{\partial}{\partial x_j}\left(\sigma_{ij}\dot{u}_i\right) - \sigma_{ij}\frac{\partial \dot{u}_i}{\partial x_j}. \qquad (2.16)$$

Using the fact that $\sigma_{ij} = \sigma_{ji}$, it can be shown that the last term on the right can be written as

$$\sigma_{ij} \frac{\partial \dot{u}_i}{\partial x_j} = \sigma_{ij} \dot{e}_{ij}. \tag{2.17}$$

Note that this does *not* imply that $\partial \dot{u}_i / \partial x_j$ and $\dot{e}_{ij}$ can be equated (for one thing, $\dot{e}_{ij}$ is symmetric whereas $\partial \dot{u}_i / \partial x_j$ is generally not). It merely states that the double sum on the left side of (2.17) equals the double sum on the right side. Hence, substitution of (2.16), with (2.17), into (2.15) yields

$$\frac{\partial}{\partial x_j} \left(\sigma_{ij} \dot{u}_i \right) - \sigma_{ij} \frac{\partial e_{ij}}{\partial t} + f_i \dot{u}_i = \frac{\partial K}{\partial t}$$

or

$$\frac{\partial K}{\partial t} + \frac{\partial W}{\partial t} = -\nabla \cdot \mathbf{I} + \mathbf{f} \cdot \frac{\partial \mathbf{u}}{\partial t}, \tag{2.18}$$

where

$$I_j = -\sigma_{ij} \frac{\partial u_i}{\partial t} \quad \text{and} \quad \frac{\partial W}{\partial t} = \sigma_{ij} \frac{\partial e_{ij}}{\partial t}. \tag{2.19}$$

As a reminder, note that the right sides of these two equations are a sum over i and a double sum over i and j, respectively.

We know what K is, but what are the physical meanings of the other terms? We demonstrate in the following that $\mathbf{I}$ is the **intensity**, or **energy flux vector**, and that W is the **potential energy density**, or more correctly, the **strain energy density**. The flux of $\mathbf{I}$ (see the definition of flux in Equation 1.37) represents the energy flowing through the surface S per unit time, e.g., the energy of a seismic wave (note that the dimensions of $\mathbf{I}$ are stress $\times$ velocity, or energy per unit area per unit time), and the energy flows in the direction of $\mathbf{I}$. We will use $\mathbf{I}$ later on to calculate the amount of energy reflected and transmitted when a seismic wave strikes an interface.

If we integrate over a volume V enclosed by a surface S and use the Gauss divergence theorem on the intensity integral, Equation (2.18) becomes

$$\frac{\partial}{\partial t} \int_V K \, dV + \frac{\partial}{\partial t} \int_V W \, dV = -\int_S \mathbf{I} \cdot \mathbf{n} \, dS + \int_V \mathbf{f} \cdot \frac{\partial \mathbf{u}}{\partial t} \, dV. \tag{2.20}$$

Note that although W appears in the second integral in Equation (2.20), we do not yet have an explicit general formula for W – we only know $(\partial W / \partial t)$. In fact, such a formula does not always exist, especially for complicated realistic materials.

Equation (2.20), or (2.18), is the mathematical expression of the **conservation of energy**. Obviously, $\int_V K \, dV$ is the total kinetic energy in the volume V. The flux of $\mathbf{I}$, i.e., $\int_S \mathbf{I} \cdot \mathbf{n} \, dS$, can be written as

$$-\int_S \mathbf{I} \cdot \mathbf{n} \, dS = -\int_S I_j n_j \, dS = \int_S \sigma_{ij} \frac{\partial u_i}{\partial t} n_j \, dS$$

$$= \int_S T_i \frac{\partial u_i}{\partial t} \, dS = \int_S \mathbf{T} \cdot \frac{\partial \mathbf{u}}{\partial t} \, dS, \qquad (2.21)$$

where (2.1) was used in going from the third to the fourth integral. Remembering that the dot product of force and velocity is the work done per unit time by the force (i.e., the power generated by the force), we see that the last integral is just the work done per unit time on the volume V by the traction $\mathbf{T}$. Similarly, the work done per unit time by the body forces is $\int_V \mathbf{f} \cdot \dot{\mathbf{u}} \, dV$, which appears in Equation (2.20). Hence, the right side of Equation (2.20) is the total amount of work done per unit time on the volume V, which must be equal to the rate of change of kinetic plus strain (potential) energy. Therefore, the quantity W must be the strain (potential) energy density.

The left side of (2.20) is $\partial \mathcal{E} / \partial t$, where $\mathcal{E}$ is the total energy in the volume V. The left side of (2.18) is $\partial E / \partial t$, where $E = K + W$ is the total energy density. This also means $\mathcal{E} = \int_V E \, dV$. Also, the effects of body forces on the propagation of seismic waves are usually negligible, in which case Equations (2.18) and (2.20) become

$$\frac{\partial E}{\partial t} = -\nabla \cdot \mathbf{I} \quad \text{and} \quad \frac{\partial \mathcal{E}}{\partial t} = -\int_S \mathbf{I} \cdot \mathbf{n} \, dS.$$

The equation on the left is called the **equation of continuity** for the energy density. All conserved physical quantities (in this case, the energy) satisfy an equation of continuity. The integral $\int_S \mathbf{I} \cdot \mathbf{n} \, dS$, which is the flux of $\mathbf{I}$, is the total net amount of energy flowing out of V through S per unit time. More specifically, if energy flows *out of* the volume V, then $\int_S \mathbf{I} \cdot \mathbf{n} \, dS$ is positive, meaning $\partial \mathcal{E} / \partial t$ is negative, i.e., the energy $\mathcal{E}$ in V decreases with time.

There is also another effect that is present, namely, the thermal heat generated by the motion of the particles. This is usually negligible in the case of seismic wave propagation, but it can be included in the conservation of energy equation, if desired, in the following way. Because heat is a conserved quantity, it satisfies a continuity equation. Consequently, we have

$$\frac{\partial H}{\partial t} = -\nabla \cdot \mathbf{q} \quad \text{and} \quad \frac{\partial \mathcal{H}}{\partial t} = -\int_S \mathbf{q} \cdot \mathbf{n} \, dS,$$

where H is the heat density (heat per unit volume), $\mathbf{q}$ is the heat flux, $\mathcal{H}$ is the total heat in the volume V, and $\int_S \mathbf{q} \cdot \mathbf{n} \, dS$ is the total net amount of heat flowing out of the volume V through the surface S per unit time. The equation on the right is obtained, as before, by integrating the equation on the left over the volume V and using Gauss' divergence theorem. Including heat results in a more general conservation of energy equation, i.e.,

$$\frac{\partial}{\partial t}(K + W + H) = -\nabla \cdot (\mathbf{I} + \mathbf{q}) + \mathbf{f} \cdot \frac{\partial \mathbf{u}}{\partial t}$$

or

$$\frac{\partial}{\partial t} \int_V (K + W + H)\, dV = -\int_S (\mathbf{I} + \mathbf{q}) \cdot \mathbf{n}\, dS + \int_V \mathbf{f} \cdot \frac{\partial \mathbf{u}}{\partial t}\, dV.$$

2.6 Strain Energy Density

We have seen that $\sum_i \sum_j \sigma_{ij}(\partial e_{ij}/\partial t)$ is the rate of change of the strain energy density W. Assume that W is a function of all the strain components. Then, using the summation convention,

$$W = W(e_{11}, \ldots, e_{33}) \quad \Longrightarrow \quad \frac{\partial W}{\partial t} = \frac{\partial W}{\partial e_{ij}} \frac{\partial e_{ij}}{\partial t} = \sigma_{ij} \frac{\partial e_{ij}}{\partial t}, \qquad (2.22)$$

where the second equation in (2.19) was used. Again, $(dW/dt) = (\partial W/\partial t)$, as $\mathbf{x}$ is independent of time t. If the two double sums are to be equal for all possible strain fields e_{ij}, then we must have

$$\sigma_{ij} = \frac{\partial W}{\partial e_{ij}}, \qquad i, j = 1, 2, 3, \qquad (2.23)$$

assuming that W is such that $\partial W/\partial e_{ij}$ is symmetric in i and j, like σ_{ij}. This equation is reminiscent of the familiar notion from basic physics that the force is the derivative of the potential energy.

To proceed further, we need a relation between σ_{ij} and e_{ij}.

2.7 Stress–Strain Relation

It is reasonable to expect that, for small stresses and strains, the stress is linearly related to the strain, just as the force acting on a simple spring is a linear function of its displacement (Hooke's law). In general, it is assumed that each stress component is a linear combination of all the strain components, i.e.,

$$\sigma_{ij} = \sum_{k=1}^{3} \sum_{l=1}^{3} c_{ijkl} e_{kl}, \qquad i, j = 1, 2, 3. \qquad (2.24)$$

This is the generalization of Hooke's law for a spring to an elastic medium, and is known as the **constitutive relation**, or the **stress–strain relation, for a linearly elastic body**. It holds, in general, for a perfectly elastic heterogeneous anisotropic medium. A heterogeneous medium is one whose physical properties vary with position – the c_{ijkl} are functions of x, y and z. In a homogeneous medium, the c_{ijkl} do not depend on position. An anisotropic medium is one whose physical properties vary with direction – in an isotropic medium, they do not.

The preceding stress–strain relation does not hold for a dissipative (anelastic) medium. In such a medium, **internal friction** causes waves to experience frequency-dependent energy loss as they propagate. However, the actual stress–strain relation for a dissipative medium, expressed in the frequency domain, has the same form as (2.24), except that the c_{ijkl} are complex and frequency dependent.

By applying the quotient rule (explained in Chapter 1) to (2.24), we may conclude that the quantities c_{ijkl} are the components of a tensor of rank 4, sometimes called the **stiffness tensor**. They are often called **elastic constants** (although they can vary with position in a heterogeneous medium, as previously mentioned), and they parametrize the medium. There are $3^4 = 81$ of them, but as we will see, only 21 of them are independent. They have the following symmetries:

$$
\begin{aligned}
c_{ijkl} &= c_{jikl} && (\text{from} \quad \sigma_{ij} = \sigma_{ji}) \\
c_{ijkl} &= c_{ijlk} && (\text{from} \quad e_{ij} = e_{ji}) \\
c_{ijkl} &= c_{klij} && (\text{from} \quad [\partial^2 W/\partial e_{ij}\partial e_{kl}] = [\partial^2 W/\partial e_{kl}\partial e_{ij}]).
\end{aligned}
\tag{2.25}
$$

To prove the first symmetry, for example, we use the fact that the stress tensor is symmetric:

$$
\sigma_{ij} = \sum_{k=1}^{3}\sum_{l=1}^{3} c_{ijkl}e_{kl} = \sigma_{ji} - \sum_{k=1}^{3}\sum_{l=1}^{3} c_{jikl}e_{kl}.
$$

If this is to hold true for all possible strain tensor fields, then we must have $c_{ijkl} = c_{jikl}$. The other two symmetries can be proven in a similar way. Other symmetry relationships can also be derived from these three.

By expanding the sums in the stress–strain relation and using the symmetries of c_{ijkl}, and the facts that $e_{kl} = e_{lk}$ and $\sigma_{ij} = \sigma_{ji}$, one can show that it can be written in matrix form:

$$
\sigma_{ij} = \sum_{k}\sum_{l} c_{ijkl}e_{kl} \qquad \text{is the same as}
$$

$$
\begin{bmatrix}
\sigma_{11} \\
\sigma_{22} \\
\sigma_{33} \\
\sigma_{23} \\
\sigma_{31} \\
\sigma_{12}
\end{bmatrix}
=
\begin{bmatrix}
c_{11} & c_{12} & c_{13} & c_{14} & c_{15} & c_{16} \\
c_{21} & c_{22} & c_{23} & c_{24} & c_{25} & c_{26} \\
c_{31} & c_{32} & c_{33} & c_{34} & c_{35} & c_{36} \\
c_{41} & c_{42} & c_{43} & c_{44} & c_{45} & c_{46} \\
c_{51} & c_{52} & c_{53} & c_{54} & c_{55} & c_{56} \\
c_{61} & c_{62} & c_{63} & c_{64} & c_{65} & c_{66}
\end{bmatrix}
\begin{bmatrix}
e_{11} \\
e_{22} \\
e_{33} \\
2e_{23} \\
2e_{31} \\
2e_{12}
\end{bmatrix},
\quad \text{i.e.,} \quad \boldsymbol{\sigma} = \mathbf{ce}.
\tag{2.26}
$$

The elements c_{ij} of the 6×6 matrix $\mathbf{c}$ are determined by the following rule: $c_{ij} = c_{klmn}$, where kl is the subscript pair on the *i*th element of $\boldsymbol{\sigma}$ (or $\mathbf{e}$), and *mn* is the subscript pair on the *j*th element of $\boldsymbol{\sigma}$ (or $\mathbf{e}$). For example, $c_{11} = c_{1111}$, $c_{23} = c_{2233}$, $c_{46} = c_{2312}$, and $c_{53} = c_{3133}$.

The symmetry $c_{klmn} = c_{mnkl}$ implies that $\mathbf{c}$ is symmetric, i.e., $c_{ij} = c_{ji}$. This means that only 21 of its 36 elements are independent. Consequently, even though there are 81 elastic constants c_{ijkl}, only 21 of them are independent – the others can be obtained from the 21. In other words, the most general anisotropic medium has 21 independent elastic constants. However, as we will see, an isotropic medium has only two independent elastic constants.

The stress–strain relation (2.24) can also be written as

$$\sigma_{ij} = \sum_{k=1}^{3}\sum_{l=1}^{3} c_{ijkl}\frac{\partial u_l}{\partial x_k} = \sum_{k=1}^{3}\sum_{l=1}^{3} c_{ijkl}\frac{\partial u_k}{\partial x_l}. \tag{2.27}$$

The proof of the first equation follows (and the proof of the second is similar). Using the definition of e_{kl}, we have, using the summation convention,

$$\sigma_{ij} = c_{ijkl}e_{kl} = \tfrac{1}{2}c_{ijkl}\frac{\partial u_k}{\partial x_l} + \tfrac{1}{2}c_{ijkl}\frac{\partial u_l}{\partial x_k} = \tfrac{1}{2}c_{ijlk}\frac{\partial u_l}{\partial x_k} + \tfrac{1}{2}c_{ijkl}\frac{\partial u_l}{\partial x_k}$$

$$= \tfrac{1}{2}c_{ijkl}\frac{\partial u_l}{\partial x_k} + \tfrac{1}{2}c_{ijkl}\frac{\partial u_l}{\partial x_k} = c_{ijkl}\frac{\partial u_l}{\partial x_k},$$

which proves the result. In the first term following the third "=" sign, the indices k and l were switched (they are just dummy summation indices and can be switched without affecting the sum), and in the first term following the fourth "=" sign, the symmetry $c_{ijkl} = c_{ijlk}$ was used.

The result can be written even more simply as

$$\sigma_{ij} = c_{ijkl}u_{l,k} = c_{ijkl}\partial_k u_l,$$

if the shorthand notations for the spatial derivatives are also used, in addition to the summation convention.

2.8 More on Strain Energy and Energy Conservation

The differential equation for the strain energy density W and its solution are, using the summation convention,

$$\frac{\partial W}{\partial e_{ij}} = \sigma_{ij} = c_{ijkl}e_{kl}, \quad W = \tfrac{1}{2}c_{ijkl}e_{ij}e_{kl} = \tfrac{1}{2}\sigma_{ij}e_{ij}. \tag{2.28}$$

To verify that $W = \tfrac{1}{2}\sum_i\sum_j\sigma_{ij}e_{ij}$ is the solution, we differentiate it with respect to strain to obtain the preceding partial differential equation for W. Using the product rule, we have, using the summation convention,

$$\frac{\partial W}{\partial e_{mn}} = \tfrac{1}{2} c_{mnkl} e_{kl} + \tfrac{1}{2} c_{ijmn} e_{ij} = \tfrac{1}{2} c_{mnkl} e_{kl} + \tfrac{1}{2} c_{mnij} e_{ij}$$

$$= \tfrac{1}{2} c_{mnkl} e_{kl} + \tfrac{1}{2} c_{mnkl} e_{kl} = c_{mnkl} e_{kl} = \sigma_{mn},$$

which verifies the solution. In the second term following the second "=" sign, a symmetry of the c-tensor was used, and in the second term following the third "=" sign, the fact that i and j are dummy indices was used.

W is generally a positive number, and is zero when there is no strain. Note that W is similar to the potential energy function of a simple spring ("$\tfrac{1}{2} k x^2$"). As previously mentioned, W can also be called the **potential energy density**. In general, the potential energy density is different from the strain energy density. For a perfectly elastic medium, they are the same, but for a dissipative, i.e., **anelastic**, medium, for instance, they are not. The equation $\partial W / \partial t = \sum_i \sum_j \sigma_{ij} (\partial e_{ij} / \partial t)$ holds for a dissipative medium, but W cannot be written down explicitly as $W = \tfrac{1}{2} \sum_i \sum_j \sigma_{ij} e_{ij}$.

2.9 Isotropy

An isotropic medium is one whose physical properties are the same in all directions. It can be shown, for an isotropic medium, that

$$c_{ijkl} = \lambda\, \delta_{ij} \delta_{kl} \; + \; \mu\, (\delta_{ik} \delta_{jl} + \delta_{il} \delta_{jk}). \tag{2.29}$$

λ and μ are known as the **Lamé constants**. If we substitute this into the constitutive relation, we get the stress–strain relation for an isotropic elastic medium:

$$\sigma_{ij} = \lambda \mathcal{D} \delta_{ij} + 2\mu e_{ij} \qquad i,j = 1, 2, 3, \tag{2.30}$$

where $\mathcal{D} = \sum_k e_{kk} = \nabla \cdot \mathbf{u}$ is the dilatation.

To prove this, we make the substitution and use the basic rule

$$\sum_{n=1}^{3} \delta_{kn} e_{np} = \sum_{n=1}^{3} \delta_{nk} e_{np} = e_{kp},$$

as $\delta_{kn} = \delta_{nk} = 0$ if $n \neq k$ and 1 if $n = k$. Substituting (2.29) into (2.24), we obtain, using the summation convention,

$$\sigma_{ij} = \lambda \delta_{ij} \delta_{kl} e_{kl} + \mu \delta_{ik} \delta_{jl} e_{kl} + \mu \delta_{il} \delta_{jk} e_{kl} = \lambda \delta_{ij} e_{kk} + \mu \delta_{ik} e_{kj} + \mu \delta_{jk} e_{ki}$$

$$= \lambda \mathcal{D} \delta_{ij} + \mu e_{ij} + \mu e_{ji} = \lambda \mathcal{D} \delta_{ij} + 2\mu e_{ij},$$

where $e_{ij} = e_{ji}$ has been used. This proves (2.30).

As shown in the preceding, the stress–strain relation can also be written in matrix form as $\sigma = \mathbf{c} \mathbf{e}$. By using either (2.29) or (2.30), it can be shown that, for a perfectly elastic isotropic medium,

$$\mathbf{c} = \begin{bmatrix} \lambda + 2\mu & \lambda & \lambda & 0 & 0 & 0 \\ \lambda & \lambda + 2\mu & \lambda & 0 & 0 & 0 \\ \lambda & \lambda & \lambda + 2\mu & 0 & 0 & 0 \\ 0 & 0 & 0 & \mu & 0 & 0 \\ 0 & 0 & 0 & 0 & \mu & 0 \\ 0 & 0 & 0 & 0 & 0 & \mu \end{bmatrix}.$$

Consider the case in which all the σ_{ij} are zero except for σ_{12} ($= \sigma_{21}$). Then

$$\sigma_{12} = 2\mu e_{12} = \mu \left(\frac{\partial u_1}{\partial x_2} + \frac{\partial u_2}{\partial x_1} \right). \qquad (2.30')$$

This case is known as a state of **simple shear**. The shear stress is directly proportional to the shear strain. The Lamé constant μ is called the **shear modulus**. Experiments have shown that the shear modulus μ is a positive number for small deformations.

The preceding stress–strain relation can also be substituted into the general expressions derived previously for the intensity components I_j and the strain energy density W, to derive expressions for these quantities for isotropic media.

2.10 Hydrostatic Stress

We have seen that the stress tensor in a fluid undergoing deformation is

$$\sigma_{ij} = -p\delta_{ij}, \qquad i,j = 1,2,3, \qquad (2.31)$$

where p is the pressure. This is called a state of **hydrostatic stress**. The normal stresses are

$$\sigma_{11} = \sigma_{22} = \sigma_{33} = -p,$$

and the shear stresses are all zero.

Consider a disturbance or deformation occurring in an isotropic solid medium that creates a state of **hydrostatic stress** (2.31) in the solid (e.g., a uniform radial pressure would create such a state). Note that if we define

$$\mathcal{S} \equiv \sum_k \sigma_{kk},$$

then (2.31) can also be written as

$$\sigma_{ij} = \tfrac{1}{3}\mathcal{S}\delta_{ij}.$$

The shear stresses in a solid are given by $\sigma_{ij} = 2\mu e_{ij}$ for $i \neq j$, and since they are all zero in a state of hydrostatic stress, we see that the shear strains in the solid are also zero in this state, with the normal strains all being equal, i.e.,

$$e_{ij} = \tfrac{1}{3}\mathcal{D}\delta_{ij}.$$

The state of hydrostatic stress would produce a change in the volume (but not the shape) of a given region V in the solid.

In general, though, shear stresses are nonzero in a solid. However, the stress field in a solid can be separated into nonshear and shear parts.

First, contract (2.30) to obtain

$$\mathcal{S} = \sum_{i=1}^{3} \sigma_{ii} = \lambda \mathcal{D} \sum_{i=1}^{3} \delta_{ii} + 2\mu \sum_{i=1}^{3} e_{ii} = 3\lambda\mathcal{D} + 2\mu\mathcal{D} \quad \Longrightarrow \quad \mathcal{S} = 3k_B\mathcal{D}, \quad (2.32a)$$

where

$$k_B \equiv \lambda + \tfrac{2}{3}\mu, \qquad (2.32b)$$

and where the normal stresses in the sum $\mathcal{S}$ are no longer equal to each other, in general.

k_B is called the **compression modulus** or the **bulk modulus** of the solid. If the solid is under hydrostatic stress, then $\mathcal{S} = -3p$, implying $k_B = -p/\mathcal{D}$. The dilatation $\mathcal{D}$ is a measure of the relative increase in volume of a small volume element due to normal strain. Therefore, k_B is normally a positive number because p and $\mathcal{D}$ normally have opposite signs. For example, if the pressure p is positive, then the volume decreases, i.e., $\mathcal{D}$ is negative. In addition, the smaller the volume decrease is, i.e., the smaller $|\mathcal{D}|$ is, the more incompressible the medium is, and the larger k_B is. Hence, k_B is a measure of the **incompressibility** of the medium. Note also that $k_B > 0 \quad \Longrightarrow \quad \lambda > -\tfrac{2}{3}\mu$.

Next, express the stress tensor in a solid as the sum of a tensor whose shear components are zero, and a **deviatoric stress tensor** σ'_{ij} containing the shear contribution:

$$\sigma_{ij} = \tfrac{1}{3}\mathcal{S}\delta_{ij} + \sigma'_{ij}. \qquad (2.33)$$

Similarly,

$$e_{ij} = \tfrac{1}{3}\mathcal{D}\delta_{ij} + e'_{ij}, \qquad (2.34)$$

where e'_{ij} is the **deviatoric strain tensor**.

Contracting these two equations shows that $\sum_i \sigma'_{ii} = \sum_i e'_{ii} = 0$.

Finally, obtain a relation between the deviatoric stress and strain tensors. Using (2.34) and (2.30) gives

$$2\mu e'_{ij} = 2\mu e_{ij} - \tfrac{2}{3}\mu\mathcal{D}\delta_{ij} = \sigma_{ij} - \lambda\mathcal{D}\delta_{ij} - \tfrac{2}{3}\mu\mathcal{D}\delta_{ij} = \sigma_{ij} - (\lambda + \tfrac{2}{3}\mu)\mathcal{D}\delta_{ij}$$

$$= \sigma_{ij} - k_B\mathcal{D}\delta_{ij} = \sigma_{ij} - \tfrac{1}{3}\mathcal{S}\delta_{ij} = \sigma'_{ij}.$$

Therefore, as an alternative to (2.30), the state of stress and strain in an isotropic solid can also be expressed by the set of equations

$$\sigma'_{ij} = 2\mu e'_{ij} \quad \text{and} \quad \mathcal{S} = 3k_B\mathcal{D}. \qquad (2.35)$$

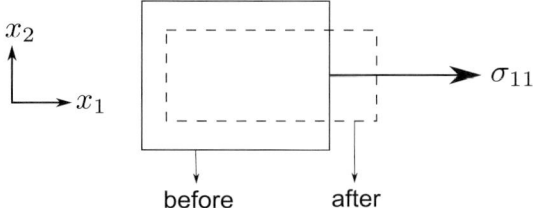

Figure 2.5 A normal stress σ_{xx} on a box.

2.11 Other Elastic Constants

Imagine a boxlike region in an isotropic medium subjected to a normal stress σ_{11} in the x direction, as shown in Figure 2.5.

From this figure, we see that $e_{11} > 0$, $e_{22} < 0$, and $e_{33} < 0$. The fact that $e_{11} = (\partial u_1 / \partial x_1)$ is positive can be seen more readily by choosing a series of points on the undeformed box with increasing x_1 values and noting that their x_1 components of displacement, i.e., their u_1-values, increase with x_1. Similarly, the fact that e_{22} and e_{33} are negative can be seen in this way as well. It can also be shown that $e_{22} = e_{33}$.

Poisson's ratio σ and **Young's modulus** Y are defined by

$$\sigma \equiv -\frac{e_{22}}{e_{11}} = -\frac{e_{33}}{e_{11}} \quad \text{and} \quad Y \equiv \frac{\sigma_{11}}{e_{11}}. \tag{2.36}$$

For an isotropic medium, it can be shown that σ and Y can be expressed in terms of λ and μ as

$$\sigma = \frac{\lambda}{2(\lambda + \mu)} \quad \text{and} \quad Y = \frac{\mu(3\lambda + 2\mu)}{\lambda + \mu}. \tag{2.37}$$

Since μ and k_B are positive, it can be shown that $Y > 0$ and $-1 < \sigma \leq \frac{1}{2}$. However, since e_{11} and e_{22} usually have opposite signs, σ can usually be further restricted to the range $0 < \sigma \leq \frac{1}{2}$. Generally, the harder the rock, the smaller the σ value, because e_{22} is smaller for harder rocks. For liquids, $\sigma = \frac{1}{2}$ because liquids cannot support a shear stress, so $\mu = 0$. Note that $\lambda = \mu$ implies $\sigma = \frac{1}{4}$ (and $Y = \frac{5}{2}\mu$). Poisson's ratio is a dimensionless number, but Young's modulus has the dimensions of force per unit area. For rocks in the continental upper crust, typical values for σ and Y are $\sigma \sim \frac{1}{4}$ and $Y \sim 60 \text{ GPa} = 60 \times 10^9 \text{ N/m}^2 = 60 \times 10^4 \text{ bar} = 59.23 \times 10^4 \text{ atm}$.

The constants λ, μ, Y, k_B, and σ can be written in terms of each other. For example, from Equation (2.37), we have

$$1 + \sigma = \frac{2(\lambda + \mu)}{2(\lambda + \mu)} + \frac{\lambda}{2(\lambda + \mu)} = \frac{3\lambda + 2\mu}{2(\lambda + \mu)} = \frac{Y}{2\mu} \implies \mu = \frac{Y}{2(1 + \sigma)}. \tag{2.38}$$

Other examples are

$$\lambda = \frac{Y\sigma}{(1+\sigma)(1-2\sigma)} = \frac{\mu(Y-2\mu)}{3\mu - Y}, \quad k_B = \frac{Y}{3(1-2\sigma)} = \frac{\mu Y}{3(3\mu - Y)},$$

$$\text{and} \quad \sigma = \frac{Y-2\mu}{2\mu}. \tag{2.39}$$

These formulas can be used to calculate values of λ, μ, and k_B from measured values of Y and σ.

If $\sigma = \frac{1}{4}$ and $Y = 60$ GPa (upper crustal rocks), then $\mu = \lambda = 24$ GPa, and $k_B = 40$ GPa.

2.12 The Equation of Motion and the Wave Equations

Consider a homogeneous medium, i.e., one in which λ and μ are constants (they do not vary with $\mathbf{x}$). We will ignore the body forces, as we first want to obtain an equation of motion only for the deformation produced by the traction $\mathbf{T}$. If we then substitute the stress–strain relation (2.30) into the equation of motion (2.14), we obtain, for $i = 1, 2, 3$,

$$\sum_j \frac{\partial \sigma_{ij}}{\partial x_j} = \lambda \sum_i \frac{\partial \mathcal{D}}{\partial x_j}\delta_{ij} + 2\mu \sum_j \frac{\partial e_{ij}}{\partial x_j} = \lambda\frac{\partial \mathcal{D}}{\partial x_i} + \mu \sum_j \frac{\partial^2 u_i}{\partial x_j^2} + \mu \frac{\partial}{\partial x_i} \sum_j \frac{\partial u_j}{\partial x_j}$$

$$= (\lambda + \mu)\frac{\partial \mathcal{D}}{\partial x_i} + \mu \nabla^2 u_i = \rho \frac{\partial^2 u_i}{\partial t^2}. \tag{2.40}$$

In vector form, this is

$$(\lambda + \mu)\nabla(\nabla \cdot \mathbf{u}) + \mu\nabla^2 \mathbf{u} = \rho \frac{\partial^2 \mathbf{u}}{\partial t^2}, \tag{2.41}$$

where $\mathcal{D} = \nabla \cdot \mathbf{u}$. This is the equation of motion for a homogeneous isotropic elastic medium. As $\nabla^2 \mathbf{u} \equiv (\nabla^2 u_1, \nabla^2 u_2, \nabla^2 u_3)$, it holds only in Cartesian coordinate frames, unless $\nabla^2 \mathbf{u}$ is evaluated using

$$\nabla^2 \mathbf{u} = \nabla(\nabla \cdot \mathbf{u}) - \nabla \times (\nabla \times \mathbf{u}),$$

in which case it holds in curvilinear coordinate frames as well (see Equation 1.83). Equivalently, if we want it to hold in curvilinear coordinate frames as well as Cartesian ones, we can use the preceding identity to replace "$\nabla^2 \mathbf{u}$" in (2.41). When we do this, Equation (2.41) becomes

$$(\lambda + 2\mu)\nabla(\nabla \cdot \mathbf{u}) - \mu\nabla \times (\nabla \times \mathbf{u}) = \rho \frac{\partial^2 \mathbf{u}}{\partial t^2}. \tag{2.42}$$

If we take the divergence of Equation (2.41), we obtain

$$(\lambda + 2\mu)\nabla^2 \mathcal{D} = \rho \frac{\partial^2 \mathcal{D}}{\partial t^2}. \tag{2.43}$$

This has the form of the 3D wave equation, i.e., $v^2 \nabla^2 \psi = \ddot{\psi}$, where v is the wave speed and ψ is the wave function. Hence, Equation (2.43) is a wave equation for the dilatation. It states that dilatational waves, also known as **compressional** waves or **longitudinal** waves or P waves, travel with a speed α, where

$$\alpha = \sqrt{\frac{\lambda + 2\mu}{\rho}}. \tag{2.44}$$

If we take the curl of equation (2.41), and use the identity $\nabla \times \nabla \psi = \mathbf{0}$, we obtain

$$\mu \nabla^2 (\nabla \times \mathbf{u}) = \rho \frac{\partial^2}{\partial t^2} (\nabla \times \mathbf{u}), \tag{2.45}$$

which also has the form of a 3D wave equation. Since $\nabla \times \mathbf{u}$ is the **rotation** (see Equation 1.36), it states that rotational waves, also known as **shear** waves, **transverse** waves, or S waves, propagate with a speed β, where

$$\beta = \sqrt{\frac{\mu}{\rho}}. \tag{2.46}$$

Note that $\beta < \alpha$, i.e., compressional waves travel faster than shear waves.

Equations (2.43) and (2.45) could have also been derived by taking the divergence and curl of Equation (2.42) and using Equations (1.29).

Equations (2.41) and (2.42) look rather complicated. However, for plane waves traveling in one direction only, they reduce to 1D wave equations. For example, for a plane wave propagating in the $\pm x_1$ direction, the three displacement components do not vary with x_2 and x_3 (i.e., $\partial u_k / \partial x_2 = \partial u_k / \partial x_3 = 0$). This also means that $\mathcal{D} = \partial u_1 / \partial x_1$. For the x component of (2.41), i.e., $i = 1$ in (2.40), we then obtain

$$(\lambda + 2\mu)\frac{\partial^2 u_1}{\partial x_1^2} = \rho \frac{\partial^2 u_1}{\partial t^2} \quad \text{i.e.,} \quad \frac{\partial^2 u_1}{\partial x_1^2} = \frac{1}{\alpha^2} \frac{\partial^2 u_1}{\partial t^2}, \tag{2.47}$$

and for $i = 2$ and $i = 3$ in (2.40) we obtain

$$\mu \frac{\partial^2 u_n}{\partial x_1^2} = \rho \frac{\partial^2 u_n}{\partial t^2} \quad \text{i.e.,} \quad \frac{\partial^2 u_n}{\partial x_1^2} = \frac{1}{\beta^2} \frac{\partial^2 u_n}{\partial t^2}, \quad n = 2, 3. \tag{2.48}$$

These are two different 1D wave equations, meaning that two types of waves can propagate. The first describes a P wave traveling with speed α in the $\pm x$ direction. Note that the wave function is u_1, the x component of displacement – the particles struck by a P wave are displaced parallel to the direction ($\pm x$) in which the P wave is traveling. The second equation describes an S wave traveling with speed β in the $\pm x$ direction. The wave function here is u_2 or u_3, meaning that the particles struck by an S wave are displaced perpendicular to the direction in which the S wave is traveling. We will see later that (2.41) also has 3D plane wave solutions.

We see there are two components to the S wave displacement. In seismic theory, the x and y axes are usually taken to be horizontal (parallel to the Earth's surface) and the z axis is vertical. Consequently, u_3 is called the **shear-vertical** or SV component and u_2 is called the **shear-horizontal** or SH component (the displacement for an S wave could be $\mathbf{u} = u_2\mathbf{e}_2 + u_3\mathbf{e}_3$). Note that the P and SV wave displacements are parallel to the xz plane, whereas the SH component is perpendicular to it. We will see later that this often means that the propagation of P and SV waves can be treated separately from the propagation of SH waves.

2.13 Elastic Constants in Terms of α and β

By using the formulas for α and β, we can express the formulas for the various elastic constants (Y, σ, etc.) in terms of α and β, or, α and β can be expressed in terms of them. Such formulas often prove useful in the analysis of seismic data. For example, α, β, and Poissons's ratio σ are related by

$$\sigma = \frac{(\beta/\alpha)^2 - \frac{1}{2}}{(\beta/\alpha)^2 - 1} \quad \text{or} \quad \left(\frac{\beta}{\alpha}\right)^2 = \frac{\frac{1}{2} - \sigma}{1 - \sigma} \tag{2.49}$$

$$\Rightarrow \quad 0 \le \frac{\beta}{\alpha} < \frac{1}{\sqrt{2}} \quad \text{for} \quad 0 < \sigma \le \frac{1}{2}. \tag{2.50}$$

These apply to a homogeneous isotropic medium.

2.14 The 1D Wave Equation

In one dimension, the wave equation is

$$\frac{\partial^2 \psi}{\partial x^2} = \frac{1}{v^2} \frac{\partial^2 \psi}{\partial t^2}. \tag{2.51}$$

This describes a wave of amplitude $\psi(x, t)$ traveling in the $\pm x$ direction with speed v. From basic physics, we know that a particular solution for ψ is any function of $x \pm vt$ (or $t \pm (x/v)$, etc.).

For example, consider $\psi = f(x - vt)$. This represents a plane wave traveling with speed v in the positive x direction. We can see this by plotting graphs of $f(x - vt)$ vs. x at increasing times t, such as those in Figure 2.6. Such graphs can be thought of as "snapshots" of the waveform in the medium at increasing times, and they show a wave pulse moving in the $+x$ direction with speed v.

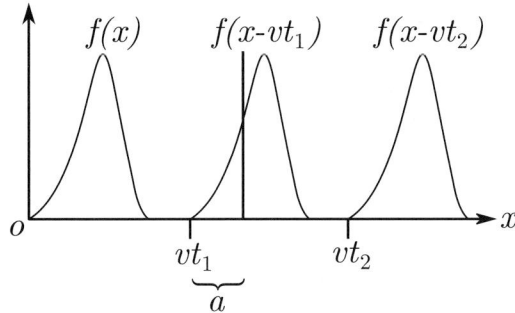

Figure 2.6 A plane wave pulse $f(x - vt)$ propagating in the $+x$ direction. The vertical line is a plane wavefront.

The wavefronts are planes normal to the x axis, and hence are described by the equation $x = $ constant. The value of the constant depends on the time t and the position of the wavefront. For example, in Figure 2.6, the equation of the wavefront is $x = vt_1 + a$.

More specifically, the position x of any given wavefront and the time t both increase as the wavefront propagates, but the **phase** of the wavefront, i.e., $x - vt$, remains constant, i.e., $x = vt + $ constant. In other words, for a given wavefront, as the time t increases, the position x of the wavefront increases just enough to keep the phase $x - vt$ constant. The wavefronts $x = $ constant are sometimes called "planes of constant phase."

The general plane wave solution to the 1D wave equation is

$$\psi = f(x - vt) + g(x + vt), \tag{2.52}$$

where f and g represent waves traveling in the positive and negative x directions, respectively. For instance, if a wave traveling in the positive x direction is reflected by a boundary, then f would represent the incident wave and g the reflected wave. The total wave field ψ is the sum of these two overlapping wave fields.

In general, wavefronts can be curved surfaces on which the phase is constant. **Rays** are lines that are perpendicular to the wavefronts (in isotropic media). They define the path of travel of a wave between the source and a receiver.

For spherical waves, the Laplacian in the 3D wave equation is written in terms of spherical coordinates. The spherical symmetry implies that $\psi = \psi(r, t)$, i.e., ψ has no angular dependence, and so the 1D wave equation becomes

$$\frac{1}{r^2} \frac{\partial}{\partial r} \left(r^2 \frac{\partial \psi}{\partial r} \right) = \frac{1}{v^2} \frac{\partial^2 \psi}{\partial t^2}. \tag{2.53}$$

The solution is

$$\psi = \frac{1}{r}f(r - vt) + \frac{1}{r}g(r + vt). \tag{2.54}$$

The term involving f represents an outgoing spherical wave, and the term in g an ingoing spherical wave. The "$1/r$" factor represents the decrease (increase) in the amplitude of the outgoing (ingoing) wave due to the spherical divergence (convergence) of the rays.

2.15 Harmonic Waves

From basic physics, we know that a harmonic cosine wave is expressed as

$$\psi = A\cos[\kappa(x - vt)] = A\cos[\kappa x - \omega t], \tag{2.55}$$

where A is the amplitude and v is the wave speed. Also $\omega = 2\pi f$ and $\kappa = 2\pi k$, where f is the frequency and k is the wave number, and where

$$\omega = v\kappa, \quad \text{i.e.,} \quad f = vk. \tag{2.56}$$

The period T is given by $T = 1/f$, and the wavelength λ is given by $\lambda = 1/k$. Furthermore, $v = \lambda f$. The wavenumber k can be thought of as a "spatial" frequency (whose units are cycles per meter, say), in which case f would be called a "temporal" frequency (with units of cycles per second, or Hz). ω is called the radial or circular or angular frequency, and κ is called the radial or circular or angular wavenumber. *A note of caution*: in many texts and papers, the English letter k is used for the angular wavenumber. We use the Greek letter κ here to be consistent with the use of the Greek letter ω for the angular frequency.

The relationships between wavelength and wavenumber, and frequency and period, can be derived in a fairly straightforward way. For example, if we examine the wave at the instant $t = 0$ (think of it as a "snapshot" or photograph of the wave at $t = 0$), we have $\psi = A\cos\kappa x$, a simple cosine function. Clearly, if x is one wavelength, that represents one cycle of the wave, i.e., $\kappa\lambda = 2\pi$, implying $\lambda = 2\pi/\kappa = 1/k$.

Often, complex exponentials are used to describe harmonic waves: since $e^{i\theta} = \cos\theta + i\sin\theta, i = \sqrt{-1}$, we may write the wave in (2.55) as

$$\psi = A e^{i(\kappa x - \omega t)}, \tag{2.57}$$

where it is understood that the real part of ψ describes the physical part of the wave. Although we have assumed so far that A is a real number, it can in fact be complex when ψ is written as a complex exponential. For example, if $A = |A|e^{i\phi}$, then the phase of the harmonic wave is changed by an amount ϕ, i.e.,

$$\psi = |A|e^{i(\kappa x - \omega t + \phi)} \implies \text{Re}(\psi) = |A|\cos(\kappa x - \omega t + \phi). \tag{2.58}$$

One must be careful when using complex exponentials. For instance, suppose we have a harmonic wave ψ given, at time $t = 0$, by

$$\psi = \cos(\kappa x), \tag{2.59}$$

and suppose that $\mathcal{E}$ is some physical quantity given by $\mathcal{E} = \psi^2$. Then, at $t = 0$,

$$\mathcal{E} = \psi^2 = \cos^2(\kappa x). \tag{2.60}$$

However, if we would use complex exponentials, we would obtain

$$\mathcal{E} = \psi^2 = \left[e^{i\kappa x} \right]^2 = e^{2i\kappa x}, \tag{2.61}$$

whose real part (the actual physical quantity) is

$$\cos(2\kappa x) = \cos^2(\kappa x) - \sin^2(\kappa x), \tag{2.62}$$

which is *not* the same as the correct answer (2.60). Whenever it is necessary to evaluate non-linear terms involving products of wave functions, such as ψ^2, one must use the real parts of the individual wave functions in forming the products to avoid the preceding error. However, if all terms in such a calculation are linear, then complex exponentials can be used throughout the calculation, and the real part of the final result gives the expression for the actual physical quantity.

Often, the **slowness** s, where $s \equiv 1/v = \kappa/\omega$, is used in the mathematical expression for a harmonic wave. Equation (2.57) then becomes

$$\psi = A\, e^{i\omega(sx-t)}. \tag{2.63}$$

In three dimensions, the wave equation is

$$\nabla^2 \psi = \frac{1}{v^2} \frac{\partial^2 \psi}{\partial t^2}, \tag{2.64}$$

and the harmonic wave solution is

$$\psi = A\, e^{i\omega(\mathbf{s}\cdot\mathbf{x}-t)} = A e^{i(\boldsymbol{\kappa}\cdot\mathbf{x}-\omega t)}, \tag{2.65}$$

where $\mathbf{s}\cdot\mathbf{x} = s_x x + s_y y + s_z z$, and $\boldsymbol{\kappa} = \omega\mathbf{s}$ is the radial wavevector. This describes a harmonic plane wave traveling in the direction of $\mathbf{s}$ (or $\boldsymbol{\kappa}$). The slowness vector $\mathbf{s}$ is perpendicular to the plane wavefronts, i.e., the planes of constant phase, which are given by $\mathbf{s}\cdot\mathbf{x} = constant$. The magnitude of $\mathbf{s}$ is $s = 1/v = \kappa/\omega$.

As we have seen, the 3D wave equation can be a vector wave equation, e.g.,

$$\nabla^2 \boldsymbol{\psi} = \frac{1}{v^2} \frac{\partial^2 \boldsymbol{\psi}}{\partial t^2},$$

in which case the harmonic wave solution is

$$\boldsymbol{\psi} = A\, e^{i\omega(\mathbf{s}\cdot\mathbf{x}-t)}\mathbf{d} = A e^{i(\boldsymbol{\kappa}\cdot\mathbf{x}-\omega t)}\mathbf{d},$$

where **d**, the **polarization vector**, is a constant vector (usually a unit vector, i.e., a vector of unit length) from which the direction of ψ can be determined. Note that if **d** is real, then Re(ψ) points alternately in the $+$**d** and $-$**d** directions, because Re(ψ) alternates between positive and negative values as **x** and t vary. Note also that **s** and **d** are generally in different directions.

It should be mentioned that some authors, instead of using the components s_x, s_y, and s_z of slowness, prefer to use **apparent velocities** $v_{(x)}$, $v_{(y)}$, and $v_{(z)}$, where

$$v_{(x)} = 1/s_x, \quad v_{(y)} = 1/s_y, \quad v_{(z)} = 1/s_z.$$

Note that

$$s_x{}^2 + s_y{}^2 + s_z{}^2 = s^2 = \frac{1}{v^2} \quad \Rightarrow \quad \frac{1}{v_{(x)}{}^2} + \frac{1}{v_{(y)}{}^2} + \frac{1}{v_{(z)}{}^2} = \frac{1}{v^2} \quad \text{and}$$

$$\mathbf{s} \cdot \mathbf{x} - t = s_x x + s_y y + s_z z - t = \frac{x}{v_{(x)}} + \frac{y}{v_{(y)}} + \frac{z}{v_{(z)}} - t.$$

This also shows that $v_{(x)}$, $v_{(y)}$, and $v_{(z)}$ are *not* the components of a vector **v** (they do not satisfy the rule that $A_x{}^2 + A_y{}^2 + A_z{}^2 = A^2$ for a vector **A**), which is why parentheses appear around the subscripts. $v_{(x)}$, $v_{(y)}$, and $v_{(z)}$ are the **apparent velocities** or **apparent wave speeds** in the x, y, and z directions, respectively. For example, imagine a plane wavefront propagating in the xz plane upward toward a flat horizontal surface (the xy plane). The slowness vector (which lies in the xz plane) makes an angle θ with the vertical. The line along which the plane wavefront intersects the surface travels in the x direction at the speed $v_{(x)} = v/\sin\theta$, where v is the speed of the wave ($v = \alpha$ or β). Similarly, the line along with the plane wavefront intersects a vertical plane (a plane parallel to the yz plane), and travels vertically at the speed $v_{(z)} = v/\cos\theta$. The proofs of these are in Appendix 2A.

2.16 3D Plane Wave Solutions of the Equation of Motion

A more general way of finding plane wave solutions is to substitute a trial plane wave for **u** into the equation of motion (EOM) and see under what conditions the plane wave solves the EOM. Either EOM, (2.41) or (2.42), can be used if one uses Cartesian coordinates. So, let the particle displacement **u** at the point **x** be

$$\mathbf{u} = A \exp\big[i\omega(s_1 x_1 + s_2 x_2 + s_3 x_3 - t)\big]\mathbf{d} = A \exp\big[i\omega(\mathbf{s} \cdot \mathbf{x} - t)\big]\mathbf{d},$$

where the amplitude A is complex in general, **s** is the slowness vector with $|\mathbf{s}| = s = 1/v$, where v is the wave speed, and **d** is the polarization vector (a constant unit vector here). Also, "exp(x)" is an alternative way of writing e^x. The polarization vector **d** points in the direction of the particle displacement **u**, and the slowness **s** points in the direction of wave propagation, or more precisely, in the direction in which the phase $\omega(\mathbf{s} \cdot \mathbf{x} - t)$ is increasing most rapidly with distance. Substitute $\mathbf{u}$ into EOM (2.41). For $\nabla \cdot \mathbf{u}$, one obtains

$$\nabla \cdot \mathbf{u} = \sum_{j=1}^{3} \frac{\partial u_j}{\partial x_j} = i\omega \sum_{j} s_j A e^{i\omega(\mathbf{s}\cdot\mathbf{x}-t)} d_j = i\omega \sum_{j=1}^{3} s_j u_j = i\omega \, \mathbf{s} \cdot \mathbf{u}.$$

Similarly, one obtains

$$\nabla(\nabla \cdot \mathbf{u}) = -\omega^2(\mathbf{s} \cdot \mathbf{u})\mathbf{s}, \quad \nabla^2\mathbf{u} = -\omega^2 s^2 \mathbf{u}, \quad \ddot{\mathbf{u}} = -\omega^2 \mathbf{u},$$
$$\text{and} \quad \nabla \times \mathbf{u} = i\omega \, \mathbf{s} \times \mathbf{u}.$$

The last result is needed only if EOM (2.42) is used. Substituting these into EOM (2.41) results in

$$(\lambda + \mu)(\mathbf{s} \cdot \mathbf{u})\mathbf{s} + (\mu s^2 - \rho)\mathbf{u} = \mathbf{0}. \tag{2.66}$$

One way to proceed is to note that there are three cases: (i) $\mathbf{s}\|\mathbf{u}$, (ii) $\mathbf{s}\perp\mathbf{u}$, and (iii) $\mathbf{s}$ and $\mathbf{u}$ are neither parallel or perpendicular to each other.

For case (i), $\mathbf{u} = a\mathbf{s}$, where a is a constant at any given time. The preceding equation then becomes

$$\left[(\lambda + \mu)s^2 + (\mu s^2 - \rho)\right]a\mathbf{s} = \mathbf{0} \quad \Longrightarrow \quad v^2 = \frac{1}{s^2} = \frac{\lambda + 2\mu}{\rho} = \alpha^2,$$

i.e., we have a plane wave traveling in the $\mathbf{s}$ direction with speed α, with the particles also being displaced parallel to the $\mathbf{s}$ direction – a P wave.

For case (ii), $\mathbf{s} \cdot \mathbf{u} = 0$ in Equation (2.66), i.e.,

$$v^2 = \frac{1}{s^2} = \frac{\mu}{\rho} = \beta^2,$$

i.e., we have a plane wave traveling in the $\mathbf{s}$ direction with speed β, with the particles also being displaced perpendicular to the $\mathbf{s}$ direction – an S wave.

For case (iii), the first and second terms of (2.66) are two nonzero vectors, not parallel or perpendicular to each other, pointing in different directions. This means that their sum cannot be zero, so there is no solution for this case.

Hence, there are two plane wave solutions, represented by cases (i) and (ii).

Another way to proceed is as follows.

If we first take the cross product of (2.66) with $\mathbf{s}$ and use $\mathbf{s} \times \mathbf{s} = \mathbf{0}$, and then take the dot product of the equation with $\mathbf{s}$ and use $\mathbf{s} \cdot \mathbf{s} = s^2$, we obtain

$$(\mu s^2 - \rho)\mathbf{u} \times \mathbf{s} = \mathbf{0} \quad \text{and} \quad \left[(\lambda + 2\mu)s^2 - \rho\right](\mathbf{u} \cdot \mathbf{s}) = 0.$$

Both of these equations must be true simultaneously. Setting $s^2 = \rho/\mu$ would make the first equation true. Then, to make the second equation true as well, it is necessary that $\mathbf{u} \cdot \mathbf{s} = 0$ (since $(\lambda+2\mu)s^2 - \rho \neq 0$). Similarly, setting $s^2 = \rho/(\lambda+2\mu)$ would make the second equation true. Then, to make the first equation true as well, it is necessary that $\mathbf{u} \times \mathbf{s} = \mathbf{0}$ (since $\mu s^2 - \rho \neq 0$). Consequently, using $v = 1/s$, we have

$$\text{either} \quad \{v^2 = \mu/\rho \quad \text{and} \quad \mathbf{u} \cdot \mathbf{s} = 0\} \quad \text{or} \quad \{v^2 = (\lambda+2\mu)/\rho \quad \text{and} \quad \mathbf{u} \times \mathbf{s} = \mathbf{0}\}.$$

The first condition corresponds to a plane shear (S) wave propagating with speed $\beta = \sqrt{\mu/\rho}$ and having a particle displacement $\mathbf{u}$ normal to the direction of wave propagation $\mathbf{s}$. The second condition corresponds to a plane compressional (P) wave propagating with speed $\alpha = \sqrt{(\lambda + 2\mu)/\rho}$ and having a particle displacement $\mathbf{u}$ parallel to the direction of wave propagation $\mathbf{s}$. Note from the preceding equations that plane P waves satisfy $\nabla \times \mathbf{u} = \mathbf{0}$ (they produce no rotation) and plane S waves satisfy $\mathcal{D} = \nabla \cdot \mathbf{u} = 0$ (they produce no dilatation). Furthermore, $\mathbf{u}$ for P waves can be written as the gradient of some scalar function ϕ, i.e., $\mathbf{u} = \nabla\phi$ (because $\nabla \times \nabla\phi = \mathbf{0}$ from Equation 1.29a), and $\mathbf{u}$ for S waves can be written as the curl of some vector function ψ, i.e., $\mathbf{u} = \nabla \times \psi$ (because $\nabla \cdot (\nabla \times \psi) = 0$ from Equation 1.29a). ϕ and ψ are called the P and S wave potentials, respectively. Potentials are discussed in greater detail in the next section.

As mentioned previously, if the shear wave is propagating in the xz plane (with the z axis perpendicular to the ground surface), i.e., if the slowness vector $\mathbf{s}$ is parallel to the xz plane, then the shear wave displacement $\mathbf{u}$ can be broken into two components, the SV component that is parallel to the xz plane, and the SH component, which is perpendicular to the xz plane (in the y direction). The P and SV displacements are then parallel to the xz plane and the SH displacement is perpendicular to it.

For a simple numerical example, let the z axis point vertically downward, and the x axis point to the right, and consider a plane wave traveling downward in the xz plane in the positive x and z directions at an angle of $\theta = 30°$ to the vertical and with a speed of $v = 2$ km/s. The slowness vector $\mathbf{s}$ points in the direction of wave propagation, hence

$$s_x = |\mathbf{s}| \sin\theta = \frac{\sin\theta}{v} = \frac{\sin(30°)}{2} = 0.25 \text{ s/km},$$

$$\text{and} \quad s_z = \frac{\cos\theta}{v} = \frac{\cos(30°)}{2} = 0.4330 \text{ s/km}.$$

Conversely, if s_x and s_z are given as in the preceding, one can determine v and θ as follows:

$$s_x{}^2 + s_z{}^2 = (0.25)^2 + (0.4330)^2 = 0.25 = (1/v^2) \quad \Rightarrow \quad v = 2 \text{ km/s},$$

$$\sin\theta = vs_x = 2(0.25) = 0.5 \quad \Rightarrow \quad \theta = 30°.$$

If the wave is a *P* wave, then particle motion is in the direction of travel of the wave, i.e., the unit polarization vector **d** is parallel to **s** meaning $\mathbf{s} \times \mathbf{d} = \mathbf{0}$, and if the wave is an *SV* wave, then particle motion is perpendicular to the direction of travel of the wave, i.e., the unit polarization vector **d** is normal to **s**, meaning $\mathbf{s} \cdot \mathbf{d} = 0$, e.g.,

$$\mathbf{d}_P = (\sin\theta, 0, \cos\theta), \quad \mathbf{d}_{SV} = (\cos\theta, 0, -\sin\theta), \quad \mathbf{d}_P \cdot \mathbf{d}_{SV} = 0.$$

It can be shown, using the rules for the dot and cross products, that for the preceding numerical example, $\mathbf{s} \times \mathbf{d} = \mathbf{0}$ for the *P* wave, and $\mathbf{s} \cdot \mathbf{d} = 0$ for the *SV* wave. If the wave is an *SH* wave, then **d** is parallel to the *y* axis.

2.17 Displacement Potentials

Consider the vector identities

$$\nabla^2 \mathbf{A} = \nabla(\nabla \cdot \mathbf{A}) - \nabla \times (\nabla \times \mathbf{A}) \quad \text{and} \quad \nabla \cdot (\nabla \times \mathbf{A}) = 0. \tag{2.67}$$

Let $\mathbf{w} \equiv \nabla^2 \mathbf{A}$, $\phi \equiv \nabla \cdot \mathbf{A}$, and $\boldsymbol{\psi} \equiv -\nabla \times \mathbf{A}$. Then these identities can be written as

$$\mathbf{w} = \nabla\phi + \nabla \times \boldsymbol{\psi} \quad \text{and} \quad \nabla \cdot \boldsymbol{\psi} = 0. \tag{2.68}$$

This is called the Helmholtz decomposition of the vector **w**. Any well-defined vector **w** can be written in this form, i.e., as the sum of a gradient and a curl. Consequently, we can write the particle displacement **u** and the body force **f** as

$$\mathbf{u} = \nabla\phi + \nabla \times \boldsymbol{\psi} \quad \text{and} \quad \mathbf{f} = \nabla\Phi + \nabla \times \boldsymbol{\Psi}. \tag{2.69}$$

Substituting this into the equation of motion (2.41) with **f** included, i.e.,

$$(\lambda + \mu)\nabla(\nabla \cdot \mathbf{u}) + \mu\nabla^2\mathbf{u} + \mathbf{f} = \rho\frac{\partial^2 \mathbf{u}}{\partial t^2}, \tag{2.70}$$

and using (2.67), yields

$$\nabla\left[(\lambda + 2\mu)\nabla^2\phi + \Phi - \rho\frac{\partial^2\phi}{\partial t^2}\right] + \nabla \times \left[\mu\nabla^2\boldsymbol{\psi} + \boldsymbol{\Psi} - \rho\frac{\partial^2\boldsymbol{\psi}}{\partial t^2}\right] = \mathbf{0}. \tag{2.71}$$

This equation could also have been derived from the equation of motion (2.42). If we take the divergence and curl of the preceding equation, and use (2.67) and $\nabla \times \nabla g = \mathbf{0}$, where *g* is an arbitrary scalar function, we obtain the following equations:

$$\nabla^2\left[(\lambda + 2\mu)\nabla^2\phi + \Phi - \rho\frac{\partial^2\phi}{\partial t^2}\right] = 0,$$

and

$$\nabla^2\left[\mu\nabla^2\boldsymbol{\psi} + \boldsymbol{\Psi} - \rho\frac{\partial^2\boldsymbol{\psi}}{\partial t^2}\right] = \nabla\left[\mu\nabla^2(\nabla\cdot\boldsymbol{\psi}) + \nabla\cdot\boldsymbol{\Psi} - \rho\frac{\partial^2}{\partial t^2}(\nabla\cdot\boldsymbol{\psi})\right]. \quad (2.72)$$

Using $\alpha^2 = (\lambda + 2\mu)/\rho$ and $\beta^2 = \mu/\rho$, these equations are satisfied if

$$\nabla^2\phi - \frac{1}{\alpha^2}\frac{\partial^2\phi}{\partial t^2} = -\frac{\Phi}{\rho\alpha^2}, \quad \nabla^2\boldsymbol{\psi} - \frac{1}{\beta^2}\frac{\partial^2\boldsymbol{\psi}}{\partial t^2} = -\frac{\boldsymbol{\Psi}}{\rho\beta^2}, \quad \nabla\cdot\boldsymbol{\psi} = 0, \quad \nabla\cdot\boldsymbol{\Psi} = 0.$$
$$(2.73)$$

Note that the first two equations are classical 3D inhomogeneous wave equations, which have well-known solutions. Hence, to solve the equation of motion for a homogeneous medium, one needs only to solve a pair of classical wave equations, and then use $\mathbf{u} = \nabla\phi + \nabla\times\boldsymbol{\psi}$ to obtain $\mathbf{u}$.

The potentials ϕ and $\boldsymbol{\psi}$ satisfy 3D wave equations that describe compressional and shear waves moving with speeds α and β, respectively. ϕ and $\boldsymbol{\psi}$ are known as the **Helmholtz displacement potentials** for compressional and shear waves, respectively. ϕ is a scalar potential function and $\boldsymbol{\psi}$ is a vector potential function.

Note that $\boldsymbol{\psi}$ is actually arbitrary to within a gradient, i.e., $\boldsymbol{\psi}' \equiv \boldsymbol{\psi} + \nabla g$, where g is an arbitrary function, would be an equally good S wave potential, because $\nabla\times\boldsymbol{\psi}' = \nabla\times\boldsymbol{\psi}$, meaning that the same $\mathbf{u}$ would be obtained regardless of which S-potential is used.

Note also that (2.72) shows that $\boldsymbol{\psi}$ could be chosen so that $\nabla\cdot\boldsymbol{\psi}$ satisfies a wave equation rather than $\nabla\cdot\boldsymbol{\psi} = 0$. However, the latter is simpler and is also sufficient. To see this, assume that $\nabla\cdot\boldsymbol{\psi}$ actually does satisfy a wave equation. Then note that $\nabla\cdot\boldsymbol{\psi}' = \nabla\cdot\boldsymbol{\psi} + \nabla^2 g$, meaning that if g is chosen so that $\nabla^2 g = -\nabla\cdot\boldsymbol{\psi}$, then $\nabla\cdot\boldsymbol{\psi}' = 0$, showing that $\boldsymbol{\psi}'$ satisfies the simpler condition. Since either $\boldsymbol{\psi}$ or $\boldsymbol{\psi}'$ is acceptable as a potential, we choose the simpler condition for the S-potential.

The condition, $\nabla\cdot\boldsymbol{\psi} = 0$, is called a **gauge** condition. It is needed because $\mathbf{u} = \nabla\phi + \nabla\times\boldsymbol{\psi}$ gives three equations (for u_x, u_y, u_z) in four unknowns (ϕ, ψ_x, ψ_y, ψ_z) – the gauge condition provides the required fourth equation.

Consider the following harmonic plane wave solutions for ϕ and $\boldsymbol{\psi}$ (neglect the body forces):

$$\phi = A\exp\left[i\omega(\mathbf{s}_\alpha\cdot\mathbf{x} - t)\right], \quad \boldsymbol{\psi} = \mathbf{B}\exp\left[i\omega(\mathbf{s}_\beta\cdot\mathbf{x} - t)\right], \quad s_\alpha = 1/\alpha, \quad s_\beta = 1/\beta.$$

ϕ represents a compressional wave traveling in the direction of $\mathbf{s}_\alpha$, and $\boldsymbol{\psi}$ represents a shear wave traveling in the direction of $\mathbf{s}_\beta$. The displacement $\mathbf{u}$ that a particle experiences due to the passage of the two waves is then given by $\mathbf{u} = \nabla\phi + \nabla\times\boldsymbol{\psi}$, which yields

$$\mathbf{u} = \mathbf{u}_\alpha + \mathbf{u}_\beta, \quad \mathbf{u}_\alpha \equiv \nabla\phi = i\omega A\mathbf{s}_\alpha\exp\left[i\omega(\mathbf{s}_\alpha\cdot\mathbf{x} - t)\right],$$
$$\mathbf{u}_\beta \equiv \nabla\times\boldsymbol{\psi} = i\omega\,\mathbf{s}_\beta\times\mathbf{B}\exp\left[i\omega(\mathbf{s}_\beta\cdot\mathbf{x} - t)\right].$$

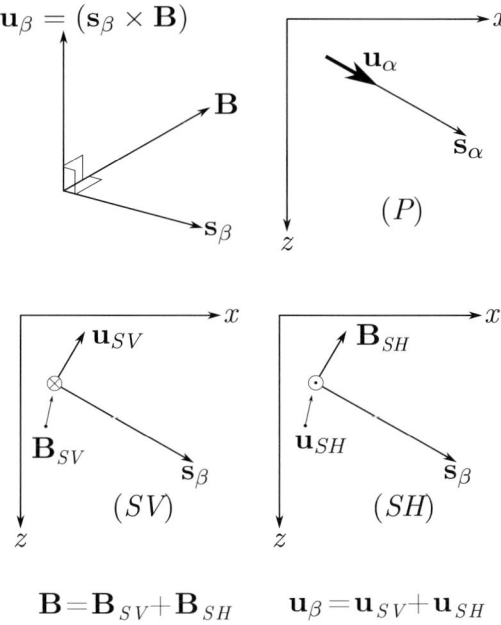

Figure 2.7 Displacements associated with potentials.

Since $\mathbf{u}_\alpha \propto \mathbf{s}_\alpha$, plane P waves cause particles in their path to move parallel to $\mathbf{s}_\alpha$, i.e., parallel to the P wave's direction of propagation. Similarly, since $\mathbf{u}_\beta \propto (\mathbf{s}_\beta \times \mathbf{B})$, meaning that $\mathbf{u}_\beta$ is perpendicular to both $\mathbf{s}_\beta$ and $\mathbf{B}$, plane S waves cause particles in their path to move perpendicular to $\mathbf{s}_\beta$, i.e., perpendicular to the S wave's direction of propagation. Furthermore,

$$\nabla \cdot \boldsymbol{\psi} = 0 \quad \Longrightarrow \quad \mathbf{s}_\beta \cdot \mathbf{B} = 0,$$

meaning that $\mathbf{s}_\beta$ is perpendicular to $\mathbf{B}$. Hence, $\mathbf{u}_\beta$, $\mathbf{s}_\beta$ and $\mathbf{B}$ are all orthogonal to each other (see Figure 2.7).

Consider an *xyz* coordinate system where the *xy* plane ($z = 0$) is the horizontal ground surface, and the vertical z axis is the depth axis. Suppose that the plane P and S waves are propagating in the *xz* plane, i.e., $\mathbf{s}_\alpha$ and $\mathbf{s}_\beta$ are parallel to the *xz* plane. This is a common situation in seismology. Then $\mathbf{u}_\alpha$ is parallel to the *xz* plane (and parallel to $\mathbf{s}_\alpha$), and $\mathbf{u}_\beta$ (which is perpendicular to $\mathbf{s}_\beta$) can be resolved into two components, one which is parallel to the *xz* plane, known as the *SV* wave component (shear vertical), and a component which is perpendicular to the *xz* plane (i.e., in the y direction), known as the *SH* wave component (shear horizontal). For the *SV* wave, $\mathbf{B}$ must point in the $\pm y$ direction, which means that $\boldsymbol{\psi} = (0, \psi, 0)$, where $\psi = B \exp\left[i\omega(\mathbf{s}_\beta \cdot \mathbf{x} - t)\right]$, i.e., the potential for the *SV* wave is, in effect, a scalar potential, just like the potential for the P wave – only a scalar wave equation needs to be solved, not a vector wave equation. A little math shows that

$$\mathbf{u}_{SV} = \nabla \times \boldsymbol{\psi}_{SV} = \nabla \times (0, \psi, 0) = \left(-\frac{\partial \psi}{\partial z}, 0, \frac{\partial \psi}{\partial x}\right). \qquad (2.74)$$

For the *SH* wave, **B** lies in the *xz* plane, meaning ψ is a vector potential, i.e., ψ = $(\psi_x, 0, \psi_z)$. However, for the *SH* case, potentials are not normally used, because the equation of motion (2.70), applied to an *SH* wave in the *xz* plane, reduces to a scalar wave equation – since $\nabla \cdot \mathbf{u} = 0$ for shear waves, and since $\mathbf{u} = (0, u, 0)$ for a plane *SH* wave, (2.70) reduces to $\mu \nabla^2 u + f = \rho \ddot{u}$. Consequently, for waves propagating parallel to the *xz* plane, the displacements of all three wave types, *P*, *SV*, and *SH*, can be obtained by solving only scalar wave equations. Because **u** for *P* and *SV* waves lies in the *xz* plane and **u** for *SH* waves is normal to the *xz* plane, the *SH* plane wave propagation problem can be treated apart from the *P-SV* problem.

Using $\nabla \times \nabla \phi = \mathbf{0}$, we see that *P* waves are irrotational waves, since $\nabla \times \mathbf{u}_\alpha = \mathbf{0}$. Using (2.67), we see that *S* waves are solenoidal waves, since $\nabla \cdot \mathbf{u}_\beta = 0$.

A more mathematically complete statement of the Helmholtz decomposition rule is Lamé's theorem: given the equation of motion for an isotropic elastic medium, if the following are true,

$$\mathbf{f} = \nabla \Phi + \nabla \times \boldsymbol{\Psi}, \quad \dot{\mathbf{u}}(\mathbf{x}, 0) = \nabla A + \nabla \times \mathbf{B}, \quad \mathbf{u}(\mathbf{x}, 0) = \nabla C + \nabla \times \mathbf{D},$$

$$\text{with} \quad \nabla \cdot \boldsymbol{\Psi} = 0, \quad \nabla \cdot \mathbf{B} = 0, \quad \nabla \cdot \mathbf{D} = 0,$$

then ϕ and ψ exist such that the following hold:

$$\mathbf{u} = \nabla \phi + \nabla \times \boldsymbol{\psi}, \quad \nabla \cdot \boldsymbol{\psi} = 0, \quad \ddot{\phi} = \frac{\Phi}{\rho} + \alpha^2 \nabla^2 \phi, \quad \ddot{\boldsymbol{\psi}} = \frac{\boldsymbol{\Psi}}{\rho} + \beta^2 \nabla^2 \boldsymbol{\psi}. \quad (2.75)$$

2.18 Helmholtz Equations

For some types of wave propagation problems, the spatial part of the solutions for ϕ and ψ are more complicated than the harmonic form, $\exp[i\omega \mathbf{s} \cdot \mathbf{x}]$, used in the preceding section. Consequently, trial solutions that are sinusoidal in time, but spatially more complicated, are attempted. For instance, assume that the *P* and *SV* wave potentials are given by $\phi = \bar{\phi}(x, y, z) \exp(-i\omega t)$ and $\psi = \bar{\psi}(x, y, z) \exp(-i\omega t)$, with similar expressions for the body force potentials. Substitution of these into the 3D wave equations for ϕ and ψ gives

$$\nabla^2 \bar{\phi} + \omega^2 s_\alpha^2 \, \bar{\phi} = -\left(\bar{\Phi}/\rho\alpha^2\right), \qquad \nabla^2 \bar{\psi} + \omega^2 s_\beta^2 \, \bar{\psi} = -\left(\bar{\Psi}/\rho\beta^2\right). \qquad (2.76)$$

These are known as the **Helmholtz equations** for $\bar{\phi}$ and $\bar{\psi}$. The plane wave solutions are, for zero body forces, for example, $\bar{\phi} = \exp(i\omega s_\alpha \cdot \mathbf{x})$ and $\bar{\psi} = \exp(i\omega \mathbf{s}_\beta \cdot \mathbf{x})\mathbf{n}$, where **n** is a unit vector. However, more complicated solutions exist as well (e.g., linear combinations of plane wave solutions – in general, Fourier series). The Helmholtz equations can also be obtained more rigorously by Fourier transforming the 3D wave equations, using the rule that

$$\mathcal{F}\{\partial\phi/\partial t\} = -i\omega\mathcal{F}\{\phi\},$$

where $\mathcal{F}$ denotes the Fourier transform (see Equation 1.150). In that case, $\bar{\phi}$ and $\bar{\psi}$ would be the forward Fourier transforms of ϕ and ψ , i.e.,

$$\bar{\phi}(x, y, z, \omega) = \int_{-\infty}^{\infty} \phi(x, y, z, t)e^{i\omega t} \, dt,$$

$$\bar{\psi}(x, y, z, \omega) = \int_{-\infty}^{\infty} \psi(x, y, z, t)e^{i\omega t} \, dt, \qquad (2.77)$$

with similar expressions existing for the body force potentials. After solutions for $\bar{\phi}$ and $\bar{\psi}$ are found, ϕ and ψ can be obtained through an inverse Fourier transform, i.e.,

$$\phi(x, y, z, t) = \frac{1}{2\pi} \int_{-\infty}^{\infty} \bar{\phi}(x, y, z, \omega)e^{-i\omega t} \, d\omega,$$

$$\psi(x, y, z, t) = \frac{1}{2\pi} \int_{-\infty}^{\infty} \bar{\psi}(x, y, z, \omega)e^{-i\omega t} \, d\omega.$$

Helmholtz equations are often simpler to work with than the original wave equations.

2.19 Near-Field and Far-Field Waves

Consider a P wave generated by a spherically symmetric point source. The wave equation for the P wave potential ϕ, with the body forces ignored, can then be written, in spherical coordinates, as

$$\nabla^2\phi = \frac{1}{r^2}\frac{\partial}{\partial r}\left(r^2\frac{\partial\phi}{\partial r}\right) = \frac{1}{\alpha^2}\frac{\partial^2\phi}{\partial t^2}.$$

The solution for an outgoing wave is

$$\phi = \frac{1}{r}f\left(t - \frac{r}{\alpha}\right),$$

where f is an arbitrary function. The displacement $\mathbf{u} = \nabla\phi = (\partial\phi/\partial r)\mathbf{e}_r$ in the case of spherical symmetry. Hence, the radial component, u_r, is given by

$$u_r = \frac{\partial\phi}{\partial r} = -\frac{1}{\alpha r}f'\left(t - \frac{r}{\alpha}\right) - \frac{1}{r^2}f\left(t - \frac{r}{\alpha}\right),$$

where $f'(\xi) \equiv df(\xi)/d\xi$. Note the $1/r$ and $1/r^2$ factors. Since the function $1/r$ dominates the function $1/r^2$ at large distances r, and since the function $1/r^2$ dominates the function $1/r$ at small distances r, the first term in u_r is called the **far-field (FF)**

wave, and the second is called the **near-field (NF) wave**. The FF wave will be much bigger than the NF wave far from the source where the receivers typically are, so for most practical purposes, the NF term can be neglected. However, for receivers close enough to the source, i.e., for waves that have not traveled very far, the NF term would have to be included in u_r. Note that the FF waveform, i.e., the shape of the FF wave, has the shape of the *derivative* of f, whereas the NF waveform has the shape of f.

2.20 Mean Values of Energy-Related Quantities

It is possible to gain some insight into the nature of harmonic wave propagation by calculating the mean (time-averaged) values of the various energy-related quantities (kinetic energy density, etc.). For example, consider a harmonic P wave

$$\mathbf{u} = A \exp[i\omega(\mathbf{s}\cdot\mathbf{x} - t)]\,\mathbf{d} = |A|\exp[i\omega(\mathbf{s}\cdot\mathbf{x} - t) + i\phi]\,\mathbf{d}, \qquad (2.78a)$$

where A is, in general, a complex amplitude, i.e., $A = |A|\exp(i\phi)$, and where $\mathbf{d}$ is the polarization vector, a constant real unit vector. For P waves, $\mathbf{d} = \alpha\mathbf{s}$ ($\Rightarrow |\mathbf{d}| = \sqrt{\mathbf{d}\cdot\mathbf{d}} = 1$). Let us replace $\mathbf{u}$ by its real part, the actual physical wave:

$$\mathbf{u} = |A|\cos[\omega(\mathbf{s}\cdot\mathbf{x} - t) + \phi]\,\mathbf{d}. \qquad (2.78b)$$

Suppose we want to calculate the mean value of the kinetic energy density K. As a reminder, the mean value $\langle g(t)\rangle$ of a function $g(t)$ over the time period $t = 0$ to $t = T$ is given by

$$\langle g(t)\rangle = \frac{\int_0^T g(t)\,dt}{\int_0^T dt} = \frac{1}{T}\int_0^T g(t)\,dt. \qquad (2.78c)$$

For a sinusoidal function, T is often taken to be the period. Substitution of Equation (2.78b) into the formula for K (Equation 2.15b) gives

$$K = \frac{1}{2}\rho\sum_i\left[\frac{\partial u_i}{\partial t}\right]^2 = \frac{1}{2}\rho\frac{\partial\mathbf{u}}{\partial t}\cdot\frac{\partial\mathbf{u}}{\partial t} = \tfrac{1}{2}\rho\omega^2|A|^2\sin^2[\omega(\mathbf{s}\cdot\mathbf{x} - t) + \phi]. \quad (2.79)$$

Hence, calculating the mean value of K involves calculating the mean value of $\sin^2[\cdots]$, which is $\frac{1}{2}$:

$$\left\langle\sin^2[\omega(\mathbf{s}\cdot\mathbf{x} - t) + \phi]\right\rangle = \frac{\int_0^T \sin^2[\omega(\mathbf{s}\cdot\mathbf{x} - t) + \phi]\,dt}{\int_0^T dt} = \frac{1}{2}, \qquad (2.80)$$

where $T = 2\pi/\omega$ is the period. We then have

$$\langle K\rangle = \tfrac{1}{4}\rho\omega^2|A|^2. \qquad (2.81)$$

In the same way, it can also be shown that the mean value of the strain energy density is $\langle W \rangle = \langle K \rangle = \frac{1}{4}\rho\omega^2|A|^2$. Hence, the mean value of the total energy density E is

$$\langle E \rangle = \langle K + W \rangle = \frac{1}{2}\rho\omega^2|A|^2. \tag{2.82}$$

Note that $\langle E \rangle$ is proportional to the square of the frequency, which states that the mean energy density increases rapidly with frequency.

The mean intensity can be calculated in the same way, using I_j in (2.19), i.e., $I_j = -\sum_i \sigma_{ij}\dot{u}_i$. The result is, after some calculation,

$$\langle \mathbf{I} \rangle = \frac{1}{2}\rho\alpha\omega^2|A|^2\mathbf{n}, \tag{2.83}$$

where $\mathbf{n}$ is a unit vector in the direction of the slowness $\mathbf{s}$. Therefore, the mean intensity is in the direction of wave propagation. For a P wave, $\mathbf{s} = (1/\alpha)\mathbf{n}$. In this case, $\langle \mathbf{I} \rangle$ can also be obtained through a simpler intuitive approach. Suppose the P wave is passing through one end of a fictitious cylindrical tube (whose axis is parallel to $\mathbf{n}$, the direction of wave propagation) with cross-sectional area a and length l. If the wave energy travels the length l in a time t, then $l = \alpha t$. Hence the mean total energy of motion in the cylinder is the mean energy density times the volume of the tube, i.e., $\langle E \rangle la$, which is equal to $\langle E \rangle \alpha ta$. Therefore, the mean intensity (the mean total energy passing through unit area in unit time) is $\langle E \rangle \alpha$, which is, from Equation (2.82), equal to $\frac{1}{2}\rho\alpha\omega^2|A|^2$. This confirms Equation (2.83). Although, in this case, the intuitive calculation of $\langle \mathbf{I} \rangle$ is much simpler than the one using I_j in Equation (2.19), the latter approach has the distinct advantage that it can be used in more complicated cases where intuition fails.

For an S wave, the same result, (2.83), for $\langle \mathbf{I} \rangle$ is obtained except that α is replaced by β.

In calculating $\langle K \rangle$ in (2.81), we had to take the real part of $\mathbf{u}$ (Equation 2.78b) first, because K in (2.79) is nonlinear in $\mathbf{u}$. However, there is a way to calculate mean values using exponentials, i.e., without first taking the real parts of the wave functions. In fact, in general it is more convenient to use complex exponentials. The calculation of time averages of products of wave functions expressed as complex exponentials is made easier by the following rule: if $\mathcal{F} = \psi_1\psi_2$, and if ψ_1, B_1, ψ_2 and B_2 are complex functions such that

$$\psi_1(\mathbf{x}, t) = B_1(\mathbf{x})\,e^{\pm i\omega t} \qquad \text{and} \qquad \psi_2(\mathbf{x}, t) = B_2(\mathbf{x})\,e^{\pm i\omega t}$$

i.e., if ψ_1 and ψ_2 can be separated into space and complex exponential time factors, then

$$\langle \mathcal{F} \rangle = \langle\, \mathrm{Re}(\psi_1)\mathrm{Re}(\psi_2)\,\rangle = \frac{1}{2}\mathrm{Re}(\psi_1\psi_2^*) = \frac{1}{2}\mathrm{Re}(\psi_1^*\psi_2), \tag{2.84}$$

where the superscript "$*$" denotes the complex conjugate.

As an example, we recalculate $\langle K \rangle$ using complex exponentials. The jth component of displacement, from (2.78a), is given by

$$u_j = A e^{i\omega(\mathbf{s}\cdot\mathbf{x}-t)} d_j, \quad \Longrightarrow \quad \frac{\partial u_j}{\partial t} = -i\omega A e^{i\theta} d_j, \quad \theta \equiv \omega(\mathbf{s}\cdot\mathbf{x} - t).$$

Therefore, using the preceding rule, we have

$$
\begin{aligned}
\langle K \rangle &= \tfrac{1}{2}\rho \left\langle \sum_j \mathrm{Re}\left[\frac{\partial u_j}{\partial t}\right] \mathrm{Re}\left[\frac{\partial u_j}{\partial t}\right] \right\rangle = \tfrac{1}{2}\rho \sum_j \left\langle \mathrm{Re}\left[\frac{\partial u_j}{\partial t}\right] \mathrm{Re}\left[\frac{\partial u_j}{\partial t}\right] \right\rangle \\
&= \tfrac{1}{4}\rho \sum_j \mathrm{Re}\left[\frac{\partial u_j}{\partial t}\left(\frac{\partial u_j}{\partial t}\right)^*\right] = \tfrac{1}{4}\rho \sum_j \mathrm{Re}\left[-i\omega A d_j e^{i\theta} i\omega A^* d_j e^{-i\theta}\right] \\
&= \tfrac{1}{4}\rho \sum_j \mathrm{Re}\left[\omega^2 |A|^2 d_j^2\right] = \tfrac{1}{4}\rho\omega^2 |A|^2,
\end{aligned}
$$

where $AA^* = |A|^2$ and $\sum_j d_j^2 = 1$. Although either of the preceding methods for computing mean values can be used, the complex exponential method, together with the preceding simplifying rule, is more convenient to use in more complicated cases, such as computing mean values in dissipative anisotropic media.

The following example shows how to calculate the mean intensity of a plane S wave, using complex exponentials. Note that the summation convention is used, as well as the comma notation for derivatives.

For S waves, the dilatation $\mathcal{D} = 0$, so $\sigma_{ij} = 2\mu e_{ij} = \mu(u_{i,j} + u_{j,i})$, with the displacement $u_k = U \exp[i\omega(s_n x_n - t)]d_k$, and $|\mathbf{d}| = 1$. Also, $\mathbf{s} \cdot \mathbf{d} = 0$ for S waves. In the following, $\eta \equiv s_n x_n - t$.

$$
\begin{aligned}
\langle I_j \rangle &= -\langle \mathrm{Re}(\sigma_{ij})\mathrm{Re}(\dot{u}_i) \rangle = -\tfrac{1}{2}\mathrm{Re}[\sigma_{ij}\dot{u}_i^*] = -\tfrac{1}{2}\mathrm{Re}[\mu u_{i,j}\dot{u}_i^* + \mu u_{j,i}\dot{u}_i^*] \\
&= -\tfrac{1}{2}\mu\mathrm{Re}[i\omega s_j U e^{i\omega\eta} d_i(-i\omega U e^{i\omega\eta} d_i)^* + i\omega s_i U e^{i\omega\eta} d_j(-i\omega U e^{i\omega\eta} d_i)^*] \\
&= -\tfrac{1}{2}\mu\mathrm{Re}[i\omega s_j U e^{i\omega\eta} d_i(i\omega U^* e^{-i\omega\eta} d_i) + i\omega s_i U e^{i\omega\eta} d_j(i\omega U^* e^{-i\omega\eta} d_i)].
\end{aligned}
$$

In the last line, it is assumed that $\mathbf{s}$ and $\mathbf{d}$ are real (meaning $s_n^* = s_n$ and $d_i^* = d_i$), which is the case for body waves in an infinite, homogeneous, perfectly elastic medium (no absorption). The radial frequency ω is, of course, also real. Hence,

$$\langle I_j \rangle = -\tfrac{1}{2}\mu\mathrm{Re}[-\omega^2 |U|^2 d_i d_i s_j - \omega^2 |U|^2 s_i d_i d_j] = \tfrac{1}{2}\mu\omega^2 |U|^2 s_j$$
$$(UU^* = |U|^2, \quad d_i d_i = |\mathbf{d}|^2 = 1, \quad s_i d_i = \mathbf{s} \cdot \mathbf{d} = 0 \text{ for S waves}).$$

Letting $\mathbf{n}$ be a unit vector in the direction of slowness $\mathbf{s}$, i.e., $\mathbf{s}=(1/\beta)\mathbf{n}$, and using $\mu = \rho\beta^2$, we have

$$\langle I_j \rangle = \tfrac{1}{2}\rho\beta\omega^2 |U|^2 n_j \quad \text{or} \quad \langle \mathbf{I} \rangle = \tfrac{1}{2}\rho\beta\omega^2 |U|^2 \mathbf{n},$$

which shows that the mean intensity, i.e., mean energy flux, is in the direction of wave propagation, i.e., in the direction of the slowness **s** (this is not true for anisotropic and anelastic media).

2.21 Inhomogeneous Media

The derivation of Equation (2.41), or (2.42), the equation of motion for a homogeneous isotropic medium, involved taking spatial derivatives of the stress–strain relation (Equation 2.30). We set $\partial\lambda/\partial x_j = 0$ and $\partial\mu/\partial x_j = 0, j = 1, 2, 3$, since the medium was homogeneous. However, in an inhomogeneous medium, also called a **heterogeneous** medium, λ and μ vary with position **x**, and so the equation of motion will include spatial derivatives of λ and μ. Equivalently, the wave speeds α and β vary with position.

Unlike homogeneous media, the deformations due to a seismic wave in an inhomogeneous medium do not separate distinctly into compressional and shear motions. There is a coupling between the two types of motion. However, if the inhomogeneity is weak, i.e., if λ and μ (or α and β) do not change much over a wavelength (the wave has a short enough wavelength or a high enough frequency), then waves that are *predominantly P*, *SV*, or *SH* can propagate. Ray paths in an inhomogeneous medium are curved rather than straight.

Theories of wave propagation in an inhomogeneous medium are generally mathematically complicated. Exact solutions to the corresponding equation of motion cannot be obtained. However, theoretical development continues because the real Earth is inhomogeneous.

2.22 Other Types of Media

So far, we have considered homogeneous, isotropic, perfectly elastic media, and in the previous section, we discussed inhomogeneity. A medium may also be **anisotropic** and **anelastic**.

Anisotropy means that the properties of a medium vary with direction. For instance, the *P* wave speed in the *x* direction in a homogeneous medium may be different from that in the *z* direction, producing ellipsoidal (rather than spherical) wavefronts.

Anelasticity means that the medium does not react to seismic waves in a perfectly elastic way. Wave energy is absorbed by the medium as the wave propagates. The waves experience frequency-dependent **attenuation** and **dispersion**. Attenuation means that the amplitude of the wave decreases as the wave propagates, due to anelasticity, and dispersion means that the wave speed varies with frequency.

Generally, the higher the frequency of the wave, the faster it travels and the more it is attenuated. These effects are caused by the phenomenon of **internal friction**. The anelasticity of the Earth is often modeled mathematically by the theory of viscoelasticity. In this theory, the stress–strain relation is more complicated, and λ, μ, etc., depend on time t. Hence, the seismic response of the medium at a given time depends on the past "history" of the medium – it is said to have a "memory."

Needless to say, mathematical complications abound in wave propagation theories for inhomogeneous, anisotropic, anelastic media.

2.23 The Acoustic Wave Equation

Acoustic waves are compressional (P) waves in a fluid or solid, hence the wave equation (2.43) for the dilatation $\mathcal{D}$ can be applied. From the equation $\sigma_{ij} = -p\delta_{ij}$ in Section 2.1, and from (2.32a), where $\mathcal{S} = \sum_k \sigma_{kk} = -3p = 3k_B\mathcal{D}$, we may conclude that the pressure p in a fluid or a homogeneous elastic solid under hydrostatic stress is given by $p = -k_B\mathcal{D}$, where $k_B = \lambda + \frac{2}{3}\mu$ is the bulk modulus. Hence, the dilatational wave equation becomes

$$\nabla^2 p = \frac{1}{\alpha^2}\frac{\partial^2 p}{\partial t^2}. \tag{2.85}$$

This is the acoustic wave equation for a solid. Note that here, p can simply be taken to be the pressure produced by the wave motion. For a fluid, $\mu = 0$ because fluids support no shear, and hence $k_B = \lambda$ and $\alpha = \sqrt{\lambda/\rho} = \sqrt{k_B/\rho}$. The acoustic wave equation has been used in the theoretical development of wave equation migration methods.

2.24 Seismic Waves Generated by a Buried Explosive Charge

A classical problem in seismology is the determination of the wave field $\mathbf{u}$ generated by an explosive source in an infinite homogeneous medium. The solution allows one to draw some conclusions about the nature of seismic waves produced by buried explosive charges in the real Earth.

For mathematical simplicity, let us consider a spherical cavity of radius a in an infinite homogeneous medium. We want to calculate the radial particle displacement at a distance r from the center of the cavity, due to an explosion in the cavity. Let the explosion be represented by a pressure pulse $p(t)$. Using spherical coordinates, the displacement is $\mathbf{u} = u_r\mathbf{e}_r + u_\theta\mathbf{e}_\theta + u_\phi\mathbf{e}_\phi$, but because of the spherical symmetry of the situation we have $u_\theta = u_\phi = 0$ and $u \equiv u_r$. The spherical symmetry also implies that the waves generated are purely compressional waves and that the P wave potential ϕ depends on r, but not on the angles θ or ϕ, i.e., $\phi = \phi(r, t)$, with $u = \partial\phi/\partial r$. (Note that the symbol ϕ is being used here for both the P wave potential and the azimuthal angle in the spherical polar coordinate system. This should not cause any confusion, as the angle ϕ will no longer appear beyond this point because of spherical symmetry.) Hence the wave equation for the P wave potential ϕ becomes

$$\nabla^2\phi = \frac{1}{r^2}\frac{\partial}{\partial r}\left(r^2\frac{\partial\phi}{\partial r}\right) = \frac{1}{\alpha^2}\frac{\partial^2\phi}{\partial t^2}. \tag{2.86}$$

There is also one boundary condition: the radial stress $\sigma_{rr} = \sigma_{rr}(r, t)$ must be equal to $-p(t)$ on the surface of the cavity, i.e.,

$$\sigma_{rr}(a, t) = -p(t). \tag{2.87}$$

It can be shown that this can be written in terms of the potential ϕ as

$$\rho\alpha^2\left[\frac{\partial^2\phi}{\partial r^2} + \left(\frac{2\sigma}{1-\sigma}\right)\frac{1}{r}\frac{\partial\phi}{\partial r}\right]_{r=a} = -p(t). \tag{2.88}$$

To find the solution $\phi(r, t)$, we Fourier transform equations (2.86) and (2.88) into the frequency domain, solve for $\bar{\phi}(r, \omega)$, and then transform back to the time domain. The result is

$$\phi(r, t) = \frac{a^3}{2\pi r\rho}\int_{-\infty}^{\infty}\int_{-\infty}^{\infty}\frac{p(\tau)\exp\left[i\omega(\tau - t_r)\right]}{a^2\omega^2 + ia\alpha\epsilon\omega - \epsilon\alpha^2}\,d\omega\,d\tau, \tag{2.89}$$

where

$$\epsilon \equiv \frac{2(1 - 2\sigma)}{1 - \sigma} \quad \text{and} \quad t_r \equiv t - \frac{r - a}{\alpha} \tag{2.90}$$

(see Grant and West, 1965, p. 37). t_r is called the **reduced time** or **retarded time**. The final step consists of specifying the mathematical form of the pressure pulse $p(t)$, working out the integrals, and then obtaining the displacement u from $u = \partial\phi/\partial r$.

A relatively simple result for u can be obtained if we assume a step function pressure pulse, i.e., $p(t) = p_0$ for $t \geq 0$, and $p(t) = 0$ for $t < 0$, where p_0 is a constant, and assume that the detector (located at r) is far from the cavity, i.e., $r \gg a$. Then, in the expression for u, we can drop the terms that are of order $1/r^2$ and higher. The result is

$$u \approx \frac{a^2 p_0\sqrt{1 - 2\sigma}}{2\mu}\frac{e^{-ct_r}}{r}\sin(\chi t_r), \tag{2.91}$$

where $t_r \geq 0$ ($u = 0$ for $t_r < 0$), and where

$$\chi \equiv \frac{\alpha}{a}\frac{\sqrt{1 - 2\sigma}}{1 - \sigma} \quad \text{and} \quad c \equiv \frac{\alpha}{a}\left(\frac{1 - 2\sigma}{1 - \sigma}\right).$$

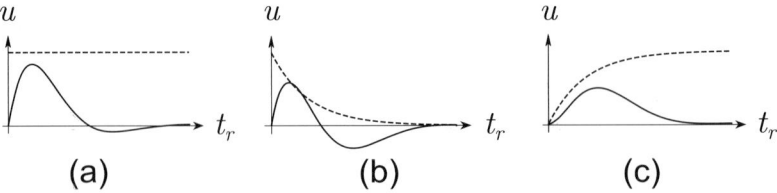

Figure 2.8 Schematic graphs of u and $p(t)$. The horizontal axis is the reduced time t_r for u and the actual time t for p. The displacement u is the solid line and the pressure p is the dashed line, where (a) $p(t) = p_0$, (b) $p(t) = p_0 e^{-bt}$, and (c) $p(t) = p_0(1 - e^{-bt})$, and where $\sigma = \frac{1}{4}$ and $b = \chi/\sqrt{2} = 2\alpha/3a$.

Note that in the expression for t_r, the second term $(r - a)/\alpha$ is the time that it takes for the P wave to go from the surface of the cavity to r. This implies that a particle at r starts to feel the pulse at time $t = (r - a)/\alpha$, i.e., at $t_r = 0$.

Equation (2.91) was originally obtained for $\lambda = \mu$ (i.e., $\sigma = \frac{1}{4}$) by Sharpe (1942a,b). Figure 2.8 shows u for three different pressure functions $p(t)$.

From the preceding results, we can draw some conclusions (Sharpe, 1942a). These conclusions can be applied to reflected signals as well, since reflection produces only amplitude and phase changes.

The fact that the amplitude is proportional to $1/\mu$ implies that as the rigidity increases, the amplitude decreases. This means that the amplitude of the signal should be greater in a medium of low rigidity (low μ) such as clay or water-saturated sand, than in a medium of high rigidity such as limestone. This agrees with field observations.

The following observation has been made in fieldwork: the preliminary "springing" of a hole with a large charge to create a large cavity results in larger amplitudes of reflected motion being obtained from subsequent small charges than would be obtained without springing. Since $u \propto a^2 p_0$, and since $p_0 \sim 1/a$ (due to the spherical divergence of the pressure wave propagating in the fluid in the hole), we see that $u \propto a$, supporting the field observation.

If one takes the Fourier transform of $u(r, t)$ to obtain the frequency spectrum, one finds that the spectrum has a peak at the frequency f_0, where

$$f_0 = \frac{\chi}{2\pi} = \left[\frac{\sqrt{1 - 2\sigma}}{2\pi(1 - \sigma)} \right] \frac{\alpha}{a}. \tag{2.92}$$

This is the dominant frequency in the waveform for u, which is an exponentially decaying sine wave. f_0 is just the frequency of the sine wave, as one might expect. Equation (2.92) implies that shots fired in a high rigidity formation should produce waves that are richer in high frequencies than shots fired in a low rigidity formation of the same σ-value, because in the high rigidity medium, α is larger and a is probably smaller as well. This agrees with field observations. For example, shots fired below the weathering layer produce signals richer in high frequencies (meaning a higher resolution) than shots fired in the weathering layer.

Another field observation is that increasing the charge size tends to decrease the resolution available in the signal (i.e., it tends to reduce the high frequency content of the spectrum). The theoretical explanation of this is as follows. The behavior of the medium in the vicinity of the shot is not linear, because of the high stress created by the explosion. The behavior becomes linear beyond a certain radius r_0 (after the stress has died down enough). Since our theory is a linear one, r_0 corresponds to "a" in our theory. Hence, increasing the charge size effectively means increasing a, which means decreasing f_0 (see Equation 2.92), in agreement with the field observations.

Figure 2.8a shows the type of signal that one might expect to be produced by a high-speed explosive, which has a short "burn time," since $p(t)$ rises very rapidly to its constant level (the rise is infinitely fast, actually). Similarly, Figure 2.8c shows what one might expect from a lower-speed explosive, which has a longer burn time. We see that, for the high-speed explosive, the response has a higher amplitude and is richer in high frequencies (there is more resolution). Again, this agrees with field observations. It also supports the previous point on charge size versus frequency, in that larger charges will tend to have longer burn times, resulting in signals with lower resolution.

Consider the approximate numerical value of f_0. Let $\sigma = \frac{1}{4}$, $\alpha = 2,000$ m/s, and $a = 1$ m. Then $f_0 \approx 300$ Hz. Such high frequencies are not observed in real seismic data because of frequency-dependent absorption (anelasticity), which drastically attenuates high frequencies. The theory presented in the preceding, which is for a perfectly elastic medium, does not account for this.

For another application of the Sharpe model, see Petten and Margrave (2012).

2.25 Seismic Waves Generated by a Directed Point Force

Consider a point force in a homogeneous medium. The force acts at a point and in a given direction and has an arbitrary time variation. The point force would produce elastic waves in the medium. The nature of these waves can be determined by solving the equation of motion, i.e., (2.41) or (2.42), with the body force term $\mathbf{f}$ included:

$$(\lambda + \mu)\nabla(\nabla \cdot \mathbf{u}) + \mu\nabla^2\mathbf{u} + \mathbf{f} = \rho\frac{\partial^2\mathbf{u}}{\partial t^2}, \qquad \text{or}$$

$$(\lambda + 2\mu)\nabla(\nabla \cdot \mathbf{u}) - \mu\nabla \times (\nabla \times \mathbf{u}) + \mathbf{f} = \rho\frac{\partial^2\mathbf{u}}{\partial t^2}. \qquad (2.93)$$

The source of the waves, i.e., the point force, is represented by the body force **f**. For example, a point force in the x direction, with amplitude A, located at the point **a**, and with time variation $s(t)$, would be represented as

$$\mathbf{f}(\mathbf{x}, t) = A\delta(\mathbf{x} - \mathbf{a})s(t)\mathbf{e}_1 \quad \text{or} \quad f_i(\mathbf{x}, t) = A\delta(\mathbf{x} - \mathbf{a})s(t)\delta_{i1}, \quad (2.94)$$

where $\delta(\mathbf{x} - \mathbf{a})$ is a 3D Dirac delta function, representing a "spike" at $\mathbf{x} = \mathbf{a}$. If the time variation is impulsive, then $s(t) = \delta(t)$.

The equation of motion can be solved using displacement potentials (see, e.g., Achenbach, 1973, p. 96, Aki and Richards, 1980, p. 70, or 2002, p. 68, and Lay and Wallace 1995, p. 323) or by Fourier transforms (see, e.g., Červený, 2001, p. 80). The Fourier transform approach has the advantage that it can be used for anisotropic media as well, whereas the displacement potential approach cannot. The solution is as follows.

Consider the standard case where $\mathbf{f} = s(t)\delta(\mathbf{x})\mathbf{e}_j$, which represents a point source in the direction $\mathbf{e}_j$ where $j = 1, 2,$ or 3, located at $\mathbf{x} = \mathbf{0}$, and with time variation $s(t)$. Then the solution is

$$u_i(\mathbf{x}, t) = \frac{(3\gamma_i\gamma_j - \delta_{ij})}{4\pi\rho r^3} \int_{r/\alpha}^{r/\beta} \tau s(t - \tau)\, d\tau + \frac{\gamma_i\gamma_j}{4\pi\rho\alpha^2 r}s\left(t - \frac{r}{\alpha}\right)$$

$$+ \frac{(\delta_{ij} - \gamma_i\gamma_j)}{4\pi\rho\beta^2 r}s\left(t - \frac{r}{\beta}\right) \equiv u_i^N + u_i^P + u_i^S, \quad i = 1, 2, 3,$$

$$(2.95)$$

where u_i is the ith component of displacement, **x** is the vector from the origin to the observation point (i.e., the location of the seismometer), $r \equiv |\mathbf{x}|$, and $\gamma \equiv \mathbf{x}/r$ is a unit vector in the direction of **x** (i.e., $\gamma_n = x_n/r$ is the direction cosine between **x** and the x_n axis). See Figure 2.9. Since the wave travels from the origin to **x**, the vectors **x** and γ are in the direction of wave propagation. To apply the formula, first

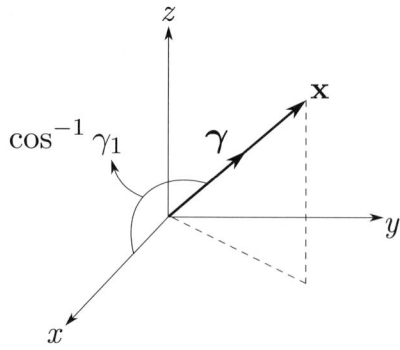

Figure 2.9 An illustration of **x**, γ, and r.

choose a value for j (to set the direction of the point force), then compute the three components of **u**.

The first term, u_i^N, is the *near-field* term. It dominates over the other two terms at small values of r. The second term, u_i^P, is the *far-field P wave* term. It is a P wave because the speed of the pulse is seen to be α, and because $\mathbf{u}^P \times \gamma = \mathbf{0}$, i.e., the displacement $\mathbf{u}^P$ is parallel to γ, i.e., the particle motion is parallel to the direction of wave propagation (which is the case for a P wave).

To prove that $\mathbf{u}^P \times \gamma = \mathbf{0}$, simply compute the three components of the cross product using (2.95). Deleting the factor $\gamma_j s(t - r/\alpha)/4\pi\rho\alpha^2 r$ because it appears in every term, and using the "is proportional to" symbol "$\propto$," gives

$$\left(\mathbf{u}^P \times \gamma\right)_1 = u_2^P\gamma_3 - u_3^P\gamma_2 \propto \gamma_2\gamma_3 - \gamma_3\gamma_2 = 0,$$

and similarly, $\quad \left(\mathbf{u}^P \times \gamma\right)_2 = 0 \quad$ and $\quad \left(\mathbf{u}^P \times \gamma\right)_3 = 0.$ $\qquad$ (2.96)

The third term, u_i^S, is the *far-field S wave* term. It is an S wave because the speed of the pulse is β and because $\mathbf{u}^S \cdot \gamma = 0$, i.e., the displacement $\mathbf{u}^S$ is perpendicular to γ, i.e., the particle motion is perpendicular to the direction of wave propagation (which is the case for an S wave). Using (2.95), the proof of $\mathbf{u}^S \cdot \gamma = 0$ goes as follows:

$$\mathbf{u}^S \cdot \gamma = \sum_n u_n^S \gamma_n \propto \sum_n (\delta_{nj} - \gamma_n\gamma_j)\gamma_n$$

$$= \sum_n \delta_{nj}\gamma_n - \gamma_j \sum_n \gamma_n\gamma_n = \gamma_j - \gamma_j = 0. \qquad (2.97)$$

It can also be shown, in the same way, that $\mathbf{u}^N \cdot \gamma \neq 0$ and $\mathbf{u}^N \times \gamma \neq \mathbf{0}$, meaning that the near-field wave is neither a P wave nor an S wave, but contains both P and S wave motion.

Note that the far-field terms decay as $1/r$ due to the geometrical spreading of the wavefronts. The near-field term appears to decay as $1/r^3$, but it actually does not decay that fast because r appears in the limits of the integral, making the integral a function of r. For example, for an impulse (or, in general, a pulse $s(t)$ of short duration), the near-field term decays roughly as $1/r^2$. Thus, it is the dominant term in u_i for small r (i.e., in the near-field). The far-field terms dominate, though, at large values of r. The far-field signals are generally more important, as most seismic measurements are made in the far-field. For values of r in between the small and the large, the waveform is a mix of the near-field and far-field waveforms.

Note that the near-field term is a convolution integral. This means that if the source pulse $s(t)$ has finite duration T (i.e., $s(t) \neq 0$ only for $0 \leq t \leq T$), then u_i^N is nonzero only in the range $(r/\alpha) \leq t \leq (r/\beta) + T$ (based on the properties of the convolution integral). To see this, apply the method outlined in Appendix 1B on the convolution of continuous signals.

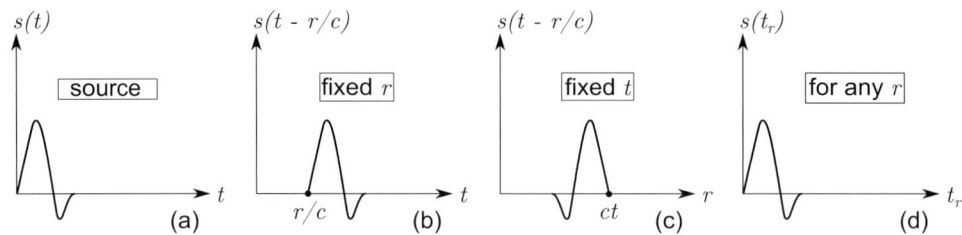

Figure 2.10 A source pulse $s(t - r/c)$ (a) at $r = 0$, (b) at a fixed nonzero r, (c) at a fixed t value, and (d) for any r, plotted against the retarded time $t_r = t - r/c$.

Note also that the terms "near-field" and "far-field" have meaning only if the source pulse is "short" enough, i.e., if the pulse duration is significantly smaller than the typical travel times, or equivalently, if the receiver distance r is large enough. Otherwise, for "long" pulses (or small r), both the near-field and far-field terms are important at all r, i.e., neither dominates.

The waveform part of the far-field terms, $s(t - r/c)$, where $c = \alpha$ or β, is plotted in Figure 2.10b as seen by a fixed observer at distance r and in Figure 2.10c as seen in a "photograph" or "snapshot" of the wave taken at time t as it propagates in the medium. If $s(t) = 0$ for $t < 0$, then r/c is the arrival time of the wave. Figure 2.10d shows (b) plotted as a function of the *reduced* or *retarded* time $t_r \equiv t - r/c$.

From the definitions of the far-field wave signals $\mathbf{u}^P$ and $\mathbf{u}^S$ in (2.95), we can see that their waveforms propagate without a change in shape (other than a loss in amplitude) because $s(t - r/c)$, where $c = \alpha$ or β, is merely $s(t)$ shifted in time by an amount r/c. The waveform does not change because the medium is nondissipative. The amplitude decays as $1/r$ due to the spherical divergence of the wavefront.

How do the near-field and far-field wave signals compare at various distances? To answer this, one could use (2.95) to compute and plot graphs of the displacement in simple cases, such as that of the point force $\mathbf{f}$ lying along an axis, with the observation points at $\mathbf{x}$ lying on that same axis, which results in a near-field wave and a far-field P wave only. Figure 2.11 shows an example for a particular signal $s(t)$ for which the integral in (2.95) can be analytically evaluated to give an exact result. There is even a case in which (2.95) can be extended to a low-loss absorbing nondispersive medium (absorption is discussed in Chapter 10) and the integral term analytically evaluated for a particular signal $s(t)$, as well as the other terms, to give an exact result (Sun, 2018).

How does the displacement, i.e., particle motion, produced by these far-field waves vary with position? A valuable way to represent the particle motions is through the use of **far-field radiation patterns**.

For example, consider the far-field P wave signals, $\mathbf{u}^P$, and consider a sphere of radius r centered on the location of the point source (i.e., on $\mathbf{x} = \mathbf{0}$), and suppose

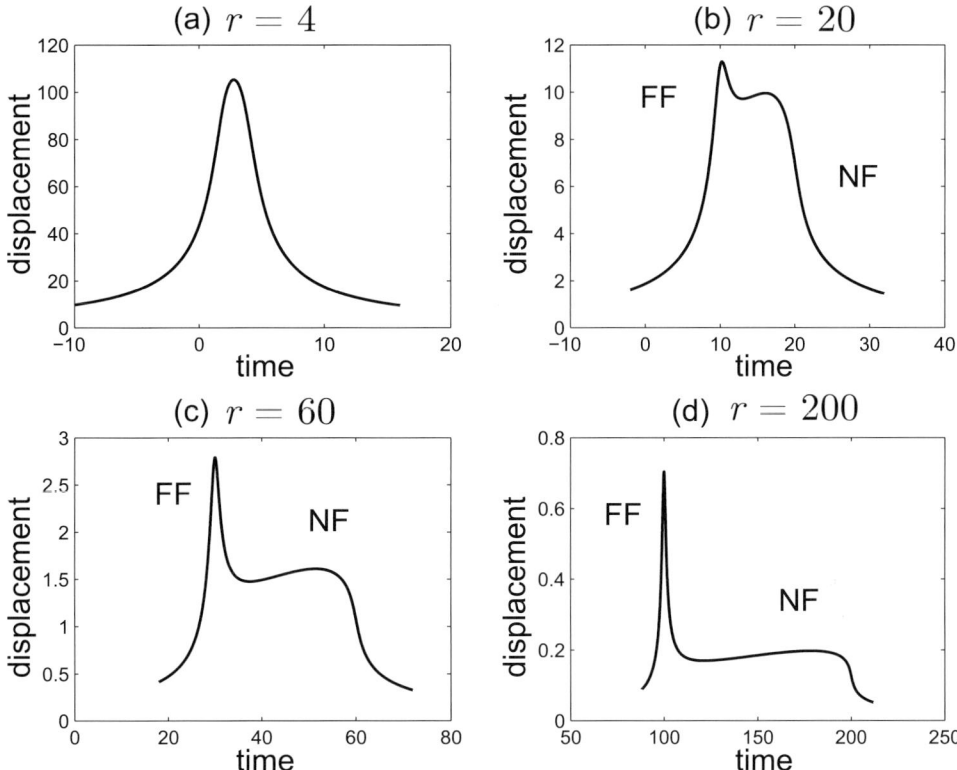

Figure 2.11 The displacement component in the direction of the body force for the case $s(t) = A[1 + (t/t_0)^2]^{-1/2}$, where t_0 and A are constants, with $\alpha = 2$ (arbitrary units), $\beta = 1$, $t_0 = 1$, $A/4\pi\rho = 500$, and for (a) $r = 4$, (b) $r = 20$, (c) $r = 60$, (d) $r = 200$. The plots show how the displacement amplitude decreases and the far-field wave dominates with increasing distance r. In (a), the near-field (NF) and the far-field (FF) signals cannot be distinguished.

there are seismometers uniformly distributed over the surface of the sphere. How would they respond to the P wave arriving at $t = r/\alpha$? If the source were a spherically symmetric explosion, they would all move in exactly the same way. Not so for our directed point force.

From (2.95), $\mathbf{u}^P = a\gamma_j\gamma$, where $a = a(r, t) \equiv s(t - r/\alpha)/(4\pi\rho\alpha^2 r)$, and γ is the radial unit vector from the source (at the origin) to the receiver (at $\mathbf{x}$). $\mathbf{u}^P$ is along γ, but at any given time t, its magnitude varies over the spherical surface of radius r as $\gamma_j = x_j/r$. Suppose the point force is in the y direction, meaning $j = 2$. Using spherical polar coordinates (see Figure 2.12a), γ and the displacement $\mathbf{u}^P$ are given by

$$\gamma = \sin\theta\cos\phi\,\mathbf{e}_1 + \sin\theta\sin\phi\,\mathbf{e}_2 + \cos\theta\,\mathbf{e}_3, \qquad \mathbf{u}^P = a\gamma_2\gamma = (a\sin\theta\sin\phi)\,\gamma.$$
$$(2.98)$$

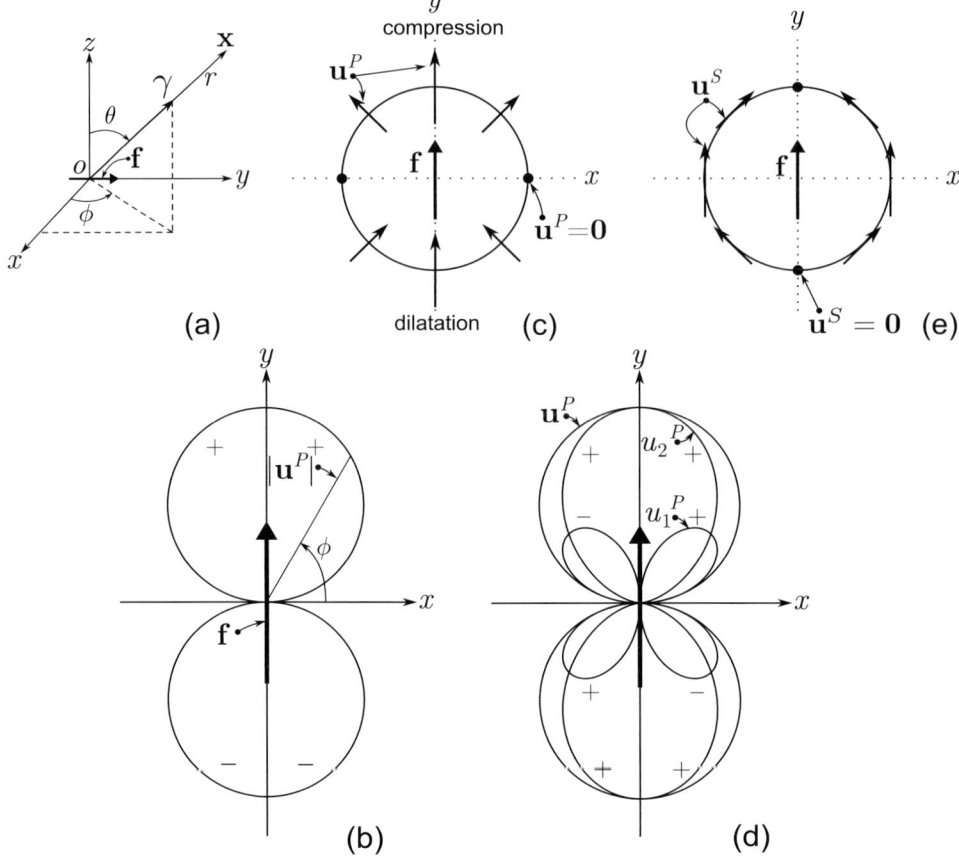

Figure 2.12 (a) Spherical polar coordinates; (b) and (d) radiation patterns for $\mathbf{u}^P$ and its components; and (c) and (e) displacement diagrams for $\mathbf{u}^P$ and $\mathbf{u}^S$, respectively.

This shows that the receivers in the *xz* plane ($\phi = 0, \pi$) do not move at all in response to the far-field *P* wave – any motion they experience would be due to the near-field wave and/or the far-field *S* wave. Consider the receivers in the *xy* plane – they lie along a circle of radius *r* centered on the origin in the *xy* plane. Their displacements are obtained by setting $\theta = \pi/2$ to obtain $\mathbf{u}^P = (a \sin \phi) \gamma$, which is plotted in Figure 2.12b, and which is called a **radiation pattern**. In this figure, the length of a radial line from the origin to the curve is the magnitude of $\mathbf{u}^P$, and the + and − signs denote its direction ($\pm \gamma$, i.e., outward or inward). Figure 2.12c, a displacement diagram, is another way to plot the radiation pattern of $\mathbf{u}^P$ – the arrows show the direction and relative magnitude of $\mathbf{u}^P$. As might be expected, the particle motion due to the far-field *P* wave is strongest along the line parallel to $\mathbf{f}$ and zero along the perpendicular line (the *x* axis), and is compressional ahead of

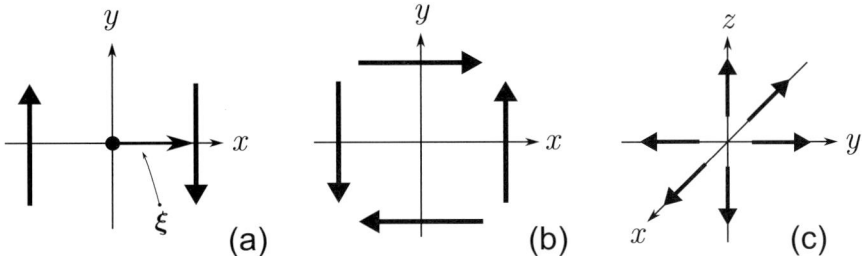

Figure 2.13 Directed point force systems for (a) a single couple, (b) a double couple without moment, and (c) a spherically symmetric explosion.

$\mathbf{f}$ and dilatational behind $\mathbf{f}$. Similarly, Figure 2.12d shows the radiation patterns of $u_1^P = a \sin\phi \cos\phi = (a/2) \sin 2\phi$ and $u_2^P = a \sin^2 \phi$ (as well as the one for $\mathbf{u}^P$ again) for $\theta = \pi/2$. The $+$ and $-$ signs denote the signs of u_1^P and u_2^P. Clearly, in the general direction of the point force (around $\phi = 90°$), the y component of motion is substantially larger than the x component (as might be expected, as the point force is in the y direction). Figure 2.12e shows the displacement diagram for the far-field S wave. As might be expected, the receivers in the "North" and "South" positions experience no transverse motion due to the far-field S wave.

If the point force is located at the vector position ξ rather than at the origin, then

$$\mathbf{f} = s(t)\delta(\mathbf{x} - \xi)\mathbf{e}_j, \quad \text{with} \quad r = |\mathbf{x} - \xi| \quad \text{and} \quad \gamma_j = (x_j - \xi_j)/r \quad \text{in (2.95). (2.99)}$$

Combinations or distributions of directed point forces, such as those shown in Figure 2.13, can be treated by summing up their individual responses. The body force for the *single couple* in Figure 2.13a, $\mathbf{f} = s(t)[\delta(\mathbf{x} - \xi) - \delta(\mathbf{x} + \xi)]\mathbf{e}_2$, was originally thought by seismologists to be a reasonable model for an earthquake source, as it appears to represent slip along a fault. However, theoretical seismograms based on it did not agree with observed ones. It turns out that a canceling moment or torque is also required. The **double couple without moment** in Figure 2.13b is now accepted as the best body force model of an earthquake source. It is composed of two single couples with opposite moments.

Equation (2.95) gives the displacement at the receiver for a single point force directed in the x_1, x_2, or x_3 directions (for $j = 1$, 2, or 3, respectively) and applied at the origin. However, typical sources can be modeled by specific combinations of point forces. For example, explosions and earthquake waves can be modeled by displacements caused by specific force *couples* (as shown in Figure 2.13). The theory for doing this can be generalized, and it leads to the important concept of the **moment tensor**. To proceed with this, first we rewrite (2.95) in a notation more suitable for generalization:

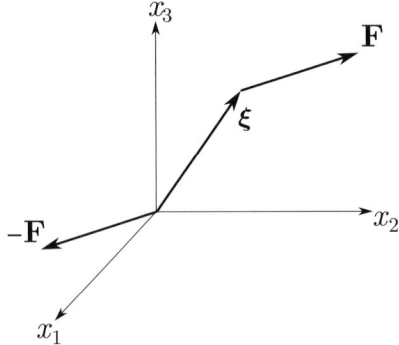

Figure 2.14 A force couple, with ξ and the force **F** pointing in arbitrary directions.

$$\mathbf{f} = s(t)\delta(\mathbf{x})\mathbf{e}_j \quad \Longrightarrow \quad u_{ij} = U_{ij}(\mathbf{x}|s) = U_{ij}^N(\mathbf{x}|s) \;+\; U_{ij}^P(\mathbf{x}|s) \;+\; U_{ij}^S(\mathbf{x}|s)$$
(2.100)

where a subscript j has been added because the terms on the right-hand side of (2.95) contain both subscripts i and j. $U_{ij}(\mathbf{x}|s)$ represents the right-hand side of (2.95), with $U_{ij}^N(\mathbf{x}|s)$, etc., being the individual terms, and "$|s$" indicates the source function used. Consider now the pair of forces, activated simultaneously at $t = 0$, shown in Figure 2.14.

We will use the **summation convention** in the calculations and formulas that follow.

The body force **f** is

$$\mathbf{f} = \mathbf{F}(t)\delta(\mathbf{x} - \xi) - \mathbf{F}(t)\delta(\mathbf{x}) = F_j(t)\delta(\mathbf{x} - \xi)\mathbf{e}_j - F_j(t)\delta(\mathbf{x})\mathbf{e}_j, \qquad (2.101)$$

where $\mathbf{F} = F_j\mathbf{e}_j$ (note the sum over j). As indicated in (2.99), if the force is applied at a point ξ rather than at the origin, then **x** must be replaced by $\mathbf{x} - \xi$ in the right-hand side of (2.95) or (2.100). Therefore, applying (2.100), the displacement u_i^C at **x** due to this pair of forces is

$$u_i^C = U_{ij}(\mathbf{x} - \xi|F_j) - U_{ij}(\mathbf{x}|F_j) = -\frac{\partial U_{ij}(\mathbf{x}|F_j)}{\partial x_k}\xi_k, \qquad (2.102)$$

(note the double sum over j and k), where the 3D Taylor series for a function $g(\mathbf{x})$ has been used, i.e.,

$$g(\mathbf{x} \pm \xi) = g(\mathbf{x}) \pm \frac{\partial g(\mathbf{x})}{\partial x_k}\xi_k + O(\xi^2). \qquad (2.103)$$

The term $O(\xi^2)$ has been omitted at the end of (2.102) because we must let $\xi \to \mathbf{0}$ to obtain a force *couple*. As we know the formula for U_{ij}, we can calculate the derivative $\partial U_{ij}/\partial x_k$. The calculation is straightforward but tedious. Using the easily proven results

$$\frac{\partial x_n}{\partial x_k} = \delta_{nk}, \qquad \frac{\partial r}{\partial x_k} = \gamma_k \qquad \text{and} \qquad \frac{\partial \gamma_n}{\partial x_k} = \frac{\delta_{nk} - \gamma_n \gamma_k}{r}, \tag{2.104}$$

and keeping only the lowest order terms, i.e., the $O(1/r)$ terms – we will examine only the far-field response here – we obtain, after some math,

$$\frac{\partial U_{ij}(\mathbf{x}|F_j)}{\partial x_k} = \frac{\partial U_{ij}^N(\mathbf{x}|F_j)}{\partial x_k} + \frac{\partial U_{ij}^P(\mathbf{x}|F_j)}{\partial x_k} + \frac{\partial U_{ij}^S(\mathbf{x}|F_j)}{\partial x_k}, \qquad \text{where} \tag{2.105a}$$

$$\frac{\partial U_{ij}^N(\mathbf{x}|F_j)}{\partial x_k} = \frac{\partial}{\partial x_k}\left[\frac{(3\gamma_i\gamma_j - \delta_{ij})}{4\pi\rho r^3}\int_{r/\alpha}^{r/\beta}\tau F_j(t-\tau)\,d\tau\right] = O\left(\frac{1}{r^2}\right), \tag{2.105b}$$

$$\frac{\partial U_{ij}^P(\mathbf{x}|F_j)}{\partial x_k} = \frac{\partial}{\partial x_k}\left[\frac{1}{4\pi\rho\alpha^2}\left(\frac{\gamma_i\gamma_j}{r}\right)F_j\left(t-\frac{r}{\alpha}\right)\right]$$

$$= -\frac{1}{4\pi\rho\alpha^3}\frac{\gamma_i\gamma_j\gamma_k}{r}F_j'\left(t-\frac{r}{\alpha}\right) + O\left(\frac{1}{r^2}\right), \tag{2.105c}$$

$$\frac{\partial U_{ij}^S(\mathbf{x}|F_j)}{\partial x_k} = \frac{\partial}{\partial x_k}\left[\frac{(\delta_{ij} - \gamma_i\gamma_j)}{4\pi\rho\beta^2 r}F_j\left(t-\frac{r}{\beta}\right)\right]$$

$$= \frac{1}{4\pi\rho\beta^3}\frac{\gamma_k(\gamma_i\gamma_j - \delta_{ij})}{r}F_j'\left(t-\frac{r}{\beta}\right) + O\left(\frac{1}{r^2}\right), \tag{2.105d}$$

where the prime $'$ denotes the derivative of F_j with respect to its argument, i.e., $F_j'(\eta) \equiv \partial F_j(\eta)/\partial \eta$.

Substituting (2.105) into (2.102) gives the displacement u_i^C due to the force couple:

$$u_i^C = u_i^P + u_i^S + O\left(\frac{1}{r^2}\right), \qquad \text{where} \tag{2.106a}$$

$$u_i^P = \frac{\gamma_i\gamma_j\gamma_k}{4\pi\rho\alpha^3 r}M_{jk}'\left(t-\frac{r}{\alpha}\right), \tag{2.106b}$$

$$u_i^S = -\frac{\gamma_k(\gamma_i\gamma_j - \delta_{ij})}{4\pi\rho\beta^3 r}M_{jk}'\left(t-\frac{r}{\beta}\right)$$

$$= -\frac{\gamma_i\gamma_j\gamma_k}{4\pi\rho\beta^3 r}M_{jk}'\left(t-\frac{r}{\beta}\right) + \frac{\gamma_k}{4\pi\rho\beta^3 r}M_{ik}'\left(t-\frac{r}{\beta}\right), \tag{2.106c}$$

and where

$$M_{jk}(t) \equiv F_j(t)\xi_k \implies M_{jk}'(t) \equiv F_j'(t)\xi_k$$

$$\implies M_{jk}'\left(t-\frac{r}{c}\right) = F_j'\left(t-\frac{r}{c}\right)\xi_k. \tag{2.107}$$

Note the double sums over j and k in (2.106), and the single sum over k in the last term of (2.106c). u_i^P and u_i^S are the far-field P wave and far-field S wave displacement produced by the force couple. Note that the waveform is the derivative of the source waveform. The complete formula for u_i^C, including the $O(1/r^2)$ near-field terms, can be found in Aki and Richards, 2002, eq. 4.29, p. 77.

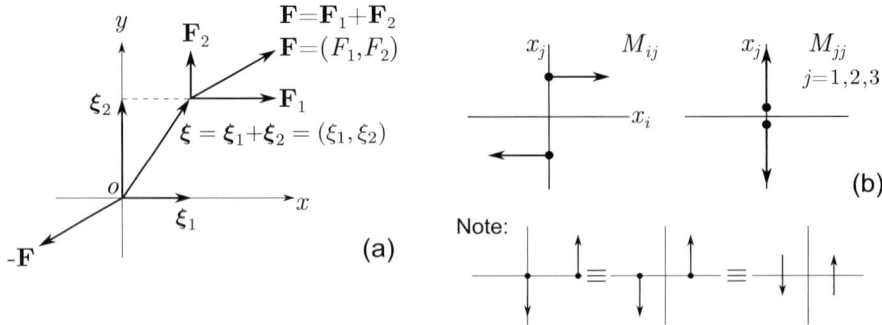

Figure 2.15 (a) Physical meaning of the moment tensor for ξ and $\mathbf{F}$ in the xy plane. $M_{12} = F_1\xi_2$ and $M_{21} = F_2\xi_1$ are moments (or torques) about O due to $\mathbf{F}_1$ and $\mathbf{F}_2$, respectively (b) Force couples corresponding to M_{ij}.

$M_{jk}, j, k = 1, 2, 3$, are the components of the **moment tensor**. From the definition of a force couple, it is understood that as $\xi_k \to 0$, $F_j \to \infty$ in such a way that $M_{jk} = F_j\xi_k$ remains finite. M_{jk} is the moment, or torque, produced by the jth component of the force $\mathbf{F}$, applied at ξ, about the origin (plus, in general, the moment produced by its opposite, the jth component of the force $-\mathbf{F}$, about the origin, which in this case is zero, as $-\mathbf{F}$ emanates from the origin). In Figure 2.15a, $M_{12} = F_1\xi_2$ is the moment about the origin O due to $\mathbf{F}_1$, and $M_{21} = F_2\xi_1$ is the moment about the origin O due to $\mathbf{F}_2$. Also, typically, $M_{ij} = M_{ji}$, because angular momentum is conserved (e.g., for earthquakes and explosions). A diagram representing the component M_{ij} is shown in Figure 2.15b. It shows that M_{ij} can be determined from the rule that the base of one force component arrow lies on the positive x_j axis and points in the positive x_i direction, and the base of the opposite force component arrow lies on the negative x_j axis and points in the negative x_i direction. The components M_{jk} can be thought of as a measure of the strength of the source.

As an example, consider the double-couple force mechanism, the body force equivalent for an earthquake source. Consider a vertical fault coinciding with the xz plane. From Figures 2.13b and 2.15b, the only nonzero components of the moment tensor are then $M_{12} = M_{21} \equiv M_0$. Applying (2.106b), and using (2.98), to obtain the far-field P wave displacement u_i^P on the ground surface (the xy plane, $\theta = \pi/2$) due to the double-couple gives

$$u_i^P = \frac{\gamma_i\gamma_1\gamma_2}{4\pi\rho\alpha^3 r}M_{12}'\left(t - \frac{r}{\alpha}\right) + \frac{\gamma_i\gamma_2\gamma_1}{4\pi\rho\alpha^3 r}M_{21}'\left(t - \frac{r}{\alpha}\right) = \frac{\gamma_i\gamma_1\gamma_2}{2\pi\rho\alpha^3 r}M_0'\left(t - \frac{r}{\alpha}\right) \tag{2.108}$$

$$= \frac{\gamma_i}{2\pi\rho\alpha^3 r}\sin^2\theta\sin\phi\cos\phi M_0'\left(t - \frac{r}{\alpha}\right) = \frac{\gamma_i\sin 2\phi}{4\pi\rho\alpha^3 r}M_0'\left(t - \frac{r}{\alpha}\right) \quad \text{for} \quad \theta = \frac{\pi}{2}. \tag{2.109}$$

Using (2.98) again, the x and y components of displacement for $\theta = \pi/2$ are

$$u_1^P = u_x^P = \frac{\cos\phi \, \sin 2\phi}{4\pi\rho\alpha^3 r} M_0{}'\left(t - \frac{r}{\alpha}\right), \quad u_2^P = u_y^P = \frac{\sin\phi \, \sin 2\phi}{4\pi\rho\alpha^3 r} M_0{}'\left(t - \frac{r}{\alpha}\right),$$

$$(2.110)$$

and hence the magnitude of the P wave displacement, i.e., the radial component of displacement, in the xy plane is

$$|\mathbf{u}^P| = |u_r^P| = \sqrt{(u_x^P)^2 + (u_y^P)^2} = \frac{|\sin 2\phi|}{4\pi\rho\alpha^3 r} \left| M_0{}'\left(t - \frac{r}{\alpha}\right)\right|, \qquad (2.111)$$

where the sign of u_r for any ϕ can be determined from u_x^P and u_y^P in (2.110). Equivalently, one may substitute into $\mathbf{u}$ the following basis vector transformation rules in the xy plane,

$$\mathbf{e}_x = (\cos\phi)\mathbf{e}_r - (\sin\phi)\mathbf{e}_\phi, \qquad \mathbf{e}_y = (\sin\phi)\mathbf{e}_r + (\cos\phi)\mathbf{e}_\phi, \qquad (2.112)$$

to obtain the P wave displacement for the double-couple as

$$\mathbf{u}^P = u_x^P \mathbf{e}_x + u_y^P \mathbf{e}_y = \frac{\sin 2\phi}{4\pi\rho\alpha^3 r} M_0{}'\left(t - \frac{r}{\alpha}\right)\mathbf{e}_r. \qquad (2.113)$$

Note that the double-couple radiation pattern for $\theta = \pi/2$ for the radial component is the same as the one for u_1^P in Figure 2.12d – in both cases, the displacement varies as $\sin 2\phi$, giving the four-leaf clover pattern. A displacement diagram for the double-couple source is shown in Figure 2.16a.

The far-field S wave displacement can be calculated from (2.106c) in a similar way (and is left as an exercise for the reader). One obtains the transverse displacement

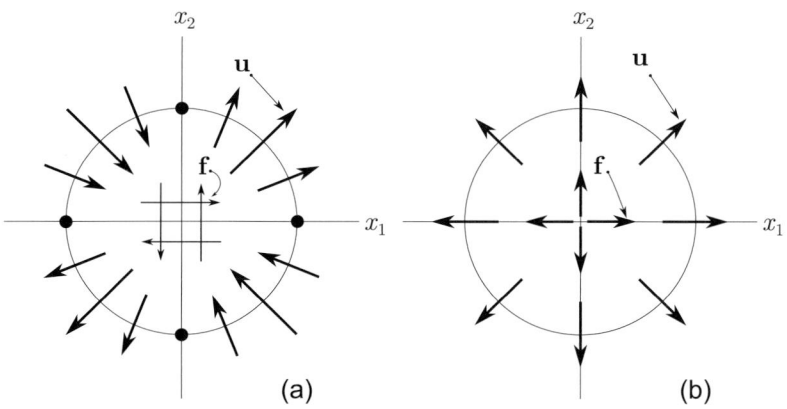

(a)   (b)

Figure 2.16 Displacement ($\mathbf{u}$) diagrams for the far-field P wave, for (a) the double-couple source mechanism (the vertical fault coincides with the xz or yz plane), and for (b) the spherically symmetric explosion (in the xy plane).

$$\mathbf{u}^S = \frac{\cos 2\phi}{4\pi\rho\beta^3 r} M_0' \left(t - \frac{r}{\beta}\right) \mathbf{e}_\phi, \tag{2.114}$$

whose radiation pattern is like the four-leaf clover pattern of u_1^P in Figure 2.12d except rotated by 45°.

As another example, consider a spherically symmetric explosion (Figure 2.13c). In this case, the only nonzero elements of the moment tensor are $M_{11} = M_{22} = M_{33} = M_0$. Applying (2.106b) to obtain the far-field P wave displacement u_i^P due to the explosion gives

$$u_i^P = \frac{\gamma_i}{4\pi\rho\alpha^3 r} \sum_{k=1}^{3} \gamma_k \gamma_k M_{kk}' \left(t - \frac{r}{\alpha}\right) = \frac{\gamma_i}{4\pi\rho\alpha^3 r} M_0' \left(t - \frac{r}{\alpha}\right), \tag{2.115}$$

or, as a vector equation,

$$\mathbf{u}^P = \frac{1}{4\pi\rho\alpha^3 r} M_0' \left(t - \frac{r}{\alpha}\right) \boldsymbol{\gamma}. \tag{2.116}$$

$\boldsymbol{\gamma}$ is a unit vector going from the origin, i.e., the center of the force system, to the observation point, i.e., it is a vector in the direction of wave propagation, meaning $\mathbf{u}$ represents pure P wave motion. As can be seen, the displacement does not vary with angle, as expected. Each point on an outgoing spherical wavefront moves in the same way. A displacement diagram for the explosion source is shown in Figure 2.16b.

2.26 Appendix 2A: Apparent Velocities

Consider a plane wave whose slowness vector or wavevector lies in the xz plane. It is traveling upward and in the positive x direction with speed v, where $v = \alpha$ or β. The direction of travel of the wave makes an angle θ to the vertical, as shown in Figure 2.17. In other words, the slowness vector $\mathbf{s}$ makes an angle θ with the vertical z axis, meaning

$$s_x = \frac{\sin\theta}{v}, \quad s_z = \frac{\cos\theta}{v} \quad \text{and} \quad s_x^2 + s_z^2 = \frac{1}{v^2}. \tag{2.117}$$

The wavefront intersects the horizontal xy plane (e.g., the surface of the Earth) along the line P and it also intersects a vertical plane perpendicular to the xz plane along the line Q. The lines P and Q are perpendicular to the xz plane. After a time Δt, the plane wavefront lies along $P'Q'$. Let L be the perpendicular distance between the two wavefronts, i.e., L is the distance the wavefront has traveled in time Δt. Then the speed $v_{(x)}$ with which P moves toward P', i.e., the **apparent wave speed** or **apparent velocity** in the x direction, is given by

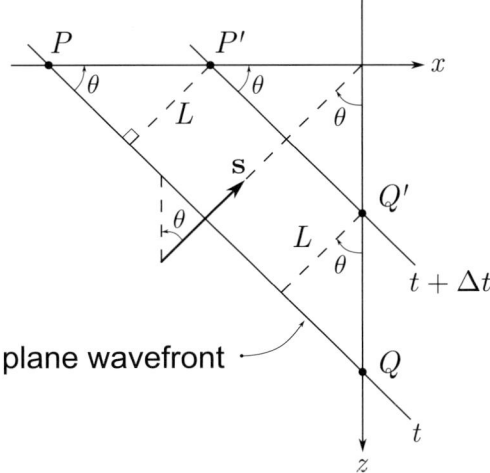

Figure 2.17 A plane wavefront propagating upward in the xz plane. The plane of the wavefront is perpendicular to the xz plane (its slowness vector lies in the xz plane).

$$v_{(x)} = \frac{\overline{PP'}}{\Delta t} = \frac{L}{\Delta t \sin \theta} = \frac{v \Delta t}{\Delta t \sin \theta} = \frac{v}{\sin \theta}. \tag{2.118}$$

Similarly, the speed $v_{(z)}$ with which Q moves toward Q', i.e., the **apparent wave speed** or **apparent velocity** in the z direction, is given by

$$v_{(z)} = \frac{\overline{QQ'}}{\Delta t} = \frac{L}{\Delta t \cos \theta} = \frac{v \Delta t}{\Delta t \cos \theta} = \frac{v}{\cos \theta}. \tag{2.119}$$

Note that

$$\frac{1}{v_{(x)}^2} + \frac{1}{v_{(z)}^2} = \frac{1}{v^2}, \tag{2.120}$$

where the rule $\sin^2 \theta + \cos^2 \theta = 1$ has been used. Note also that

$$s_x = \frac{1}{v_{(x)}} \quad \text{and} \quad s_z = \frac{1}{v_{(z)}}. \tag{2.121}$$

Clearly, $v_{(x)}$ and $v_{(z)}$ are not the components of a vector $\mathbf{v}$ because (2.120) is not the rule that the components of vectors satisfy. In other words, $v_{(x)}^2 + v_{(z)}^2 \neq v^2$. This is why parentheses appear around the subscripts.

Note that if the wave is traveling vertically upward, then $\theta = 0$, giving an infinite apparent velocity in the x direction, i.e., $v_{(x)} \rightarrow \infty$, meaning that the wavefront strikes the points along P and P' (and all points on the surface) simultaneously (as a vertically traveling wavefront would indeed do), and giving $v_{(z)} = v$, i.e., the apparent velocity in the vertical direction is just the speed v of the wave, as can be seen from Figure 2.17 with $\theta = 0$.

Exercises

1. (a) Show that e_{ij} and η_{ij} are tensors.
   (b) Show that $\mathcal{D} = \sum_{i=1}^{3} e_{ii}$, the dilatation, is a scalar invariant.
   (c) Show that the stress–strain relation has the same form in all coordinate systems (rotated with respect to one another), i.e. show that if $\sigma_{ij} = \lambda\delta_{ij}\mathcal{D} + 2\mu e_{ij}$, then $\sigma'_{ij} = \lambda\delta_{ij}\mathcal{D}' + 2\mu e'_{ij}$. (First show σ_{ij} is a tensor. Also see Equation 1.115.)

2. From the stress–strain relation of Equation (2.30), obtain e_{ij} as a function of σ_{ij}.

3. Consider a hypothetical material in which the stress components σ_{ij} are *quadratic* in the strain components e_{ij}, i.e.,

$$\sigma_{ij} = \sum_{k,l,m,n} v_{ijklmn}\, e_{kl}\, e_{mn}, \tag{1}$$

and assume v_{ijklmn} is given by

$$v_{ijklmn} = a\delta_{ij}\delta_{kl}\delta_{mn} + b(\delta_{ik}\delta_{jl}\delta_{mn} + \delta_{im}\delta_{jn}\delta_{kl}) + c(\delta_{il}\delta_{jm}\delta_{kn} + \delta_{in}\delta_{jk}\delta_{lm}), \tag{2}$$

where a, b, and c are constants.

   (a) Substitute (2) into (1) to obtain the stress–strain relation.
   (b) Write σ_{11} and σ_{12} out in full in terms of the derivatives $\partial u_i/\partial x_j$. (See, for example, Equation 2.30.)

4. An iron sphere 1 meter in diameter is placed at the bottom of a deep ocean trench 10 km below the water surface. Calculate the dilatation of the sphere due to compression. By how much does its radius decrease due to submersion and compression? In doing your calculation, verify that you can neglect the pressure difference between the top and the bottom of the sphere. The pressure-depth equation is $p = p_0 + \rho g z$, where z is the depth below the water surface, $p_0 = 1.013 \times 10^5$ N/m^2 = atmospheric pressure at sea level, $\rho = 1.0$ g/cm^3 = water density, $g = 9.8$ m/s^2 = acceleration of gravity. Young's modulus and Poisson's ratio for iron are $Y = 2.0 \times 10^{11}$ N/m^2 and $\sigma = 0.29$.

5. Let $\sigma_{11} > 0$, and $\sigma_{22} = \sigma_{33} = 0$ (see Figure 2.5).

   (a) By substituting the preceding conditions into the stress–strain relation, show that $e_{22} = e_{33}$, and verify $\sigma = \lambda/2(\lambda + \mu)$.
   (b) By adding the three equations for σ_{11}, σ_{22}, and σ_{33}, derive the equation $Y = \mu(3\lambda + 2\mu)/(\lambda + \mu)$.

6. (a) Substitute the stress–strain relation (2.30) into Equation (2.28) for the strain energy density W to obtain a formula for W in terms of $\mathcal{D}$ and e_{ij}.
   (b) Go further and express W in terms of the components of displacement.
   (c) Use your result in part (a) to express W in terms of $\mathcal{D}$ and e'_{ij} (the deviatoric strain tensor).
   (d) What is W for a liquid? Express W in terms of the dilatation $\mathcal{D}$ and also in terms of the pressure p.
   (e) What is W for the 1D case in which $u_1 = u(x, t)$, $u_2 = u_3 = 0$?

7. Show that $\psi = (A/r)\exp\left[-i\omega(t - r/v)\right]$ is a solution to the spherical wave equation (2.53).

8. Show that $\psi = (A/r)f(t - r/v)$ is a solution to the spherical wave equation (2.53), where f is any function of $t - r/v$.

9. The particle displacement $\mathbf{u}$ for an *SH* wave is $\mathbf{u} = U\exp\left[i\omega(\mathbf{s} \cdot \mathbf{x} - t)\right]\mathbf{e}_y$, where the slowness $\mathbf{s}$ is in the direction of propagation of the wave, and U can, in general, be a complex number. Calculate the mean intensity $\langle \mathbf{I} \rangle$ of the wave.

10. Consider a plane P wave propagating in the x direction in an infinite homogeneous isotropic medium. The particle displacement due to the wave is $\mathbf{u} = u(x, t)\mathbf{e}_x$.

    (a) What are the components σ_{ij} of the stress tensor?
    (b) If the wave propagates only in the x direction, why are σ_{yy} and σ_{zz} nonzero? In other words, explain the physical significance of σ_{yy} and σ_{zz}.

11. The particle displacement $\mathbf{u} = A\cos\left[\omega(z/v - t)\right]\mathbf{e}_x$, where A is a constant, is produced by a seismic plane wave.

    (a) What type of wave is this: P, SV, or SH?
    (b) Calculate the components σ_{ij} of the stress tensor, the energy flux $\mathbf{I}$, the potential energy density W, and the kinetic energy density K.
    (c) Compare your result for the stress σ_{ij} with the displacement $\mathbf{u}$. What can you deduce from the comparison?
    (d) Using the results of part (b), check to see if the energy conservation equation, namely $\partial(K + W)/\partial t = -\nabla \cdot \mathbf{I}$, is satisfied.
    (e) Calculate formulas for the means (time averages) of σ_{ij}, $\mathbf{I}$, W, and K.
    (f) If $A = 1\,\mathrm{mm}$, the density is $2\,\mathrm{g/cm^3}$, the speed is $2\,\mathrm{km/s}$ and the frequency is $30\,\mathrm{Hz}$, how much energy passes through an area of one square meter on the plane $z = $ constant in one cycle? Give your answer in ergs $(\mathrm{g \cdot cm^2/s^2})$. For how long could this amount of energy power a $60\,\mathrm{W}$ light bulb (if energy could be transferred)?

12. For a harmonic P wave, show that $W = K$.

13. (a) Substitute the stress–strain relation for an isotropic medium into the general formulas for the intensity $I_j (j = 1, 2, 3)$ and the strain energy density W, to obtain formulas for I_j and W for an isotropic medium.

    (b) What do your formulas for I_j and W reduce to for a general plane P wave traveling in the z direction, i.e., $\mathbf{u} = u(z, t)\mathbf{e}_z$?

    (c) What do they reduce to if $u(z, t) = A \cos[\omega(z/\alpha - t)]$, i.e., if the P wave is harmonic?

    (d) Obtain the mean values of your results in part (c).

14. A certain physical scalar quantity $\mathcal{B}$ is given by

$$\mathcal{B} = \mu \left[\mathcal{D}^2 + 2 \sum_{i=1}^{3} \sum_{j=1}^{3} \left(\frac{\partial u_i}{\partial x_j} \right)^2 \right],$$

where $\mathcal{D}$ is the dilatation. Calculate $\langle \mathcal{B} \rangle$, the mean value of $\mathcal{B}$, for the *SH* wave $\mathbf{u} = U \exp[i\omega(\mathbf{s} \cdot \mathbf{x} - t)]\mathbf{e}_y$ traveling in the xz plane. U is, in general, complex. Use Equation (2.84).

15. The jth component of a certain physical vector quantity $\mathbf{E}$ is given by

$$E_j - - \left[(\lambda + \mu) \sum_{i=1}^{3} \frac{\partial u_i}{\partial x_i} \dot{u}_j + \mu \sum_{i=1}^{3} \frac{\partial u_j}{\partial x_i} \dot{u}_i \right].$$

The particle displacement $\mathbf{u}$ for an *SH* wave is $\mathbf{u} = U \exp[i\omega(\mathbf{s} \cdot \mathbf{x} - t)]\mathbf{e}_y$, where the slowness $\mathbf{s}$ is in the direction of propagation of the wave, and U is, in general, a complex number. Calculate $\langle \mathbf{E} \rangle$ (the mean value of $\mathbf{E}$) for the *SH* wave.

16. A certain physical scalar quantity $\mathcal{S}$ is given by

$$\mathcal{S} = \frac{1}{2}(\lambda + \mu) \sum_{i=1}^{3} \sum_{j=1}^{3} \frac{\partial u_i}{\partial x_i} \frac{\partial u_j}{\partial x_j} + \frac{1}{2}\mu \sum_{i=1}^{3} \sum_{j=1}^{3} \left(\frac{\partial u_i}{\partial x_j} \right)^2.$$

The particle displacement $\mathbf{u}$ for a certain *SH* wave is $\mathbf{u} = U \exp[i\omega(\mathbf{s} \cdot \mathbf{x} - t)]\mathbf{e}_y$. Calculate the formula for the mean (time average) of $\mathcal{S}$.

17. The jth component of a certain physical vector quantity $\mathbf{J}$ is given by

$$J_j = 2\mu \left(\sum_{i=1}^{3} \frac{\partial u_i}{\partial x_i} \dot{u}_j - \sum_{i=1}^{3} \frac{\partial u_j}{\partial x_i} \dot{u}_i \right).$$

The particle displacement **u** for an *SH* wave is $\mathbf{u} = U \exp[i\omega(\mathbf{s} \cdot \mathbf{x} - t)]\mathbf{e}_y$, where the slowness **s** is in the direction of propagation of the wave, and U is, in general, a complex number. Calculate $\langle \mathbf{J} \rangle$ (the mean value of **J**).

18. The *j*th component of a certain physical vector quantity **E** is given by

$$E_j = -(\lambda + \mu) \sum_{i=1}^{3} \frac{\partial u_i}{\partial x_i} \dot{u}_j - \mu \sum_{i=1}^{3} \frac{\partial u_i}{\partial x_j} \dot{u}_i .$$

   Calculate the mean value of **E** for a plane harmonic *P* wave traveling in the positive *z* direction.

19. $I_j = -\sum_{i=1}^{3} \sigma_{ij}(\partial u_i / \partial t)$ is the *j*th component of the intensity, or energy flux.

   (a) Express I_j for an isotropic medium in terms of the components of **u** only (substitute the formulas for σ_{ij}, $\mathcal{D}$, and e_{ij}). Compare your result with the quantity E_j, where

$$E_j = -(\lambda + \mu) \sum_{i=1}^{3} \frac{\partial u_i}{\partial x_i} \frac{\partial u_j}{\partial t} - \mu \sum_{i=1}^{3} \frac{\partial u_i}{\partial x_j} \frac{\partial u_i}{\partial t} .$$

   Are I_j and E_j equal? Can you guess where E_j comes from?

   (b) The mean intensity $\langle \mathbf{I} \rangle$ for a plane harmonic *P* wave traveling in an arbitrary direction in an isotropic medium is given by (2.83). Calculate $\langle \mathbf{E} \rangle$ for such a wave as well, and compare your result with $\langle \mathbf{I} \rangle$. Comment on what you find.

20. Derive Equations (2.42) through (2.46).

21. Read Section 2.24, entitled "Seismic Waves Generated by a Buried Explosive Charge," and answer the following questions.

   (a) The variable t_r in (2.90) is called the *reduced time* or *retarded time*. Note that the arrival time of the pulse always corresponds to $t_r = 0$, no matter what r is. Consider Figure 4.5, in which the *x-t* graph shows the travel time curves corresponding to the events on a basic exploration seismic shot record for a one-layer model. Suppose these curves were plotted against t_r, where $t_r \equiv t - x/v_2$, rather than against t. Sketch the resulting *x-t_r* graph and comment on it. In crustal and global seismology, seismograms are often plotted against reduced time, for practical reasons. What is the practical advantage in doing so?

   (b) True or false? Explain your answers. Assume $p(t) = p_0$.

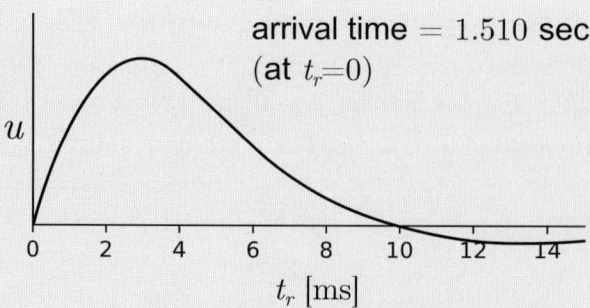

Figure 2.18 See exercise 21c.

   i. In a given medium, signals with lower frequency content tend to have higher amplitudes.

   ii. When going from a higher to a lower rigidity medium, the charge size should be reduced if one wants to prevent the resolution from being lowered.

   iii. When going from a higher to a lower rigidity medium, a higher-speed explosive should be used if one wants to prevent the resolution from being lowered.

(c) A nuclear device is buried at a depth of 100 m, detonated, and recorded by a surface seismometer 5 km from the "epicenter." The P wave signal is shown in Figure 2.18. The medium parameters ($\alpha, \beta, \rho, \sigma$, etc.) are not known. Assume $p(t) = p_0$. Estimate the size of the underground cavity created by the explosion.

22. Consider a solid body experiencing a static deformation. Suppose that the stress tensor for this deformation is given by

$$\sigma = \begin{pmatrix} \sigma_{11} & \sigma_{12} & \sigma_{13} \\ \sigma_{21} & \sigma_{22} & \sigma_{23} \\ \sigma_{31} & \sigma_{32} & \sigma_{33} \end{pmatrix} = \begin{pmatrix} 4 & 3 & 0 \\ 3 & 2 & 0 \\ 0 & 0 & 0 \end{pmatrix}$$

Note that since the normal and shear stresses in the x_3 direction, i.e., σ_{31}, σ_{32}, and σ_{33}, are all zero, this tensor describes a body experiencing normal and shear stresses parallel to the x_1-x_2 plane only (i.e., the x-y plane).

As discussed in the text at the end of Section 2.1, there are three mutually orthogonal planes in the body across which there are no shear stresses, only normal stresses. These normal stresses are the principal stresses, and the planes are the principal planes. The directions normal to the principal planes are the directions of the principal axes.

(a) Compute the values of the principal stresses (τ_1, τ_2, and τ_3) and the normal unit vectors along the principal axes ($\mathbf{n}_1$, $\mathbf{n}_2$, and $\mathbf{n}_3$) by solving the eigenvalue problem $\boldsymbol{\sigma}\mathbf{n} = \tau\mathbf{n}$.

(b) The principal axis frame can be obtained by a rotation of the original frame. Determine the rotation matrix for doing this. Sketch a diagram showing the original frame and the principal axis frame. What is the stress tensor in the principal axis frame?

Some information you may need: If you need to review how to calculate eigenvalues and eigenvectors, read through the numerical example in Section 9.2 (or refer to a linear algebra text).

$$ax^2 + bx + c = 0 \quad \Rightarrow \quad x = \frac{-b \pm \sqrt{b^2 - 4ac}}{2a}$$

$$\det \begin{pmatrix} a & b & c \\ d & e & f \\ g & h & i \end{pmatrix} = a \det \begin{pmatrix} e & f \\ h & i \end{pmatrix} - b \det \begin{pmatrix} d & f \\ g & i \end{pmatrix} + c \det \begin{pmatrix} d & e \\ g & h \end{pmatrix}.$$

$$\det \begin{pmatrix} a & b \\ c & d \end{pmatrix} = ad - bc, \quad \det(\mathbf{A}) = \det(\mathbf{A}^T).$$

23. The components A'_{ij} of a tensor $\mathbf{A}$ in a frame of reference that is rotated about the x_3 axis by an angle ϕ (see, e.g., Figure 1.10b) can be calculated from

$$A'_{ij} = \sum_{k=1}^{3} \sum_{\ell=1}^{3} R_{ik} R_{j\ell} A_{k\ell}, \quad \text{i.e.,} \quad \mathbf{A}' = \mathbf{R}\mathbf{A}\mathbf{R}^T.$$

Consider the box in Figure 2.1, and suppose that the traction $\mathbf{T}$ on the x_2 face (the face normal to the x_2 axis) is parallel to the x_2 direction, and that the tractions on all the other faces are zero. What is the stress tensor $\boldsymbol{\sigma}$ in this case? Calculate the components σ'_{ij} of the stress tensor in a frame of reference that is rotated about the x_3 axis by an angle ϕ (see, e.g., Figure 1.10b). How does $\boldsymbol{\sigma}'$ compare with $\boldsymbol{\sigma}$? Is symmetry preserved upon rotation? Is the value of the contraction of $\boldsymbol{\sigma}$ preserved upon rotation (i.e., is the contraction invariant)? What is $\boldsymbol{\sigma}'$ for $\phi = 90°$? Is it what you expect? Why?

24. As can be seen in Equation (2.26), the stress–strain relation (Hooke's law) can be written as a matrix equation, i.e., $\boldsymbol{\sigma} = \mathbf{ce}$, or in indicial notation (using the summation convention) $\sigma_i = c_{ij}e_j$, where $\boldsymbol{\sigma}$ and $\mathbf{e}$ are the column vectors

$$\boldsymbol{\sigma} = \left[\sigma_1, \sigma_2, \sigma_3, \sigma_4, \sigma_5, \sigma_6\right]^T = \left[\sigma_{11}, \sigma_{22}, \sigma_{33}, \sigma_{23}, \sigma_{31}, \sigma_{12}\right]^T$$

and $\mathbf{e} = \left[e_1, e_2, e_3, e_4, e_5, e_6\right]^T = \left[e_{11}, e_{22}, e_{33}, 2e_{23}, 2e_{31}, 2e_{12}\right]^T,$

where T signifies the transpose.

(a) Show that the equation of motion can be written in matrix form as $\mathbf{Dce}$ $+ \mathbf{f} = \rho\ddot{\mathbf{u}}$, where $\mathbf{u} = (u_x, u_y, u_z)^T$ and $\mathbf{f} = (f_x, f_y, f_z)^T$, and where $\mathbf{D}$ is a matrix whose elements are either zeros or partial derivative operators (e.g., $\partial/\partial x$). Derive $\mathbf{D}$. Also, write this matrix equation of motion in indicial form.

(b) Write the strain-displacement relation $e_{ij} = \frac{1}{2}(u_{i,j} + u_{j,i})$ as a matrix equation (using $\mathbf{u}$ and $\mathbf{e}$ as defined previously). Then substitute it into the matrix equation you derived in (a) to obtain the final form of the matrix equation of motion. Express this last equation in indicial form as well. Also, consider the matrix operating on $\mathbf{u}$ in this last equation. Show that if the elements of this matrix were numbers rather than derivative operators (as would be the case after a 3D Fourier space transform), then this matrix is symmetric (hints: $\mathbf{A} = \mathbf{A}^T \iff \mathbf{A}$ is symmetric; $(\mathbf{AB})^T = \mathbf{B}^T\mathbf{A}^T$).

(c) Calculate the elements of $\mathbf{c}$ for an isotropic medium.

25. For a point body force in the x direction, i.e., $\mathbf{f} = \delta(\mathbf{x})s(t)\mathbf{e}_x$, calculate $\mathbf{u}$ (the particle displacement) along the direction of the force, i.e., along the x axis (apply Equation 2.95). For the waveform $s(t)$, use

(a) The Dirac delta function $\delta(t)$
(b) The Heaviside step function $H(t)$ ($= 0$ for $t < 0$, and 1 for $t > 0$)
(c) $\exp(-t/T)H(t)$
(d) $B(t) = H(t) - H(t - T) = \{1, \ 0 \le t \le T; \ 0, \ \text{otherwise}\}$, where T is a positive constant giving a measure of the duration of the waveform. $B(t)$ is sometimes called a "boxcar" function.

Plot or sketch graphs of how u_x varies with time for different distances from the source.

26. For a point body force in the x direction, i.e., $\mathbf{f} = s(t)\delta(\mathbf{x})\mathbf{e}_x$, calculate $\mathbf{u}$ (the particle displacement) along the direction of the force, i.e., along the x axis (apply Equation 2.95). Use $s(t) = 1/[1 + (t/T)^2]$, where T is a positive constant giving a measure of the duration of $s(t)$, so that the integral for the near-field term can be done analytically. For the values $T = 1$, $\alpha = 2$, and $\beta = 1$ (arbitrary units), plot four graphs of u_x against time t for each

of the following distances (from the source): $r = 5$, 10, 50, and 100. Scale the graphs so that the waveforms can be clearly seen. Also, plot the same four graphs again showing the near-field and far-field curves separately. Comment on the graphs. Note: ignore the fact that $s(t)$ is noncausal – we are mainly interested in the waveforms here.

3

Reflection and Transmission of Plane Waves

When a plane wave impinges upon a flat boundary, some of it is reflected by the boundary and some of it is transmitted through the boundary. In this chapter, we discuss how the relative amplitudes of the reflected and transmitted plane waves can be computed from knowledge of the physical conditions that must be satisfied at the boundary. We also discuss the phase differences that can exist between the reflected and transmitted waves and the incident wave.

3.1 A Normally Incident P Wave Pulse

Consider a flat boundary $z = 0$ (the xy plane) separating two elastic media with parameters ρ_1, λ_1, and μ_1 (for $z < 0$), and ρ_2, λ_2, and μ_2 (for $z > 0$). Consider a compressional plane wave pulse in medium 1 traveling in the positive z direction toward the boundary. When it hits the boundary, it will be partially reflected by the boundary and partially transmitted through it. All particle motion in this situation will be in the $+z$ or $-z$ directions, i.e., u_x and u_y are zero. Let the incident, reflected and transmitted waveforms for the displacement, be

$$u_z^{(I)}(z,t) = f\left(t - \frac{z}{\alpha_1}\right), \quad u_z^{(R)}(z,t) = g\left(t + \frac{z}{\alpha_1}\right), \quad u_z^{(T)}(z,t) = h\left(t - \frac{z}{\alpha_2}\right). \quad (3.1)$$

These functions all satisfy the equation of motion for 1D plane waves, which is the 1D wave equation (2.47) with x_1 replaced by z and u_1 by u_z for this case. At the boundary, these three waves are related by certain physical conditions, known as **boundary conditions**. They allow us to determine g and h, given f. In this problem, the boundary conditions state that at the interface $z = 0$, the normal displacement u_z and the normal pressure σ_{zz} due to all the wave motion on side 1 are equal to u_z and σ_{zz} due to all the wave motion on side 2, i.e., u_z and σ_{zz} must be continuous across the boundary. This is often expressed by stating that the two media are in **welded contact**. Expressed mathematically, the conditions are

112

$$\left[u_z^{(1)}\right]_{z=0} = \left[u_z^{(2)}\right]_{z=0} \quad \text{and} \quad \left[\sigma_{zz}^{(1)}\right]_{z=0} = \left[\sigma_{zz}^{(2)}\right]_{z=0} \tag{3.2}$$

where the superscripts "(1)" and "(2)" refer to medium 1 and medium 2.

From the stress–strain relation (2.30), remembering that $u_x = u_y = 0$ here, we obtain

$$\sigma_{zz} = \rho\alpha^2 \frac{\partial u_z}{\partial z}. \tag{3.3}$$

Hence, since $u_z^{(1)} = u_z^{(I)} + u_z^{(R)}$ and $u_z^{(2)} = u_z^{(T)}$, the boundary conditions become

$$f(t) + g(t) = h(t), \tag{3.4}$$

$$-Z_1 f'(t) + Z_1 g'(t) = -Z_2 h'(t), \tag{3.5}$$

where the prime denotes the derivative of the function with respect to its argument, and where Z_1 and Z_2 are the **acoustic impedances** of media 1 and 2 respectively, given by $Z_1 = \rho_1\alpha_1$ and $Z_2 = \rho_2\alpha_2$. Equation (3.5) can be integrated to give

$$Z_1 f(t) - Z_1 g(t) = Z_2 h(t). \tag{3.6}$$

One could add an integration constant C to (3.6), but if $f(t) = 0$ for all t (no incident pulse), then $g(t) = 0$ and $h(t) = 0$ for all t as well, which implies $C = 0$. Solving Equations (3.4) and (3.6) for g and h gives

$$g(t) = -Rf(t) \quad \text{and} \quad h(t) = Tf(t) \tag{3.7}$$

where

$$R = \frac{Z_2 - Z_1}{Z_2 + Z_1} \quad \text{and} \quad T = \frac{2Z_1}{Z_2 + Z_1}. \tag{3.8}$$

R is the relative amplitude of the reflected pulse, and is called the **particle displacement reflection coefficient**. T is the relative amplitude of the transmitted pulse, and is called the **particle displacement transmission coefficient**.

As they stand, Equations (3.7) relate the waves at $z = 0$. However, we can generalize them to other values of z by realizing that t in these equations can take any value between $-\infty$ and ∞ (although the reflected and transmitted pulses do not physically exist until after the incident wave strikes the boundary, the functions f, g, and h are defined for all times in the mathematical sense). In particular, we can write

$$g\left(t + \frac{z}{\alpha_1}\right) = -Rf\left(t + \frac{z}{\alpha_1}\right) = -Rf\left(t - \frac{(-z)}{\alpha_1}\right), \tag{3.9}$$

$$h\left(t - \frac{z}{\alpha_2}\right) = Tf\left(t - \frac{z}{\alpha_2}\right) = Tf\left(t - \frac{(\alpha_1/\alpha_2)z}{\alpha_1}\right), \tag{3.10}$$

or, using Equations (3.1), we obtain

$$u_z^{(R)}(z,t) = -Ru_z^{(I)}(-z,t), \tag{3.11}$$

$$u_z^{(T)}(z,t) = Tu_z^{(I)}\left(\frac{\alpha_1}{\alpha_2}z,t\right). \tag{3.12}$$

Much information about the reflected and transmitted waves is contained in these equations.

Firstly, consider Equation (3.11). Since the reflected pulse exists for $z < 0$, Equation (3.11) states, for instance, that

$$u_z^{(R)}(-2,t) = -Ru_z^{(I)}(+2,t),$$

i.e., that the value of the reflected pulse at $z = -2$ (see the leftmost pulse in Figure 3.1b) is equal to $-R$ times the value of incident pulse at $z = +2$ (the dashed line in Figure 3.1b). Since the incident pulse does not actually exist for $z > 0$ (which is why it is dashed in Figure 3.1b), we must think of $u_z^{(I)}(2,t)$ as the value it *would* have had at $z = +2$ if there had been no boundary putting an end to medium 1. Hence, Equation (3.11) states that the reflected pulse is the mirror image of the incident pulse, but a mirror image whose amplitude is reduced from 1 to R, and also whose displacement polarity is reversed if $R > 0$.

In Figure 3.1, the incident displacement $u_z^{(I)}$ is positive. In other words, when the incident wave strikes a particle in its path, the particle moves in the $+z$ direction (see the heavy dots in Figure 3.1a), i.e., the wave *pushes* the particle away from itself. More generally, the wave *compresses* the regions of matter it encounters.

If the **first motion** of a particle is a push, we will refer to it as a **compression**. If the first motion of a particle is a *pull*, i.e., if the wave pulls the particle into itself, or more generally, if the wave *expands* the regions of matter it encounters, we will refer to it as a **rarefaction** (sometimes also called a **dilatation**). A compression can be thought of as a positive pressure acting on a particle, and a rarefaction as a negative pressure (or a tension).

In Figure 3.1, the reflected displacement $u_z^{(R)}$ is negative, since $R = +\frac{1}{3}$ (see Equation (3.11)). In other words, when the reflected wave strikes a particle in its path, the particle moves in the $-z$ direction (see Figure 3.1b), i.e., the reflected wave also *pushes* the particle away from itself, meaning the first motion due to the reflected wave is also a compression.

In general, Equations (3.8) and (3.11) show that for normally incident P waves, *if $Z_2 > Z_1$ (i.e., if $R > 0$), a compression reflects as a compression* – there is no polarity reversal, i.e., no **phase change**, in the pressure (although there is one in the displacement u_z).

Similarly, *if $Z_2 < Z_1$ (i.e., if $R < 0$), a compression reflects as a rarefaction* – there *is* a polarity reversal, i.e., a phase change, in the pressure (but not in the displacement u_z).

More generally, one can say that if Z_{inc} and Z_{tra} are the acoustic impedances in the incidence and transmission media, then for normally incident *P* waves, if $Z_{tra} > Z_{inc}$, a compression reflects as a compression, and if $Z_{tra} < Z_{inc}$, a compression reflects as a rarefaction.

Suppose a standard geophone and a hydrophone are placed in the incidence medium in Figure 3.1. The standard geophone measures u_z (actually, it measures $\partial u_z / \partial t$, but this can be ignored in the following argument). The incident wave displaces the geophone in the $+z$ direction and the reflected wave displaces it in the $-z$ direction. On the other hand, the hydrophone measures pressure. Therefore, both the incident and reflected waves have the same effect on the hydrophone – it is compressed by both of them. Hence, the geophone trace would show a polarity reversal, whereas the hydrophone trace would not (see the traces in Figure 3.1).

In Figure 3.1a, it is assumed that the source of the waves is to the left of the geophone and hydrophone (the open square and circle) in the incidence medium. However, if these two receivers were to the left of the source (analogous to a common situation in exploration seismology), then the incident wave, i.e., the direct wave (a compression) striking the geophone would push it to the left, i.e., in the negative z direction, as does the reflected wave, in which case the two pulses on the geophone trace would both kick downward, i.e., they would have the same polarity.

Equation (3.12) states that the value of the transmitted pulse at some $z > 0$ is the value that the incident pulse would have had at $\alpha_1 z / \alpha_2$, except the amplitude is T, not 1. For instance, in Figure 3.1b, where $\alpha_1 / \alpha_2 = \frac{1}{2}$ and $T = \frac{2}{3}$, the value of the transmitted pulse (the rightmost pulse in Figure 3.1b) at $z = 4$ is T times the value of the fictitious incident pulse (the dashed line) at $z = 2$. Hence, the transmitted pulse experiences a spatial scale change. In other words, although the waveform is the same, its spatial width is greater than that of the incident wave if $\alpha_2 > \alpha_1$, or smaller if $\alpha_2 < \alpha_1$, a well-known experimental fact. For an incident harmonic wave, this means that the wavelength λ of the wave changes upon transmission: if $\alpha_2 > \alpha_1$, then $\lambda_2 > \lambda_1$ by the same amount, i.e., the wavelength of the transmitted wave is greater than the wavelength of the incident wave by a factor of α_2 / α_1, and vice versa. This can also be seen from the harmonic wave formula for the *P* wave speed, $\alpha = \lambda f$, since the frequency f does not change upon transmission or reflection (as discussed later in this section).

Note that there is no polarity reversal in either the displacement u_z or the pressure, because $T > 0$. For example, a compression always transmits as a compression.

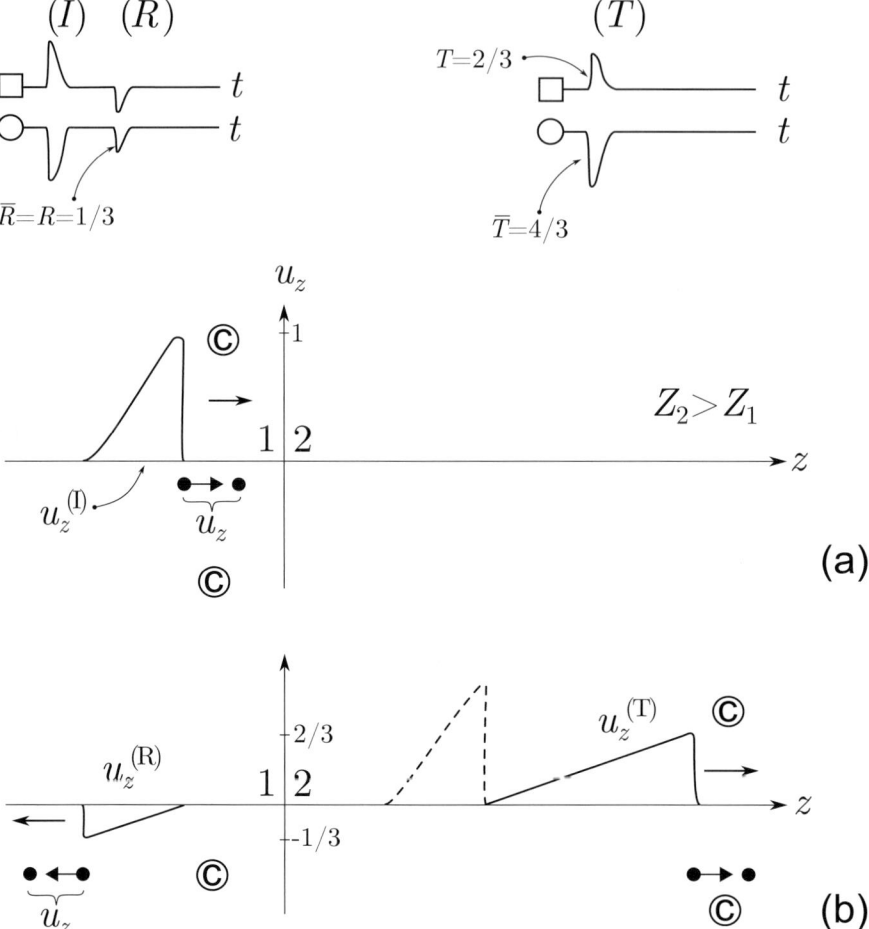

Figure 3.1 Reflection and transmission of a compressional pulse at a boundary ($z = 0$) for $\alpha_2 = 2\alpha_1$ and $\rho_1 = \rho_2$ (a) before and (b) after reflection. $R = \frac{1}{3}$ and $T = \frac{2}{3}$. The heavy dots represent particles in the medium, and the arrows on the dots give the directions of $\mathbf{u}^{(I)}$ and $\mathbf{u}^{(R)}$. The circled "c" denotes a compression. Also shown schematically are the seismic time traces recorded by a standard geophone (square) and a standard hydrophone (circle) in the two media, with the polarities drawn in accordance with the Society of Exploration Geophysicists (SEG) convention, which states that, at a given time, for a standard geophone, if the z component of displacement of the geophone is positive, i.e., if $u_z > 0$, then the corresponding point on the geophone trace has a positive value, and for a standard hydrophone, if the wave impinging upon the hydrophone is a compression, then the corresponding point on the hydrophone trace has a negative value.

Since the time t experiences no scale change, the temporal widths of the three waveforms are the same (which is also a well-known experimental fact) – if receivers were placed to record all three waves, then the waveforms of the three waves would all have the same width on the seismic traces. For a harmonic wave, this means that the frequency f does not change upon reflection or transmission. This is also why we represented the waveforms in Equation (3.1) as functions of $t \pm z/\alpha_j$ rather than $z \pm \alpha_j t$, i.e., to associate scale changes with space but not time.

The preceding results can also be seen by analyzing the pressure pulses rather than the displacement pulses. From (3.3) and the previous equations, we can obtain the following equations, which are analogous to Equations (3.11) and (3.12):

$$\sigma_{zz}^{(R)}(z,t) = \overline{R}\sigma_{zz}^{(I)}(-z,t), \tag{3.13}$$

$$\sigma_{zz}^{(T)}(z,t) = \overline{T}\sigma_{zz}^{(I)}\left(\frac{\alpha_1}{\alpha_2}z,t\right), \tag{3.14}$$

where

$$\overline{R} = R = \frac{Z_2 - Z_1}{Z_2 + Z_1} \qquad \text{and} \qquad \overline{T} = \frac{Z_2}{Z_1}T = \frac{2Z_2}{Z_2 + Z_1}. \tag{3.15}$$

$\overline{R}$ is the relative amplitude of the reflected pressure pulse, and is often called the **pressure reflection coefficient**. $\overline{T}$ is the relative amplitude of the transmitted pressure pulse, and is often called the **pressure transmission coefficient**. Hence, once again, if $Z_2 < Z_1$, then $\overline{R} < 0$ and the reflected and incident pressure pulses have opposite polarities, meaning that an incident compression reflects as a rarefaction, etc. $\overline{R}$ and $\overline{T}$ are used in the modeling or analysis of seismic traces recorded by hydrophones, which measure the pressure σ_{zz}, whereas R and T are used in the modelling or analysis of traces recorded by geophones, which measure the particle displacement $\mathbf{u}$ (standard geophones measure u_z, whereas three-component (3-C) geophones measure all three components of $\mathbf{u}$). Actually, geophones measure the particle velocity $\partial \mathbf{u}/\partial t$, as mentioned earlier, but the reflection and transmission coefficients for particle velocity are the same as those for particle displacement, since the time t is not scaled upon reflection or transmission. We could prove this more rigorously by obtaining equations analogous to Equations (3.11) and (3.12) for $\partial u_z/\partial t$, rather than u_z – we would find that these equations are identical to Equations (3.11) and (3.12) except that u_z is replaced by $\partial u_z/\partial t$.

The preceding results also apply to normally incident transverse waves, e.g., shear waves in a solid, or a wave on a string. For example, consider a thin string attached to a thick rope. Then an incident wave on the thin string experiences a polarity reversal in displacement upon reflection, a familiar result from elementary physics.

Suppose that the boundary $z = 0$ is the Earth's surface (land or sea) and that medium 2, the transmission medium, is the air (or ideally, a vacuum). Then there is essentially no transmitted wave to speak of, since the acoustic impedance of air is about four orders of magnitude less than that of the rock. In other words, $Z_2 \ll Z_1$, meaning that we can assume $Z_2 \approx 0$. The boundary is a **free surface**, which means that the stress must be zero at the surface. This is the only boundary condition – there is none for the displacement. Hence, the boundary condition is given by Equation (3.5) or (3.6) with $h = 0$, and the solution is $g(t) = f(t)$, which implies that $R = -1$. In physical terms, this means that there is no amplitude reduction or polarity reversal in the displacement upon reflection from a free surface, but that there *is* a polarity reversal in pressure upon reflection, i.e., an incident compression reflects as a rarefaction. The result that $R = \overline{R} = -1$ could have been obtained from R in Equation (3.8) by setting $Z_2 = 0$.

3.2 More on Polarity Reversals

First, let us consider how standard geophones and hydrophones respond to wave motion. Referring to Figure 3.2, the spring coil component is the inertial element (it doesn't move when waves make the case move). The relative motion between the coil and the magnet generates an electromotive force (EMF) in the coil, with the EMF voltage being proportional to the *velocity* of the coil. The sign of the EMF, which determines the polarity of the pulse on the geophone trace, depends on the direction of relative motion of the coil. For instance, an upgoing compression will

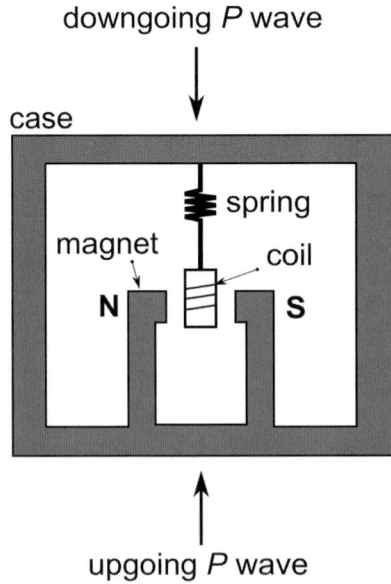

Figure 3.2 A schematic cross-section of a standard geophone.

stretch the spring, resulting in one polarity, whereas a downgoing compression will compress the spring resulting in the opposite polarity.

On the other hand, a hydrophone responds to pressure, so that on a hydrophone trace, a pulse produced by an upgoing compression would have the same polarity as a pulse produced by a downgoing compression. However, an upgoing or downgoing rarefaction would produce a pulse with the opposite polarity.

Consider the case of a **ghost reflection** on land seismic data, which typically has a polarity opposite to that of the associated primary reflection (see Figure 3.3a). From the preceding discussions of polarity reversals, we can see why that is so: the primary is a compression whose leading front pushes up on the geophone, whereas the ghost is a rarefaction whose leading front pulls down on the geophone.

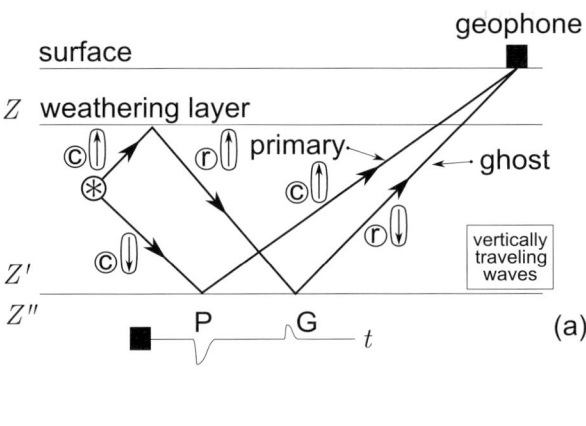

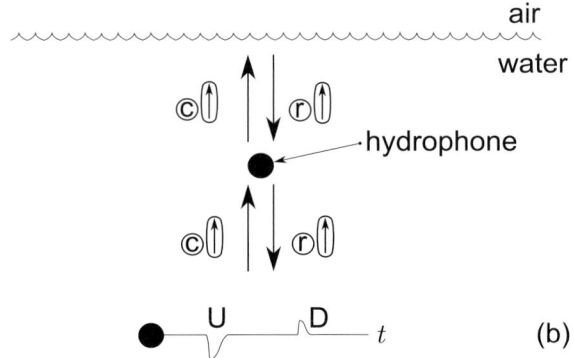

Figure 3.3 Examples of polarity reversals on (a) land data, and (b) marine data. For the ghost reflection, the acoustic impedances are such that $Z < Z' < Z''$, and the ray paths are vertical or near-vertical, but are drawn at a slant for easier viewing. The circled arrows give the direction of the displacement vector **u**. A circled "c" denotes a compression, and a circled "r" denotes a rarefaction. If the source is close enough to the base of the weathering layer, the primary and ghost overlap on the trace.

Similarly, in the case of an upgoing marine seismic wave being reflected by the sea surface, the upgoing compression comes down as a rarefaction, resulting in a hydrophone trace containing two pulses of opposite polarity (see Figure 3.3b).

A few final comments on polarity:

- If the z axis is chosen to point downwards into the Earth, then the SEG polarity convention for standard geophones and hydrophones can be summarized as follows: *a geophone and a hydrophone will record a downgoing vertically traveling P wave with opposite polarities but an upgoing vertically traveling P wave with the same polarities*. This difference has been used in marine seismology – geophone and hydrophone data sections can be multiplied together to separate downgoing and upgoing energy.

- The SEG polarity convention for standard hydrophones is mentioned in the caption of Figure 3.1. However, it should be noted that in practice, hydrophones are often wired with the opposite convention, so that an impinging compression corresponds to a positive trace value.

- A polarity convention has been proposed for multicomponent geophones (Brown, Stewart, and Lawton, 2002). It is a natural extension of the one for standard geophones, i.e., if the x_i component, $i = 1, 2, 3$, of displacement of the geophone is positive, i.e., if $u_i > 0$, then the corresponding point on the geophone trace has a positive value.

3.3 Reflection and Transmission of Harmonic Plane Waves

In seismic wave theory, it is generally easier to work with harmonic plane waves of the form $\exp\left[i\omega(\mathbf{s}\cdot\mathbf{x} - t)\right]$ rather than arbitrary unknown waveforms (any functions of $\mathbf{s}\cdot\mathbf{x} - t$). In particular, reflection and transmission coefficients can be computed more easily in this way. In principle, there is no loss of generality, since a plane wave pulse can be represented as a superposition of harmonic plane waves. Harmonic waves are sometimes called **steady-state** waves.

As an example, we treat again the problem of the normally incident P wave discussed earlier. Consider a plane harmonic P wave normally incident upon an interface at $z = 0$ (the xy plane), producing reflected and transmitted plane harmonic P waves.

The particle displacement $\mathbf{u}$ produced by each of the three waves can be written in the form

$$\mathbf{u} = U\,e^{-i\omega(t-\mathbf{s}\cdot\mathbf{x})}\mathbf{d}, \qquad (3.16)$$

where U is the amplitude of the wave, $\mathbf{s}$ is the slowness vector (which points in the direction of propagation of the wave), and $\mathbf{d}$ is a constant unit vector, often called the **polarization vector** (see Chapter 2). **d** is parallel to the straight line containing **u**. Note that **u** oscillates back and forth along this line as **x** and t vary, whereas **d** is fixed on this line. For a plane P wave, **u** oscillates along a straight line parallel to **s** (in this case, the $\pm z$ direction), pointing in one direction and then in the opposite direction as **x** and/or t change. One of these two directions is chosen as the direction of the constant unit vector **d**. Which one? The convention we use for a P wave is that **d** points in the direction of propagation of the P wave, i.e., $\mathbf{d} = \mathbf{s}/|\mathbf{s}|$. We will refer to this as the *positive* direction of particle motion or displacement for a P wave. We also use this convention for nonnormally incident P waves (see Section 3.4). As for the amplitude U, we let it, in general, be a complex number to allow for phase changes (see Equation 2.58).

Let A_I, A_R, and A_T be the amplitudes of the incident, reflected and transmitted P waves, respectively.

The displacement produced by the incident wave is

$$\mathbf{u}^{(I)} = A_I \exp\left[-i\omega(t - \mathbf{s}^{(I)}\cdot\mathbf{x})\right]\mathbf{d}^{(I)}, \quad \mathbf{s}^{(I)} = (0,\,0,\,1/\alpha_1), \quad \mathbf{d}^{(I)} = (0,\,0,\,1).$$

As $\mathbf{x} = (x, y, z)$, the term $t - \mathbf{s}^{(I)}\cdot\mathbf{x}$ becomes $t - z/\alpha_1$ (see Equation 3.1).

Similarly, the displacement produced by the reflected wave is

$$\mathbf{u}^{(R)} = A_R \exp\left[-i\omega(t - \mathbf{s}^{(R)}\cdot\mathbf{x})\right]\mathbf{d}^{(R)}, \quad \mathbf{s}^{(R)} = (0,\,0,\,-1/\alpha_1), \ \mathbf{d}^{(R)} = (0,\,0,\,-1).$$

The displacement produced by the transmitted wave is

$$\mathbf{u}^{(T)} = A_T \exp\left[-i\omega(t - \mathbf{s}^{(T)}\cdot\mathbf{x})\right]\mathbf{d}^{(T)}, \quad \mathbf{s}^{(T)} = (0,\,0,\,1/\alpha_2), \quad \mathbf{d}^{(T)} = (0,\,0,\,1).$$

As can be seen, our convention for choosing **d** means that $\mathbf{d} = \mathbf{e}_z$ for the incident and transmitted waves, and $\mathbf{d} = -\mathbf{e}_z$ for the reflected wave. The slowness vector for each wave points in the direction of travel of the wave.

As these are plane P waves, the particles are displaced in the $\pm z$ direction only, i.e., $u_x = u_y = 0$. Consequently, from the preceding we obtain

$$u_z^{(I)} = A_I \exp\left[-i\omega\left(t - \frac{z}{\alpha_1}\right)\right], \tag{3.17a}$$

$$u_z^{(R)} = -A_R \exp\left[-i\omega\left(t + \frac{z}{\alpha_1}\right)\right], \tag{3.17b}$$

$$u_z^{(T)} = A_T \exp\left[-i\omega\left(t - \frac{z}{\alpha_2}\right)\right]. \tag{3.17c}$$

For the normal stress, i.e., pressure, we apply Equation (3.3). For a given wave, we have

$$u_z = U \exp\left[-i\omega(t - s_x x - s_y y - s_z z)\right]d_z \implies$$

$$\sigma_{zz} = \rho\alpha^2 \frac{\partial u_z}{\partial z} = \rho\alpha^2 i\omega s_z u_z = \rho\alpha^2 i\omega\left(\pm\frac{1}{\alpha}\right)u_z = \pm i\omega Z u_z,$$

where $Z \equiv \rho\alpha$ and where the $\pm$ sign is chosen if the wave is propagating in the $\pm z$ direction. Note that the pressure is $90°$ out of phase with the displacement (since $i = e^{i\pi/2}$). Hence, we obtain

$$\sigma_{zz}^{(I)} = i\omega Z_1 A_I \exp\left[-i\omega\left(t - \frac{z}{\alpha_1}\right)\right], \tag{3.18a}$$

$$\sigma_{zz}^{(R)} = i\omega Z_1 A_R \exp\left[-i\omega\left(t + \frac{z}{\alpha_1}\right)\right], \tag{3.18b}$$

$$\sigma_{zz}^{(T)} = i\omega Z_2 A_T \exp\left[-i\omega\left(t - \frac{z}{\alpha_2}\right)\right]. \tag{3.18c}$$

Substitution of Equations (3.17) and (3.18) into the boundary conditions (Equations 3.2) gives

$$A_I e^{-i\omega t} - A_R e^{-i\omega t} = A_T e^{-i\omega t}, \tag{3.19a}$$

$$i\omega Z_1 A_I e^{-i\omega t} + i\omega Z_1 A_R e^{-i\omega t} = i\omega Z_2 A_T e^{-i\omega t} \tag{3.19b}$$

which can be simplified to give

$$1 - R = T \quad \text{and} \quad Z_1 + Z_1 R = Z_2 T, \tag{3.20a}$$

where $R = A_R/A_I$ is the relative amplitude of the reflected wave, and $T = A_T/A_I$ is the relative amplitude of the transmitted wave. The solutions of Equations (3.20a) are just Equations (3.8), as expected, i.e.,

$$R = \frac{Z_2 - Z_1}{Z_2 + Z_1} \quad \text{and} \quad T = \frac{2Z_1}{Z_2 + Z_1}. \tag{3.20b}$$

Equations (3.17) with $A_R = RA_I$ and $A_T = TA_I$ are simply Equations (3.11) and (3.12) applied to this case. Equations (3.17b) and (3.18b), i.e.,

$$u_z^{(R)} = -RA_I \exp\left[-i\omega\left(t + \frac{z}{\alpha_1}\right)\right] \quad \text{and} \quad \sigma_{zz}^{(R)} = i\omega Z_1 RA_I \exp\left[-i\omega\left(t + \frac{z}{\alpha_1}\right)\right],$$

show that if $Z_2 > Z_1$ (i.e., $R > 0$), there is a polarity reversal in u_z, but not in the pressure $\sigma_{zz}^{(R)}$, as we have already discovered.

Similarly, Equations (3.18) can clearly be written in the form of Equations (3.13) and (3.14) with $\overline{R}$ and $\overline{T}$ defined by Equations (3.15).

Note that each pressure wave in (3.18) can be written generically as

$$\sigma_{zz} = \overline{A} \exp\left[-i\omega\left(t \pm \frac{z}{\alpha}\right)\right], \quad \overline{A} = i\omega Z A,$$

where $\bar{A}$ is the amplitude of the pressure wave and A is the amplitude of the corresponding displacement wave. Hence, the pressure reflection and transmission coefficients can also be easily derived in terms of the displacement coefficients:

$$\bar{R} = \frac{\bar{A}_R}{\bar{A}_I} = R \quad \text{and} \quad \bar{T} = \frac{\bar{A}_T}{\bar{A}_I} = \frac{Z_2}{Z_1} T,$$

which agrees with (3.15).

We see that using harmonic waves, rather than general waveforms, leads to the same formulas for the reflection and transmission coefficients and the same conclusions about polarity reversals.

3.4 Reflection and Transmission of Plane Waves at Nonnormal Incidence

Consider the general case of a seismic plane wave that impinges nonnormally upon a flat geological interface separating two rock layers. To obtain the reflection and transmission coefficients, we must again solve the equations describing the boundary conditions. In seismology, the so-called **welded contact boundary conditions** are usually used. They state that the displacement $\mathbf{u}$ and traction $\mathbf{T}$ on one side of the interface, and $\mathbf{u}$ and $\mathbf{T}$ on the other side of the interface, match (are equal) at the boundary. In other words, the three components of $\mathbf{u}$ and the three components of $\mathbf{T}$ are continuous across the boundary. These conditions give six equations whose solutions are the relative amplitudes of the reflected and transmitted waves. Instead of $\mathbf{u}$, the particle velocity $\partial \mathbf{u}/\partial t$ could also be used, since the incident, reflected, and transmitted waves all have the same frequency. However, it is usually $\mathbf{u}$ that is used in the boundary conditions.

Assume the flat boundary is a plane that is perpendicular to the z (depth) axis. Often $z = 0$ (the xy plane) is used. The three components of displacement are u_x, u_y, and u_z. The three components of traction can be obtained from the formula $T_i = \sum_j \sigma_{ij} n_j$, where $\mathbf{n}$ is a unit vector normal to the interface – in this case, $\mathbf{n} = (0, 0, 1)$. Hence,

$$T_x = \sigma_{xz}, \quad T_y = \sigma_{yz}, \quad T_z = \sigma_{zz}.$$

The boundary condition for the normal component of displacement u_z then states that the wave motion causes no normal separation of the rock layers to occur, which is certainly physically reasonable. The boundary conditions for the tangential components of displacement, u_x and u_y, state that the wave motion also causes no slippage of one rock layer across the other along the boundary to occur. Similarly, the boundary condition for $\mathbf{T}$ states that the normal stress component, σ_{zz}, and the tangential stress components, σ_{xz} and σ_{yz}, are continuous across the boundary, meaning that there are no unbalanced stresses at the interface, which is physically reasonable for two solid layers in welded contact. The continuity of **T** can be demonstrated with a *Gaussian pill box* argument – see, e.g., Stein and Wysession (2003, pp. 51–52).

Assume that the incident wave is propagating in the *xz* plane. If it is a *P* or *SV* wave, it will cause the particles of the medium to move in the *xz* plane, but there will be no *y* component of particle motion. This means that the particle motion due to the reflected and transmitted waves will also be in the *xz* plane, with no *y* component. Similarly, if the incident plane wave is an *SH* wave, particles will move in the *y* direction with no component of motion in the *xz* plane, and the reflected and transmitted waves will produce particle motion in the *y* direction only, as well. In other words, an incident *P* or *SV* wave will generate both *P* and *SV* reflected and transmitted waves, but no *SH* waves, and an incident *SH* wave will generate only *SH* reflected and transmitted waves, but no *P* or *SV* waves. This means that the six boundary conditions can be separated into two groups: four of the equations, i.e., the ones for u_x, u_z, σ_{xz}, and σ_{zz}, apply to the *P-SV* case, and the remaining two, i.e., the ones for u_y and σ_{yz}, apply to the *SH* case. The four *P-SV* equations can be solved for the relative amplitudes of the four reflected and transmitted *P* and *SV* waves, and the two *SH* equations can be solved for the relative amplitudes of the two reflected and transmitted *SH* waves.

Consider first the *SH* case. Figure 3.4 shows the relevant geometry and notation. Each wave has a displacement $\mathbf{u} = U \exp[i\omega(\mathbf{s}\cdot\mathbf{x} - t)]\mathbf{d}$, where the amplitude U and slowness **s** are different for each wave. We let $\mathbf{d} = \mathbf{e}_y$ for each wave, hence we only need the *y* component of **u** for *SH* waves:

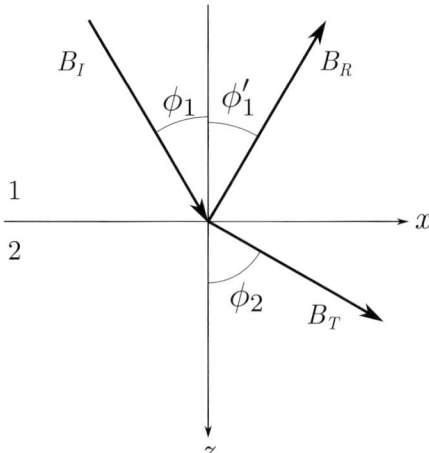

Figure 3.4 Reflection and transmission of an *SH* wave. B_I, B_R, and B_T are the amplitudes (U) of the waves. The positive direction of particle motion is the *y* direction (perpendicular to and out of the page).

$$u_y = U \exp\left[i\omega(\mathbf{s} \cdot \mathbf{x} - t)\right].$$

From the stress–strain relation (Equation 2.30), we obtain

$$\sigma_{yz} = \mu \frac{\partial u_y}{\partial z} = i\omega s_z \rho \beta^2 U \exp\left[i\omega(\mathbf{s} \cdot \mathbf{x} - t)\right], \tag{3.21}$$

where $\mu = \rho\beta^2$ has been used. For the incident, reflected, and transmitted waves, $s_z = \pm|\mathbf{s}|\cos\phi$ is given by

$$s_z^{(I)} = \frac{\cos\phi_1}{\beta_1}, \quad s_z^{(R)} = -\frac{\cos\phi_1'}{\beta_1}, \quad \text{and} \quad s_z^{(T)} = \frac{\cos\phi_2}{\beta_2}. \tag{3.22}$$

These are to be substituted into Equation (3.21) for each wave.

We can now apply the boundary conditions, which state that u_y and σ_{yz} must be continuous across the boundary, which we take to be $z = 0$. We obtain

$$\left[u_y^{(I)} + u_y^{(R)}\right]_{z=0} = \left[u_y^{(T)}\right]_{z=0}, \qquad \left[\sigma_{yz}^{(I)} + \sigma_{yz}^{(R)}\right]_{z=0} = \left[\sigma_{yz}^{(T)}\right]_{z=0}.$$

Noting that $\mathbf{s} \cdot \mathbf{x} = s_x x$ for all waves at the boundary (since $s_y = 0$), these become

$$B_I \exp\left(i\omega s_x^{(I)} x\right) + B_R \exp\left(i\omega s_x^{(R)} x\right) = B_T \exp\left(i\omega s_x^{(T)} x\right), \tag{3.23a}$$

$$\begin{aligned} \rho_1\beta_1 B_I \exp\left(i\omega s_x^{(I)} x\right) \cos\phi_1 &- \rho_1\beta_1 B_R \exp\left(i\omega s_x^{(R)} x\right) \cos\phi_1' \\ &= \rho_2\beta_2 \cos\phi_2 B_T \exp\left(i\omega s_x^{(T)} x\right), \end{aligned} \tag{3.23b}$$

where the factor $\exp(-i\omega t)$ has been canceled from each term in each equation, and where the factor $i\omega$ has been canceled from each term in the second equation. If we solve these equations as they stand, the solutions B_R/B_I and B_T/B_I will be functions of x, which is physically unrealistic. To eliminate the x dependence, we must equate all the horizontal slowness components, i.e., we set $s_x^{(I)} = s_x^{(R)} = s_x^{(T)} \equiv s_x$, which implies

$$\frac{\sin\phi_1}{\beta_1} = \frac{\sin\phi_1'}{\beta_1} = \frac{\sin\phi_2}{\beta_2} = s_x \equiv p, \tag{3.24}$$

where p is called the **ray parameter**. Note that this is just **Snell's law of reflection and refraction**. It implies that $\phi_1' = \phi_1$, as is expected. The complex exponentials in Equations (3.23) are now all the same and can be canceled from each term, eliminating the x dependence. Equations (3.23) then become

$$B_I + B_R = B_T, \qquad \rho_1\beta_1(B_I - B_R)\cos\phi_1 = \rho_2\beta_2 B_T \cos\phi_2, \tag{3.25}$$

whose solutions for the unknowns B_R and B_T are

$$R_S = \frac{B_R}{B_I} = \frac{\rho_1\beta_1\cos\phi_1 - \rho_2\beta_2\cos\phi_2}{\rho_1\beta_1\cos\phi_1 + \rho_2\beta_2\cos\phi_2}, \quad T_S = \frac{B_T}{B_I} = \frac{2\rho_1\beta_1\cos\phi_1}{\rho_1\beta_1\cos\phi_1 + \rho_2\beta_2\cos\phi_2}. \tag{3.26}$$

R_S and T_S are the particle displacement reflection and transmission coefficients for the *SH* case. It should be pointed out that the same formulas would have been obtained if the z axis had been pointing in the opposite direction (upward).

Suppose the incident wave is in medium 2. If we do the calculation for this case, we find that the reflection and transmission coefficients can be obtained from the preceding ones simply by switching the 1s and 2s.

It should also be pointed out, for clarity, that although we arrived at Snell's law in (3.24) by eliminating the x dependence in the boundary conditions (3.23), it is not necessary to apply the boundary conditions to derive Snell's law. Simpler geometrical derivations for Snell's law can be found in elementary textbooks. For example, it can be derived by equating the apparent wavelengths along the boundary for the incident, reflected, and transmitted waves, or by applying Fermat's principle of least time (it can be shown that of all the possible paths a ray could follow in going from a source point in the incidence medium to a receiver point in the transmission medium, the path that has the least travel time is also the one that obeys Snell's law).

Consider now the *P-SV* case. Figure 3.5 shows the relevant geometry and notation. In the figure, it is anticipated (correctly), based on the preceding *SH* results, that the angle of incidence equals the angle of reflection for waves of the same type.

The boundary conditions can be derived in the same way that they were for the *SH* case. The derivation is, however, considerably longer.

Note that there are two possible incident waves in medium 1: the incident wave is either a *P* wave or an *SV* wave. Instead of doing two sets of calculations for the reflection and transmission coefficients, one set for each incident wave, we can save ourselves some work by including both incident waves in one set of calculations. In the resulting equations, we then simply set one of the incident wave amplitudes equal to zero.

The particle displacement $\mathbf{u}$ produced by each wave has the form $\mathbf{u} = U \exp[i\omega(\mathbf{s} \cdot \mathbf{x} - t)]\mathbf{d}$, where the amplitude U, slowness $\mathbf{s}$ and polarization vector $\mathbf{d}$ are different for each wave. $U = A_I$ for the incident P wave, $U = A_R$ for the reflected P wave, etc. (see Figure 3.5). $\mathbf{s}$ points in the direction of travel of the wave, and $|\mathbf{s}|$ is the reciprocal of the wave's speed.

Since $\mathbf{u}$ oscillates back and forth, one of the two oscillation directions of $\mathbf{u}$ is chosen as the direction of $\mathbf{d}$. We use the following convention for choosing $\mathbf{d}$: for a *P* wave, $\mathbf{d}$ points in the direction of propagation of the wave, i.e., in the direction of $\mathbf{s}$; and for an *SV* wave, $\mathbf{d}$ isperpendicular to $\mathbf{s}$, with $d_x > 0$ (see Figure 3.5).

Various conventions have been employed in the literature for choosing $\mathbf{d}$. The sign convention for $\mathbf{d}$ adopted in the previous paragraph has the advantage that one obtains the same results for the reflection and transmission coefficients, regardless of whether the z axis is taken to point up or down.

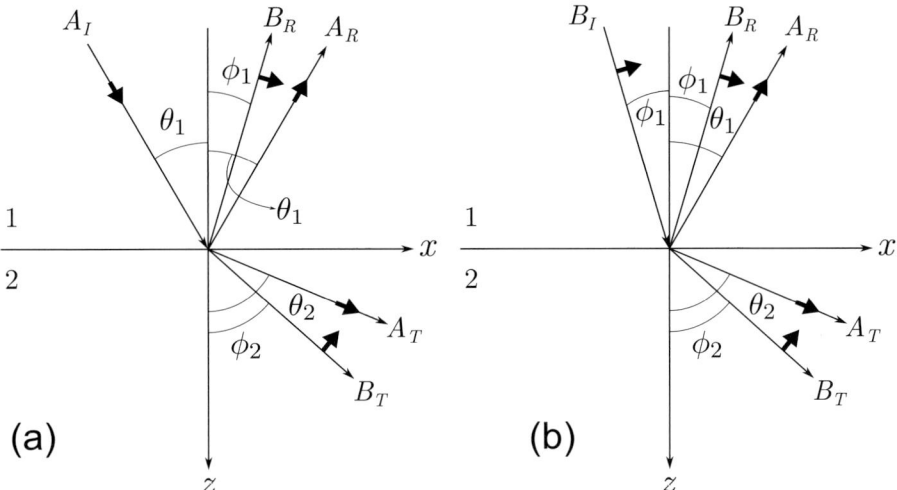

Figure 3.5 Reflection and transmission of an incident (a) *P* wave and (b) *SV* wave. A_I, A_R, and A_T are the amplitudes of the *P* waves, and B_I, B_R, and B_T are the amplitudes of the *SV* waves. The figures incorporate Snell's law of reflection. The boldface arrows are the unit vectors **d** for each wave, which define the positive direction of particle motion. The slowness vector **s** for each wave is along the ray for the wave.

From Figure 3.5, we obtain the following expressions for the slowness and polarization vectors of the waves:

$$A_I: \quad \mathbf{s} = \left(\frac{\sin\theta_1}{\alpha_1}, 0, \frac{\cos\theta_1}{\alpha_1}\right), \quad \mathbf{d} = (\sin\theta_1, 0, \cos\theta_1)$$

$$B_I: \quad \mathbf{s} = \left(\frac{\sin\phi_1}{\beta_1}, 0, \frac{\cos\phi_1}{\beta_1}\right), \quad \mathbf{d} = (\cos\phi_1, 0, -\sin\phi_1)$$

$$A_R: \quad \mathbf{s} = \left(\frac{\sin\theta_1}{\alpha_1}, 0, -\frac{\cos\theta_1}{\alpha_1}\right), \quad \mathbf{d} = (\sin\theta_1, 0, -\cos\theta_1)$$

$$B_R: \quad \mathbf{s} = \left(\frac{\sin\phi_1}{\beta_1}, 0, -\frac{\cos\phi_1}{\beta_1}\right), \quad \mathbf{d} = (\cos\phi_1, 0, \sin\phi_1)$$

$$A_T: \quad \mathbf{s} = \left(\frac{\sin\theta_2}{\alpha_2}, 0, \frac{\cos\theta_2}{\alpha_2}\right), \quad \mathbf{d} = (\sin\theta_2, 0, \cos\theta_2)$$

$$B_T: \quad \mathbf{s} = \left(\frac{\sin\phi_2}{\beta_2}, 0, \frac{\cos\phi_2}{\beta_2}\right), \quad \mathbf{d} = (\cos\phi_2, 0, -\sin\phi_2)$$

As for the *SH* case, we must equate all the s_xs to ensure that the reflection and transmission coefficients do not vary with x). This gives Snell's law of reflection and refraction:

$$p = \frac{\sin\theta_1}{\alpha_1} = \frac{\sin\theta_2}{\alpha_2} = \frac{\sin\phi_1}{\beta_1} = \frac{\sin\phi_2}{\beta_2}. \tag{3.27}$$

Note that $\theta_n > \phi_n$, $n = 1, 2$ (see Figure 3.5) because $\alpha_n > \beta_n$. The welded-contact boundary conditions are that u_x, u_z, σ_{xz}, and σ_{zz} are continuous across the interface. These four conditions lead to four equations for the unknown amplitudes of the reflected and transmitted waves.

As an example, the derivation of the third equation, for the continuity of the shear stress component σ_{xz}, is outlined here. The others are left as an exercise for the reader.

The displacement for each wave has the form $\mathbf{u} = U \exp[i\omega(\mathbf{s} \cdot \mathbf{x} - t)]\mathbf{d}$, meaning that σ_{xz} for each wave has the form

$$\sigma_{xz} = 2\mu e_{xz} = \mu \left(\frac{\partial u_x}{\partial z} + \frac{\partial u_z}{\partial x} \right) = i\omega\mu(s_z u_x + s_x u_z)$$
$$= i\omega \exp\left[i\omega(\mathbf{s} \cdot \mathbf{x} - t)\right] \mu U \left[s_z d_x + s_x d_z\right].$$

The boundary condition states that the total shear stress σ_{xz} in medium 1 at the interface $z = 0$ is equal to that of medium 2 at $z = 0$, i.e.,

$$\left[\sigma_{xz}(A_I) + \sigma_{xz}(B_I) + \sigma_{xz}(A_R) + \sigma_{xz}(B_R)\right]_{z=0} = \left[\sigma_{xz}(A_T) + \sigma_{xz}(B_T)\right]_{z=0}.$$

Next we substitute, for each term, the appropriate expression for σ_{xz}, using the appropriate expressions for μ, U, s_x, s_z, d_x, and d_z. We note that the quantity $i\omega \exp[i\omega(\mathbf{s} \cdot \mathbf{x} - t)] = i\omega \exp[i\omega(s_x x - t)]$ at the interface, and that this quantity is the same for each term (because the s_xs are all the same) and hence will cancel out of the equation – so we can leave it out at the start. We obtain

$$\mu_1 A_I \left[\left(\frac{\cos\theta_1}{\alpha_1} \right) \sin\theta_1 + \left(\frac{\sin\theta_1}{\alpha_1} \right) \cos\theta_1 \right]$$
$$+ \mu_1 B_I \left[\left(\frac{\cos\phi_1}{\beta_1} \right) \cos\phi_1 + \left(\frac{\sin\phi_1}{\beta_1} \right)(-\sin\phi_1) \right]$$
$$+ \mu_1 A_R \left[\left(-\frac{\cos\theta_1}{\alpha_1} \right) \sin\theta_1 + \left(\frac{\sin\theta_1}{\alpha_1} \right)(-\cos\theta_1) \right]$$
$$+ \mu_1 B_R \left[\left(-\frac{\cos\phi_1}{\beta_1} \right) \cos\phi_1 + \left(\frac{\sin\phi_1}{\beta_1} \right)(\sin\phi_1) \right]$$
$$= \mu_2 A_T \left[\left(\frac{\cos\theta_2}{\alpha_2} \right) \sin\theta_2 + \left(\frac{\sin\theta_2}{\alpha_2} \right) \cos\theta_2 \right]$$
$$+ \mu_2 B_T \left[\left(\frac{\cos\phi_2}{\beta_2} \right) \cos\phi_2 + \left(\frac{\sin\phi_2}{\beta_2} \right)(-\sin\phi_2) \right].$$

Using $\mu_j = \rho_j \beta_j^2$, $j = 1, 2$, $p = (\sin\theta_j)/\alpha_j = (\sin\phi_j)/\beta_j$, and $\cos^2\phi_j - \sin^2\phi_j = 1 - 2\sin^2\phi_j = 1 - 2\beta_j^2 p^2$ gives

$$2\rho_1 \beta_1^2 p(A_I - A_R) \cos\theta_1 + \rho_1 \beta_1(1 - 2\beta_1^2 p^2)(B_I - B_R)$$
$$= 2\rho_2 \beta_2^2 p A_T \cos\theta_2 + \rho_2 \beta_2(1 - 2\beta_2^2 p^2)B_T,$$

which is the equation for the boundary condition expressing the continuity of σ_{xz} across the interface.

The remaining three boundary conditions can be obtained in a similar way.

Altogether, Equations (3.28a), (3.28b), (3.28c), and (3.28d) are the boundary conditions for u_x, u_z, σ_{xz}, and σ_{zz} respectively:

$$(A_I + A_R) \sin\theta_1 + (B_I + B_R) \cos\phi_1 = A_T \sin\theta_2 + B_T \cos\phi_2 \tag{3.28a}$$

$$(A_I - A_R) \cos\theta_1 - (B_I - B_R) \sin\phi_1 = A_T \cos\theta_2 - B_T \sin\phi_2 \tag{3.28b}$$

$$2\rho_1\beta_1^2 p(A_I - A_R) \cos\theta_1 + \rho_1\beta_1(1 - 2\beta_1^2 p^2)(B_I - B_R)$$
$$= 2\rho_2\beta_2^2 pA_T \cos\theta_2 + \rho_2\beta_2(1 - 2\beta_2^2 p^2)B_T \tag{3.28c}$$

$$\rho_1\alpha_1(1 - 2\beta_1^2 p^2)(A_I + A_R) - 2\rho_1\beta_1^2 p(B_I + B_R) \cos\phi_1$$
$$= \rho_2\alpha_2(1 - 2\beta_2^2 p^2)A_T - 2\rho_2\beta_2^2 pB_T \cos\phi_2. \tag{3.28d}$$

Equations (3.28a) through (3.28d) are known as the **Zoeppritz equations**. They are a system of four linear equations in four unknowns (the four amplitudes of the reflected and transmitted waves). They can be solved to obtain the mathematical formulas for the reflection and transmission coefficients, which are considerably more complicated than those for the *SH* case.

For an incident *P* wave (Figure 3.5a), we set $B_I = 0$ in (3.28a) through (3.28d) and solve the equations for the relative amplitudes, i.e., the particle displacement reflection and transmission coefficients:

$$R_{PP} = \frac{A_R}{A_I}, \quad R_{PS} = \frac{B_R}{A_I}, \quad T_{PP} = \frac{A_T}{A_I}, \quad T_{PS} = \frac{B_T}{A_I}.$$

Similarly, for an incident *SV* wave, we set $A_I = 0$ in (3.28a) through (3.28d) and solve for

$$R_{SP} = \frac{A_R}{B_I}, \quad R_{SS} = \frac{B_R}{B_I}, \quad T_{SP} = \frac{A_T}{B_I}, \quad T_{SS} = \frac{B_T}{B_I}.$$

The reflection and transmission coefficients for *P* and *SV* waves incident from medium 1 are (Aki and Richards, 1980, 2002, chapter 5):

$$\begin{aligned}
R_{PP} &= \left[(b\xi_1 - c\xi_2)F - (a + d\xi_1\eta_2)Hp^2\right]D^{-1} \\
R_{PS} &= -2\xi_1 p(\alpha_1/\beta_1)(ab + cd\xi_2\eta_2)D^{-1} \\
T_{PP} &= 2\rho_1\xi_1(\alpha_1/\alpha_2)FD^{-1} \\
T_{PS} &= 2\rho_1\xi_1(\alpha_1/\beta_2)pHD^{-1} \\
R_{SP} &= -2\eta_1 p(\beta_1/\alpha_1)(ab + cd\xi_2\eta_2)D^{-1} \\
R_{SS} &= -\left[(b\eta_1 - c\eta_2)E - (a + d\xi_2\eta_1)Gp^2\right]D^{-1} \\
T_{SP} &= -2\rho_1\eta_1 p(\beta_1/\alpha_2)GD^{-1} \\
T_{SS} &= 2\rho_1\eta_1(\beta_1/\beta_2)ED^{-1}
\end{aligned} \tag{3.28e}$$

where

$$a = \gamma_2 - \gamma_1, \quad b = \gamma_2 + \chi_1 p, \quad c = \gamma_1 + \chi_2 p, \quad d = 2(\rho_2 \beta_2{}^2 - \rho_1 \beta_1{}^2),$$

$$\chi_l = 2\rho_l \beta_l{}^2 p, \quad \gamma_l = \rho_l(1 - 2\beta_l{}^2 p^2), \quad \xi_l = \frac{\cos \theta_l}{\alpha_l}, \quad \eta_l = \frac{\cos \phi_l}{\beta_l}, \quad l = 1, 2,$$

$$E = b\xi_1 + c\xi_2, \quad F = b\eta_1 + c\eta_2, \quad G = a - d\xi_1 \eta_2,$$

$$H = a - d\xi_2 \eta_1, \quad D = EF + GHp^2.$$

The formulas for P and SV waves incident from medium 2 can be obtained by switching subscripts 1 and 2 in the preceding formulas.

As a check, one finds that when the Zoeppritz equations, or the preceding formulas for R_{PP} and T_{PP}, are applied to the case of a normally incident P wave, the reflection and transmission coefficients in (3.8) are obtained, as expected.

The Zoeppritz equations (3.28a) through (3.28d) can also be written in a convenient matrix form as

$$\mathbf{M U}_{RT} = \mathbf{N U}_I, \tag{3.28f}$$

where

$$\mathbf{M} = \begin{bmatrix} -\alpha_1 p & -\cos \phi_1 & \alpha_2 p & \cos \phi_2 \\ \cos \theta_1 & -\beta_1 p & \cos \theta_2 & -\beta_2 p \\ \chi_1 \cos \theta_1 & \beta_1 \gamma_1 & \chi_2 \cos \theta_2 & \beta_2 \gamma_2 \\ -\alpha_1 \gamma_1 & \chi_1 \cos \phi_1 & \alpha_2 \gamma_2 & -\chi_2 \cos \phi_2 \end{bmatrix}$$

$$\mathbf{N} = \begin{bmatrix} \alpha_1 p & \cos \phi_1 & -\alpha_2 p & -\cos \phi_2 \\ \cos \theta_1 & -\beta_1 p & \cos \theta_2 & -\beta_2 p \\ \chi_1 \cos \theta_1 & \beta_1 \gamma_1 & \chi_2 \cos \theta_2 & \beta_2 \gamma_2 \\ \alpha_1 \gamma_1 & -\chi_1 \cos \phi_1 & -\alpha_2 \gamma_2 & \chi_2 \cos \phi_2 \end{bmatrix}$$

$$\mathbf{U}_{RT} = \begin{bmatrix} A_R \\ B_R \\ A_T \\ B_T \end{bmatrix}, \quad \mathbf{U}_I = \begin{bmatrix} A_I \\ B_I \\ 0 \\ 0 \end{bmatrix}.$$

The last two columns of $\mathbf{N}$ are not needed to produce the equations in (3.28), i.e., the last two columns of $\mathbf{N}$ could be filled with zeros, because the last two entries of $\mathbf{U}_I$ are zero. The last two columns of $\mathbf{N}$ are needed if the two zeros in $\mathbf{U}_I$ are replaced by the amplitudes of the two possible waves that could be incident from the lower medium, as explained four paragraphs down.

As can be seen, the second and third rows of the matrix $\mathbf{N}$ are identical to the second and third rows of $\mathbf{M}$, whereas the first and fourth rows of $\mathbf{N}$ are the negative of the first and fourth rows of $\mathbf{M}$.

If one uses potentials instead of displacements to derive the boundary conditions, e.g., ϕ(refl. P) $= A'_R \exp[i\omega(\mathbf{s}^{(RP)} \cdot \mathbf{x} - t)]$, etc., then the resulting set of four equations involving the potential amplitudes A'_R, ... are called the **Knott equations**.

If needed, pressure coefficients (hydrophones measure pressure, not $\mathbf{u}$) can be worked out. The general harmonic plane wave expression for the pressure p (not to be confused with the ray parameter p in the preceding equations) for a given wave is $p = -\sigma_{zz} = V \exp[i\omega(\mathbf{s} \cdot \mathbf{x} - t)]$. The pressure amplitude V can be obtained by calculating σ_{zz} from $\mathbf{u} = U \exp[i\omega(\mathbf{s} \cdot \mathbf{x} - t)]\mathbf{d}$ (using the stress–strain relation) for a particular wave and comparing to obtain V. Then the PP pressure reflection coefficient, for instance, would be V_R/V_I.

In deriving the Zoeppritz equations, for convenience we included both of the two possible incident waves in medium 1 in the calculation (even though only one or the other is incident from a physical viewpoint). We could also have included the two possible incident waves in medium 2 as well. In that case, the two lower elements of $\mathbf{U}_I$ would no longer be zero, but rather the amplitudes of the two upgoing incident waves in medium 2 (matrices $\mathbf{M}$ and $\mathbf{N}$ would remain the same). However, if this is done, we would need to use a more descriptive notation for the amplitudes and coefficients. For example, the symbol A_I for the amplitude of the incident P wave, or the symbol R_{PP} for the PP reflection coefficient, does not tell us in which medium, 1 or 2, the incident wave is located. Similarly, the upgoing P wave in medium 1, whose amplitude we have denoted by A_R because it is a reflected wave, could also be a transmitted wave if the incident wave is in medium 2. A better notation would be $\mathbf{U}_I = [A_1, B_1, A_2, B_2]^T$ for the incident wave amplitudes; $\mathbf{U}_{RT} = [A'_1, B'_1, A'_2, B'_2]^T$ for the reflected and transmitted wave amplitudes; and P_2P_2, P_1S_2, etc., for the reflection and transmission coefficients, or R_{PP2}, T_{PS1}, etc., (with the subscripts 1 and 2 denoting the incidence medium). P_1S_2 would indicate a P wave incident in medium 1 and scattered SV wave in medium 2, i.e., P_1S_2 would be a transmission coefficient. Similarly, P_1P_1 would be the P to P reflection coefficient in medium 1. Another possibility is to use overhead acute or grave accents to indicate the directions in which the waves are traveling, e.g., $\mathbf{U}_I = [\grave{P}_1, \grave{S}_1, \acute{P}_2, \acute{S}_2]^T$ for the incident wave amplitudes; $\mathbf{U}_{RT} = [\acute{P}_1, \acute{S}_1, \grave{P}_2, \grave{S}_2]^T$ for the reflected and transmitted wave amplitudes; and $\acute{P}\grave{P}$, $\grave{P}\grave{S}$, etc., for the reflection and transmission coefficients (Aki and Richards, 1980, 2002, chapter 5).

Welded-contact boundary conditions were used in the preceding in deriving the Zoeppritz equations, i.e., it was assumed that u_x, u_z, σ_{xz}, and σ_{zz} are continuous across the interface. However, some types of interfaces, such as certain fractures, joints and faults, may not be in perfect welded-contact. In such cases, the two stress components may still be continuous across the interface, but not the two displacement components u_x and u_z. Consider, for instance, the following set of four boundary conditions:

$$u_{x2} - u_{x1} = c_x\sigma_{xz}, \quad u_{z2} - u_{z1} = c_z\sigma_{zz}, \quad \sigma_{xz1} = \sigma_{xz2}, \quad \sigma_{zz1} = \sigma_{zz2},$$

where the subscripts 1 and 2 refer to media 1 and 2, respectively, and where c_x and c_z are constants called the **specific compliances** (they are zero for the welded-contact case). Formulas for the reflection and transmission coefficients can also be derived for this more general case (Chaisri and Krebes, 2000). Unlike the welded-contact case, the coefficients are functions of the frequency.

3.5 Reflection Off a Free Surface

Referring to Figure 3.5, suppose that medium 2 is a vacuum. In that case, the interface $z = 0$ is a "stress-free" surface. There are no transmitted waves. The air–ground interface is a close approximation to a stress-free surface – transmitted waves exist in the air, but they are very weak.

Consider the *P-SV* case. Suppose, for example, that the incident wave is a *P* wave. The free-surface boundary conditions are that σ_{xz} and σ_{zz} in the solid (medium 1) must be zero at the interface $z = 0$, i.e., at the surface. These are expressed by setting the left-hand sides of Equations (3.28c) and (3.28d) equal to zero (with $B_I = 0$). This gives

$$2\rho\beta^2 p(A_I - A_R)\cos\theta - \rho\beta(1 - 2\beta^2 p^2)B_R = 0 \qquad (3.29a)$$

$$\rho\alpha(1 - 2\beta^2 p^2)(A_I + A_R) - 2\rho\beta^2 pB_R\cos\phi = 0, \qquad (3.29b)$$

where the subscript "1" has been dropped because we are dealing with waves only in one medium. The solutions of these equations for the two reflection coefficients $R_{PP} = A_R/A_I$ and $R_{PS} = B_R/A_I$ are

$$R_{PP} = \frac{4\beta^3 p^2\cos\theta\cos\phi - \alpha(1 - 2\beta^2 p^2)^2}{4\beta^3 p^2\cos\theta\cos\phi + \alpha(1 - 2\beta^2 p^2)^2}, \qquad (3.30a)$$

$$R_{PS} = \frac{4\beta\alpha p(1 - 2\beta^2 p^2)\cos\theta}{4\beta^3 p^2\cos\theta\cos\phi + \alpha(1 - 2\beta^2 p^2)^2}. \qquad (3.30b)$$

Similar results can be obtained for an incident *SV* wave.

For an incident *SH* wave, the boundary condition is that $\sigma_{yz} = 0$ at the boundary $z = 0$, which is expressed by setting the left-hand side of the second of Equations (3.25) equal to zero. We then have,

$$\rho\beta(B_I - B_R)\cos\phi = 0, \quad \implies \quad R_S = \frac{B_R}{B_I} = 1. \qquad (3.31)$$

This could have also been obtained by setting $\rho_2 = 0$ in Equation (3.26).

Imagine an upgoing plane *P* wave striking the ground surface $z = 0$. A geophone located on the surface would be affected not only by the upcoming incident wave, but also by the two downgoing reflected waves. This is called the **free surface effect**. More precisely, the displacement $\mathbf{u}^{(G)}$ of the geophone would be given by

$$
\begin{aligned}
\mathbf{u}^{(G)} &= \mathbf{u}^{(IP)} + \mathbf{u}^{(RP)} + \mathbf{u}^{(RS)} \\
&= A_I e^{i\omega(px-t)} [\mathbf{d}^{(IP)} + R_{PP}\mathbf{d}^{(RP)} + R_{PS}\mathbf{d}^{(RS)}] \\
&\equiv A_I e^{i\omega(px-t)} \mathbf{d}^{(G)}.
\end{aligned}
\tag{3.32a}
$$

This effect must be taken into account when computing synthetic seismograms or when doing careful quantitative interpretations of geophone signals. Graphs showing how $d_x^{(G)}$ and $d_z^{(G)}$, called **surface conversion coefficients**, vary with incidence angle for different β/α values can be found in Červený and Ravindra (1971, figure 2.10). For an incident P wave, $d_x^{(G)}$ starts at zero at $\theta = 0°$, increases smoothly to a maximum value, and decreases to zero at $\theta = 90°$, and $d_z^{(G)}$ starts at a value of 2 at $\theta = 0°$, and generally decreases smoothly to zero at $\theta = 90°$.

In the case of a normally incident P wave, there is no reflected SV wave, and $p = 0$, and $R_{PP} = -1$ (as indicated in the last paragraph of Section 3.1). Hence, taking the z direction upward,

$$
\begin{aligned}
\mathbf{d}^{(G)} &= [\mathbf{d}^{(IP)} + R_{PP}\mathbf{d}^{(RP)} + R_{PS}\mathbf{d}^{(RS)}] = [\mathbf{d}^{(IP)} - \mathbf{d}^{(RP)}] \\
&= [\mathbf{e}_z - (-\mathbf{e}_z)] = 2\mathbf{e}_z \quad \Rightarrow \quad \mathbf{u}^{(G)} = 2A_I e^{-i\omega t}\mathbf{e}_z,
\end{aligned}
\tag{3.32b}
$$

i.e., the geophone is displaced upward by an amount equal to twice the amplitude of the incident wave. This is the free surface effect for normal incidence. Because $R_{PP} = -1$, an incident compression, for example, reflects as a rarefaction, or equivalently, the two displacements, u_z^{IP} and u_z^{RP}, are in phase (they are both upward) and therefore interfere constructively to produce a wave of amplitude $2A_I$. This can also be predicted from the results of experiments involving waves on strings, studied in elementary physics courses. Consider a horizontal string connected to a thin vertical rod by a light metallic ring around the rod, so that the string can slide up and down the rod via the metal ring. Consider a wave on the string traveling toward the rod. If the wave has an amplitude A, then when it reaches the rod, the metal ring attached to the end of the string slides up the rod to a height of $2A$, then comes back down, producing a reflected wave (with no polarity reversal). See, e.g., French (1971, figure 8.1b).

3.6 Critical Angles

Consider the reflection and transmission of SH waves, and suppose that, in Figure 3.4, we have $\beta_2 > \beta_1$. Then Snell's law, Equation (3.24), implies that $\phi_2 > \phi_1$ (as in Figure 3.4). Hence, there is a critical value ϕ_{1c} of ϕ_1 at which

$\phi_2 = 90°$. At this angle, the transmitted wave travels along the boundary, rather than into medium 2. The angle ϕ_{1c} is known as the **critical angle** and is given by

$$\phi_{1c} = \sin^{-1}(\beta_1/\beta_2). \tag{3.33}$$

If $\phi_1 > \phi_{1c}$, then Snell's law implies that $\sin \phi_2$ is a real number that is greater than 1, which means that ϕ_2 is a complex angle, i.e., $\phi_2 = 90° + ia$, where a is a real number. The transmitted wave travels along the boundary for $\phi_1 \geq \phi_{1c}$.

For example, suppose $\beta_1 = 1$ km/s and $\beta_2 = 2$ km/s. Then $\phi_{1c} = 30°$. So for an incident wave with $\phi_1 = 40° > \phi_{1c}$, we have

$$\sin \phi_2 = (\beta_2/\beta_1) \sin \phi_1 = (2/1) \sin 40° = 1.29 > 1$$

meaning ϕ_2 is complex. Fortunately, we usually do not need to calculate a, the imaginary part of ϕ_2.

If $\beta_2 < \beta_1$, then there is no critical angle. Since $\phi_2 < \phi_1$ in this case, ϕ_2 is always less than $90°$ for all values of ϕ_1.

Using Equations (3.22) and (3.24), the displacement of the transmitted wave is given by

$$\mathbf{u}^{(T)} = B_T \exp\left[i\omega(\mathbf{s} \cdot \mathbf{x} - t)\right]\mathbf{e}_y = B_T \exp\left[i\omega\left(px + \frac{\cos \phi_2}{\beta_2}z - t\right)\right]\mathbf{e}_y, \tag{3.34}$$

which is a wave traveling downward in the positive x and z directions for **subcritical** incidence angles ($\phi_1 < \phi_{1c}$). Let us examine the nature of the transmitted wave for **supercritical** incidence angles ($\phi_1 \geq \phi_{1c}$). In that case, we know that $\sin \phi_2 > 1$, which implies that

$$\cos \phi_2 = \sqrt{1 - \sin^2 \phi_2} = \pm i\sqrt{\sin^2 \phi_2 - 1}$$

$$= i\sqrt{\frac{\beta_2^2}{\beta_1^2} \sin^2 \phi_1 - 1} = i\sqrt{\beta_2^2 p^2 - 1} \qquad (\phi_1 \geq \phi_{1c}, \ \omega > 0). \tag{3.35}$$

In words, $\cos \phi_2$ is a purely imaginary number ($i = \sqrt{-1}$) for supercritical incidence angles. We chose the "+" sign in Equation (3.35) for reasons discussed later. Substituting Equation (3.35) into Equation (3.34) gives

$$\mathbf{u}^{(T)} = B_T \exp\left[-\frac{\omega}{\beta_2}\sqrt{\beta_2^2 p^2 - 1}\, z\right]\exp\left[i\omega(px - t)\right]\mathbf{e}_y \qquad (\phi_1 \geq \phi_{1c}, \ \omega > 0). \tag{3.36}$$

This is a plane wave traveling along the boundary in the positive x direction with speed $1/p = \beta_1/\sin \phi_1$ (which varies from β_2 at $\phi_1 = \phi_{1c}$ to β_1 at $\phi_1 = 90°$), and whose amplitude of particle motion (which is in the y direction) decays exponentially in the positive z direction. It is known as an **inhomogeneous wave** or **evanescent wave**. If we had chosen the "−" sign for $\cos \phi_2$ in Equation (3.35), the amplitude would *grow* with distance from the boundary, rather than decay, which is physically unacceptable in this case. Also, the choice of the "+" sign assumes that $\omega > 0$. If $\omega < 0$ (which occurs in the inverse Fourier transform), then we would choose the "−" sign to obtain amplitude decay. Requiring the amplitude to decay with distance from the interface is sometimes called the **radiation condition**.

Since all three waves have the same horizontal component of slowness, p, the three plane wavefronts of the incident, reflected and transmitted waves meet at a single point on the interface that travels along the boundary in the positive x direction with speed $1/p = \beta_1/\sin\phi_1$.

What happens to the reflection and transmission coefficients for $\phi_1 > \phi_{1c}$? Substituting Equation (3.35) into Equation (3.26) for R_S gives

$$R_S = \frac{\rho_1\beta_1\cos\phi_1 - i\rho_2\beta_2\sqrt{(\beta_2/\beta_1)^2\sin^2\phi_1 - 1}}{\rho_1\beta_1\cos\phi_1 + i\rho_2\beta_2\sqrt{(\beta_2/\beta_1)^2\sin^2\phi_1 - 1}}, \tag{3.37a}$$

where $\phi_1 \geq \phi_{1c}$. This shows that for *SH* waves, R_S is a *complex number* for supercritical incidence angles, and hence has a magnitude and phase, i.e., $R_S = |R_S|\exp(i\psi)$. Note that

$$R_S = \frac{A - iB}{A + iB}, \quad A = \rho_1\beta_1\cos\phi_1, \quad B = \rho_2\beta_2\sqrt{(\beta_2/\beta_1)^2\sin^2\phi_1 - 1} \tag{3.37b}$$

which means that the magnitude of R_S is 1 for supercritical incidence angles, as can be seen with the following calculation:

$$|R_S| = \left|\frac{A - iB}{A + iB}\right| = \frac{|A - iB|}{|A + iB|} = \frac{\sqrt{A^2 + B^2}}{\sqrt{A^2 + B^2}} = 1. \tag{3.37c}$$

To obtain the phase angle ψ of R_S, one must work out the real and imaginary parts, as follows:

$$R_S = \frac{A - iB}{A + iB} = \frac{(A - iB)(A - iB)}{(A - iB)(A + iB)} = \left(\frac{A^2 - B^2}{A^2 + B^2}\right) + i\left(\frac{-2AB}{A^2 + B^2}\right) \tag{3.37d}$$

$$\Rightarrow \quad \psi = \tan^{-1}\left[\frac{\text{Im}(R_S)}{\text{Re}(R_S)}\right] = \tan^{-1}\left[\frac{-2AB}{A^2 - B^2}\right], \tag{3.37e}$$

where one must take care to choose the correct branch of the arctangent ($\tan^{-1}$) function, as explained in the next paragraph. Note that $|R_S|$ could also have been calculated from (3.37d) as

$$|R_S| = \sqrt{\left[\text{Re}(R_S)\right]^2 + \left[\text{Im}(R_S)\right]^2} = 1$$

but the calculation in (3.37c) is less work. *P-SV* reflection and transmission coefficients can also be either real or complex, and when they are complex, their magnitudes are not generally equal to 1.

Suppose a reflection or transmission coefficient is given by the complex number $a + ib$ at some angle of incidence. Its phase angle ψ is then given by $\psi = \tan^{-1}(b/a)$. One must take care to use the correct branch of the arctangent function in computing the phase angle. Many calculators or tables give only the principal branch values, which lie between $-90°$ and $+90°$. If a and b are both positive, say, then ψ lies on the principal branch. However, if $a < 0$ and $b > 0$, say, then ψ lies between $+90°$ and $+180°$, meaning that the branch just above the principal branch must be used, i.e., $180°$ must be added to the principal branch value. For instance, suppose that $R_S = -0.6 + 0.8i$ at some angle of incidence. Then $|R_S| = 1$, and $\psi = \tan^{-1}(0.8/-0.6) \approx -53° + 180° = 127°$.

It is not hard to see that the general rule to follow is: *if* $\mathrm{Re}(R_S) < 0$, *then add* $\pm 180°$ *to the principal branch value of* ψ.

Consider the waveform $A \exp(-i\omega t)$, which has the form of the time part of our reflected and transmitted waves. If A is complex, i.e., $A = |A| \exp(i\psi)$, then the waveform becomes $|A| \exp[-i(\omega t - \psi)]$. Hence, it *lags* $\exp(-i\omega t)$ in time, i.e., it is *delayed* in time, if $\psi > 0$. Therefore, ψ is called the **phase-delay** or **phase-lag**. If $\psi < 0$, the waveform *leads* $\exp(-i\omega t)$ in time, hence $-\psi$ would then be a **phase-advance** or **phase-lead**. For instance, $\psi = 23°$ represents a phase-delay of $23°$, and $\psi = -23°$ represents a phase-advance of $23°$. Of course, since sinusoidal waves are infinite in length or duration, one might also think of $\psi = -23°$ representing a phase-delay of $+337°$. The numerical example of the previous paragraph indicates that the reflected wave lags the incident wave in time, i.e., $\psi = 127°$ is a phase-delay of $127°$.

The two concepts of phase-delay and phase-advance are more informative and fundamental than the concept of phase alone. The value of the "phase" is ambiguous, since it changes sign if we use the opposite sign convention to express plane waves, i.e., if we use $\exp[i\omega(t - \mathbf{s}\cdot\mathbf{x})]$ (often used in engineering texts) as opposed to $\exp[i\omega(\mathbf{s}\cdot\mathbf{x} - t)]$ (often used in physics texts), which is what we have been using. On the other hand, the phase-delay and phase-advance do not change in sign when the wave convention is changed. To see this, note that each way of expressing a plane wave can be obtained from the other by replacing i with $-i$. This means that, referring to the previous numerical example, using the opposite convention would result in $R_S = -0.6 - 0.8i$, which gives a "phase" of $-127°$ (as opposed to $+127°$ in our convention). Now, let us examine the phase-delay in both cases. In the sign convention used in this book, we have

$$R_S \exp(-i\omega t) = |R_S| \exp(+i\,127°) \exp(-i\omega t) = |R_S| \exp\left[-i(\omega t - 127°)\right]$$

indicating that the phase-delay is $+127°$. In the opposite sign convention, we have

$$R_S \exp(+i\omega t) = |R_S| \exp(-i\,127°) \exp(+i\omega t) = |R_S| \exp\left[+i(\omega t - 127°)\right]$$

indicating that the phase-delay is again $+127°$, as with our convention.

Another way to see that the phase-delay and phase-advance are independent of the sign convention is to repeat the analysis between Equations 3.35 and 3.37 with the two sign conventions for expressing the waves.

Here is an example of how to compute an *SH* reflection coefficient for a precritical and postcritical incidence angle. Let $\beta_1 = 1$ km/s, $\beta_2 = 2$ km/s, $\rho_1 = 2.1$ g/cm^3, $\rho_2 = 2.2$ g/cm^3. Then the critical angle is

$$\phi_{1c} = \sin^{-1}(\beta_1/\beta_2) = 30°.$$

Let $\phi_1 = 20°$, which is a subcritical incidence angle. Then the transmitted wave angle ϕ_2 can be obtained from Snell's law as

$$\phi_2 = \sin^{-1}\left[(\beta_2/\beta_1)\sin\phi_1\right] = 43.1602° \quad \Rightarrow \quad \cos\phi_2 = 0.7294$$

or $\cos\phi_2$ can be computed from (3.35) as

$$\cos\phi_2 = \sqrt{1 - (\beta_2/\beta_1)^2 \sin^2\phi_1} = 0.7294.$$

Substituting these values into the formula for the *SH* reflection coefficient R_S in (3.26) gives

$$R_S = |R_S|\exp(i\psi) = -0.2385 \quad \Rightarrow \quad |R_S| = 0.2385, \quad \psi = 180°$$

where ψ is the phase-lag, i.e., phase-delay.

Next let $\phi_1 = 40°$, which is a supercritical incidence angle. Then the transmitted wave is an evanescent wave, as previously discussed – it travels parallel to the interface, and its amplitude decays exponentially away from the interface into the transmission medium. In this case, $\sin\phi_2 > 1$, and assuming that the angular frequency ω is positive,

$$\cos\phi_2 = +i\sqrt{(\beta_2/\beta_1)^2 \sin^2\phi_1 - 1} = 0.8079i.$$

Equation (3.37b) gives

$$A = 1.6087, \quad B = 3.5548.$$

Note that it is not necessary to convert the numerical values of the densities and velocities so that the distance and mass units are the same (e.g., meters and kilograms) because the conversion factors would cancel out of all the terms in R_S (3.26) or (3.37b). The same applies to the precritical incidence angle calculation of the previous paragraph. Substituting these values of A and B into (3.37c,d,e), or (3.37b), gives

$$R_S = -0.6600 - 0.7512i, \quad \Rightarrow \quad |R_S| = 1, \quad \psi = -131.30°,$$

where ψ represents a phase-lag of $-131.30°$, or physically, a phase-lead of $131.30°$, as ψ is negative. To get ψ, it is necessary to subtract $180°$ from the principal branch value (48.70°) of the arctangent in (3.37e) because $\text{Re}(R_S) < 0$. One can also see from the complex value of R_S that R_S lies in quadrant III of the complex plane, and hence must have a phase angle ψ between $-90°$ and $-180°$.

Getting back to critical angles, the *P-SV* case has similar features. In Figure 3.5b, the wave speeds are such that $\beta_1 < \alpha_1$ and $\beta_1 < \beta_2 < \alpha_2$ (see Snell's law), meaning that *three* critical angles exist for the incident *SV* wave – one for the reflected *P* wave, one for the transmitted *P* wave, and one for the transmitted *SV* wave. Evanescent *P* and *SV* plane waves also travel along the interface and experience amplitude decay away from the interface, although their particle motion is somewhat more complicated.

Figures 3.6 and 3.7 are examples of plots of some of the *P-SV* reflection and transmission coefficients in (3.28e). In Figure 3.6, the P_1P_2 critical angle is $\theta_{1c} = \sin^{-1}(\alpha_1/\alpha_2) = 44.0°$. Note that a cusp occurs in the coefficients at θ_{1c}. However, not all cusps are associated with critical angles (see the P_1S_1 coefficient, in which the first cusp occurs merely because P_1S_1 changes sign). In Figure 3.7, the curves

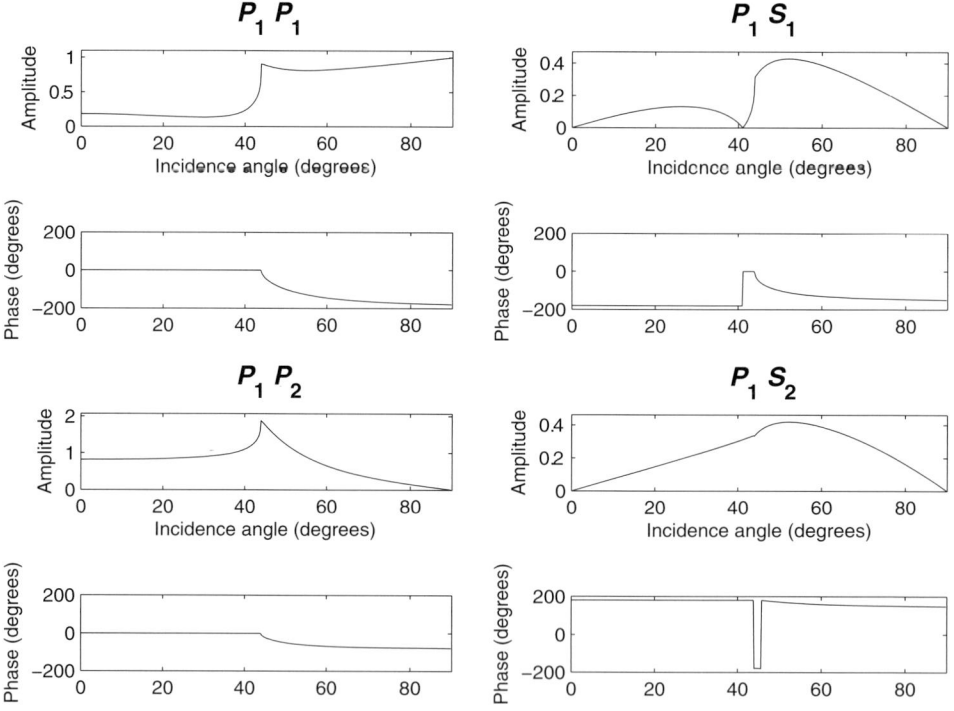

Figure 3.6 Particle displacement reflection and transmission coefficients (relative displacement amplitudes) for an incident *P* wave in medium 1. For each coefficient, the upper graph is the magnitude and the lower one is the phase-*lag* (i.e., phase-*delay*). For layer 1, the *P* and *S* wave speeds, in km/s, are 2.5 and 1.44, respectively, and for layer 2 they are 3.6 and 2.08. The density is $2\,\text{g/cm}^3$ in each layer.

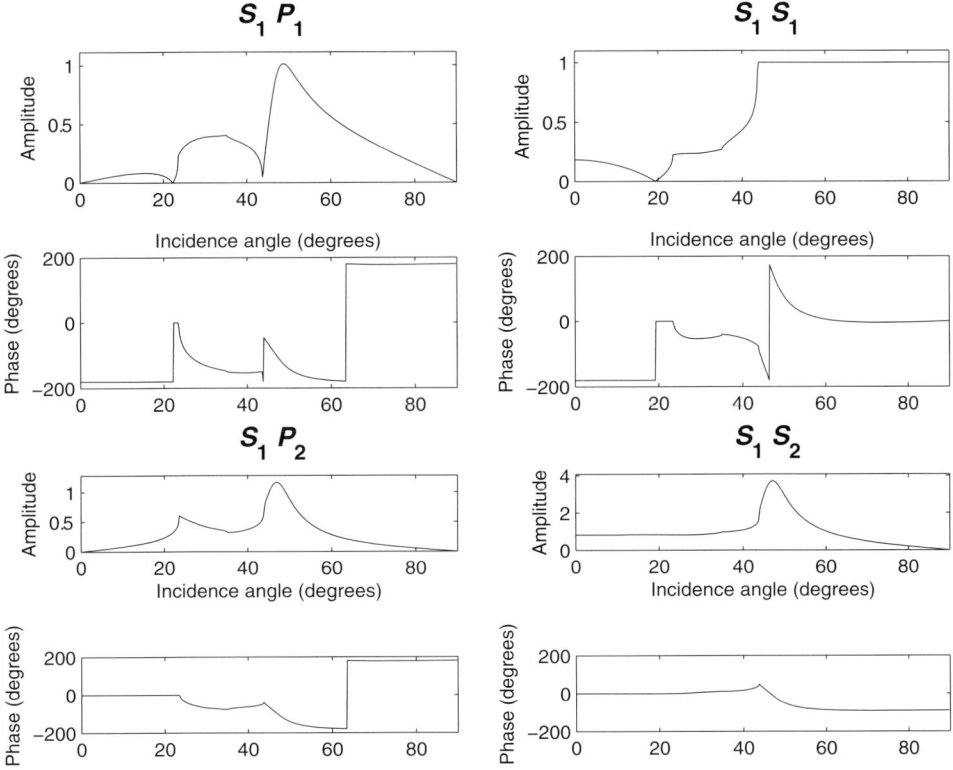

Figure 3.7 The same as Figure 3.6, except for an incident SV wave in medium 1.

appear more complicated because there are three critical angles, not just one as for an incident P wave. The three critical angles are the $S1P2$ critical angle $\phi_{1c} = \sin^{-1}(\beta_1/\alpha_2) = 23.6°$, the $S1P1$ critical angle $\phi_{1c} = \sin^{-1}(\beta_1/\alpha_1) = 35.2°$, and the $S1S2$ critical angle $\phi_{1c} = \sin^{-1}(\beta_1/\beta_2) = 43.8°$.

Figure 3.8 is another example of R_{PP} in (3.28e). It shows that the R_{PP} curves can get more complicated when another critical angle is introduced.

The variation of amplitude with angle of incidence can also be measured experimentally in various realistic cases and compared with theory (see, e.g., Bouzidi and Schmitt, 2012) to confirm the validity of the formulas. Plane wave reflection coefficients can also be inverted to obtain geological information about the subsurface, although in some cases a more careful approach involving the full wavefield based on spherical waves is required for accuracy (see, e.g., Dettmer, Dosso and Holland, 2007).

3.7 Energy Reflection and Transmission Coefficients

Referring to the reflection and transmission coefficients R and T given by Equation (3.8) for a normally incident P wave, note that if $Z_2 \ll Z_1$, then $T \approx 2$ (and

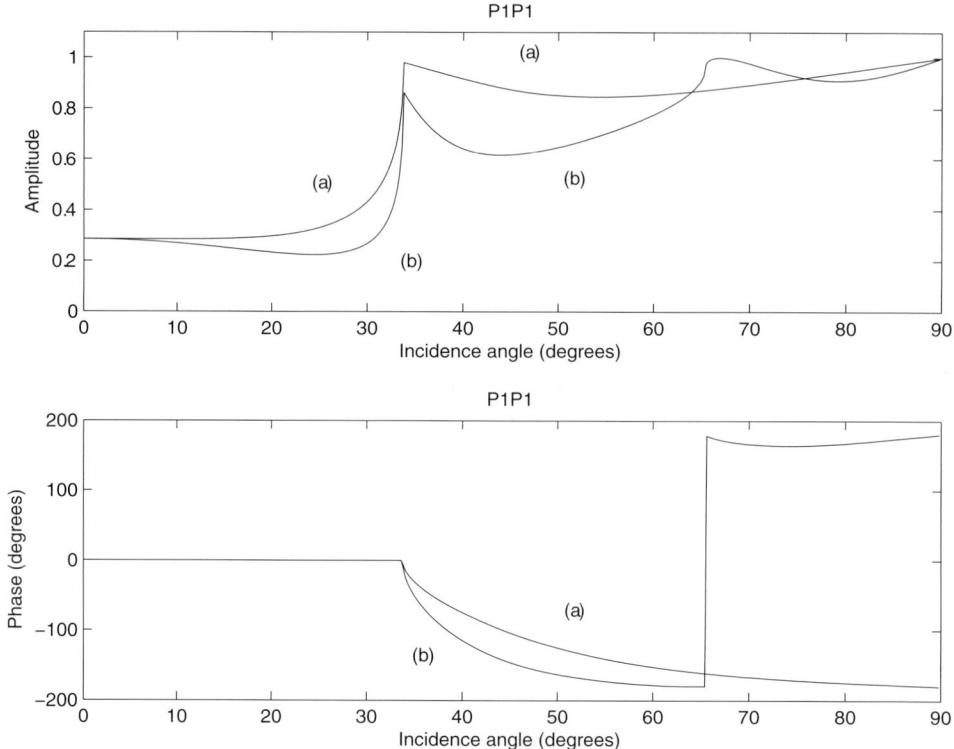

Figure 3.8 Graphs of the P_1P_1 reflection coefficient R_{PP} for $\rho_1 = \rho_2 = 2$ g/cm^3, $\alpha_1 = 2$ km/s, $\beta_1 = 1.35$ km/s, $\alpha_2 = 3.6$ km/s, and (a) $\beta_2 = 1.8$ km/s, i.e., $\beta_2 < \alpha_1$, and (b) $\beta_2 = 2.2$ km/s, i.e., $\beta_2 > \alpha_1$. The P_1P_2 critical angle is 33.75°, and the P_1S_2 critical angle (for $\beta_2 = 2.2$ km/s) is 65.38°. The phase graph is the phase-*lag*, i.e., phase-*delay*.

$R \approx -1$), i.e., the amplitude of the transmitted wave is *twice* that of the incident wave. Does this violate the law of conservation of energy? Doesn't T always have to be less than or equal to one in magnitude? The answer to both these questions is no. One must remember that R and T are particle displacement coefficients. To relate wave reflection and transmission to energy conservation, we must use coefficients that are measures of relative energy content, rather than relative particle displacement. For a normally incident P wave, the **energy reflection coefficient** $\mathcal{R}$ and the **energy transmission coefficient** $\mathcal{T}$ are such coefficients.

$\mathcal{R}$ is defined as the ratio of the magnitude of the mean intensity of the reflected wave to that of the incident wave. Using Equation (2.83), i.e., $\langle \mathbf{I} \rangle = \frac{1}{2}\rho\alpha\omega^2|A|^2\mathbf{n}$, where $\mathbf{n}$ is a unit vector in the direction of the slowness $\mathbf{s}$, it is given by

$$\mathcal{R} = \frac{\frac{1}{2}\rho_1\alpha_1\omega^2|A_R|^2}{\frac{1}{2}\rho_1\alpha_1\omega^2|A_I|^2} \quad \Rightarrow \quad \mathcal{R} = \left|\frac{A_R}{A_I}\right|^2 = |R|^2 = \left(\frac{Z_2 - Z_1}{Z_2 + Z_1}\right)^2. \tag{3.38}$$

Similarly, $\mathcal{T}$ is defined as the ratio of the magnitude of the mean intensity of the transmitted wave to that of the incident wave. From Equation (2.83), it is given by

$$\mathcal{T} = \frac{\frac{1}{2}\rho_2\alpha_2\omega^2|A_T|^2}{\frac{1}{2}\rho_1\alpha_1\omega^2|A_I|^2} \quad \Rightarrow \quad \mathcal{T} = \left(\frac{Z_2}{Z_1}\right)\left|\frac{A_T}{A_I}\right|^2 = \left(\frac{Z_2}{Z_1}\right)|T|^2 = \frac{4Z_1Z_2}{(Z_2+Z_1)^2}.$$
(3.39)

Note that $0 \leq \mathcal{R} \leq 1, 0 \leq \mathcal{T} \leq 1$, and that $\mathcal{R} + \mathcal{T} = 1$. Hence, energy is conserved. If Z_1 and Z_2 are relatively far apart in value, i.e., if $Z_2 \gg Z_1$ or $Z_2 \ll Z_1$, i.e., if there is a high acoustic impedance contrast, then $\mathcal{R} \approx 1$ and $\mathcal{T} \approx 0$ (i.e., $\mathcal{T} \ll 1$), meaning that most of the incident energy is reflected and very little is transmitted. Note that if $Z_2 \ll Z_1$, then $T \approx 2$, but $\mathcal{T} \approx 0$.

Note also that $\mathcal{R}$ is the fraction of the incident energy that is reflected, and $\mathcal{T}$ is the fraction of the incident energy that is transmitted.

Energy reflection and transmission coefficients can also be worked out for non-normally incident waves. For example, for an incident P wave, it can be shown that the conservation of energy is expressed by

$$\mathcal{R}_{PP} + \mathcal{R}_{PS} + \mathcal{T}_{PP} + \mathcal{T}_{PS} = 1,$$
(3.40)

where

$$\mathcal{R}_{PP} = \left|\frac{A_R}{A_I}\right|^2, \quad \mathcal{R}_{PS} = \left(\frac{\rho_1\beta_1\cos\phi_1}{\rho_1\alpha_1\cos\theta_1}\right)\left|\frac{B_R}{A_I}\right|^2,$$
(3.41)

$$\mathcal{T}_{PP} = \left(\frac{\rho_2\alpha_2\cos\theta_2}{\rho_1\alpha_1\cos\theta_1}\right)\left|\frac{A_T}{A_I}\right|^2, \quad \mathcal{T}_{PS} = \left(\frac{\rho_2\beta_2\cos\phi_2}{\rho_1\alpha_1\cos\theta_1}\right)\left|\frac{B_T}{A_I}\right|^2.$$
(3.42)

Note that the amplitude ratios in (3.41) and (3.42) are just the particle displacement reflection and transmission coefficients, i.e., $A_R/A_I = R_{PP}, B_R/A_I = R_{PS}, A_T/A_I = T_{PP}$, and $B_T/A_I = T_{PS}$. Note also that the numerators contain the parameters associated with the reflected or transmitted wave and the denominators contain those associated with the incident wave. Similar formulas exist for an incident SV wave and an incident SH wave. Each of these energy coefficients can be expressed as

$$\left|\frac{\langle\mathbf{I}\rangle'\cdot\mathbf{n}}{\langle\mathbf{I}\rangle\cdot\mathbf{n}}\right|_{z=0} \quad \text{or} \quad \left|\frac{\langle I_z'\rangle}{\langle I_z\rangle}\right|_{z=0},$$
(3.43)

where $\mathbf{n}$ is a unit vector normal to the boundary, $\mathbf{I}$ is the intensity vector, the numerator is the normal component of the mean intensity of the reflected or transmitted wave, and the denominator is that of the incident wave. Each energy coefficient vanishes when the reflected or transmitted wave associated with it becomes evanescent, because $\langle I_z'\rangle = 0$ (note also from Equation 3.42 that $\mathcal{T}_{PP} = 0$ when $\theta_2 = 90°$ and $\mathcal{T}_{PS} = 0$ when $\phi_2 = 90°$). For instance, $\mathcal{T}_{PP}$ in Equation (3.42) is zero when θ_1 is greater than the critical angle of incidence for the transmitted P wave, even though the displacement coefficient T_{PP} is not zero (see Figure 3.6). T_{PP} is a measure of the amplitude of particle displacement in the transmitted wave, whereas $\mathcal{T}_{PP}$ is a measure of the energy flowing into the transmission medium. Also, for this supercritical θ_1, the reflection coefficient $R_{PP} = A_R/A_I$ is complex, but $\mathcal{R}_{PP}$ is still real (as it must be), because it is the squared magnitude of a complex number.

Let us now consider the energy partitioning problem somewhat more carefully by considering the normal component of the intensity or energy flux.

Equations (3.40) and (3.43) suggest that I_z, the z component of the intensity, is continuous across the boundary $z = 0$. This is true, and can be seen by applying the formula $I_j = -\sum_i \sigma_{ij} \dot{u}_i$ (see Equation 2.19) for $j = 3$, i.e., $j = z$:

$$I_z^{(P\text{-}SV)} = -(\sigma_{xz} \dot{u}_x + \sigma_{zz} \dot{u}_z) \qquad \text{and} \qquad I_z^{(SH)} = -\sigma_{yz} \dot{u}_y. \qquad (3.44)$$

The welded-contact boundary conditions state that u_x, u_y, u_z, σ_{xz}, σ_{yz}, and σ_{zz} are continuous across the interface $z = 0$. These are precisely the quantities that appear in the formulas for I_z. Therefore, I_z must also be continuous across the interface (which makes sense, from a physical viewpoint).

Note that this also implies that I_x and I_y are not necessarily continuous across the interface $z = 0$.

The continuity of I_z also has subtle implications for the conservation of energy. Consider for instance the *SH* case. An incident plane *SH* wave in medium 1 produces a reflected *SH* wave in medium 1 and a transmitted *SH* wave in medium 2. The total displacement in medium 1 is the sum of the incident and reflected displacements, i.e., $u_y^{(1)} = u_y^{(I)} + u_y^{(R)}$. Similarly, since each stress component is linear in the components of displacement, we have $\sigma_{yz}^{(1)} = \sigma_{yz}^{(I)} + \sigma_{yz}^{(R)}$. Continuity of I_z across the interface then means

$$\left[-\sigma_{yz}^{(1)} \dot{u}_y^{(1)} \right]_{z=0} = \left[-\sigma_{yz}^{(2)} \dot{u}_y^{(2)} \right]_{z=0} \implies$$

$$\left[-\left(\sigma_{yz}^{(I)} + \sigma_{yz}^{(R)}\right)\left(\dot{u}_y^{(I)} + \dot{u}_y^{(R)}\right) \right]_{z=0} = \left[-\sigma_{yz}^{(T)} \dot{u}_y^{(T)} \right]_{z=0}. \qquad (3.45)$$

This becomes, after expanding the products and using the previous formula for I_z,

$$\left[I_z^{(I)} + I_z^{(R)} + C_z^{(IR)} \right]_0 = \left[I_z^{(T)} \right]_0 \quad \text{where} \quad C_z^{(IR)} \equiv -\left(\sigma_{yz}^{(I)} \dot{u}_y^{(R)} + \sigma_{yz}^{(R)} \dot{u}_y^{(I)} \right), \quad (3.46)$$

and where the subscript "0" means that the corresponding quantity is to be evaluated at $z = 0$. The quantity $C_z^{(IR)}$ describes an interaction of the incident stress field with the reflected particle velocity field and the reflected stress with the incident particle velocity. Its mean value, $\langle C_z^{(IR)} \rangle$, is zero for perfectly elastic media, but nonzero for dissipative media, i.e., for anelastic or viscoelastic media (see, e.g., Borcherdt, 1977, 2009; Krebes 1983).

Note also that the preceding calculation shows that the total intensity (normal component) in medium 1 is not in general the sum of the incident and reflected intensities, i.e., $I_z^{(1)} \neq I_z^{(I)} + I_z^{(R)}$. This is because the intensity is *quadratic*, not linear, in the displacement components u_j. This is also generally the case for I_x and I_y.

Taking the mean value of the equation on the left in (3.46) and rearranging gives

$$\mathcal{R} + \mathcal{T} + \mathcal{C} = 1 \quad \text{where} \quad \mathcal{R} = -\frac{\left\langle I_z^{(R)} \right\rangle_0}{\left\langle I_z^{(I)} \right\rangle_0}, \quad \mathcal{T} = \frac{\left\langle I_z^{(T)} \right\rangle_0}{\left\langle I_z^{(I)} \right\rangle_0}, \quad \mathcal{C} = -\frac{\left\langle C_z^{(IR)} \right\rangle_0}{\left\langle I_z^{(I)} \right\rangle_0}.$$

$$(3.47)$$

The equation on the left expresses the conservation of energy in the *SH* reflection-transmission problem. A minus sign is used in the definition of $\mathcal{R}$ because $I_z^{(R)}$ and $I_z^{(I)}$ have opposite signs. $\mathcal{R}$ is the *energy reflection coefficient*. A fraction $\mathcal{R}$ of the normal energy flux in the incident wave is reflected. $\mathcal{T}$ is the *energy transmission coefficient*. A fraction $\mathcal{T}$ of the normal energy flux in the incident wave is transmitted. $\mathcal{C}$ is the *energy interaction coefficient*. A fraction $\mathcal{C}$ of the normal energy flux in the incident wave is used up in the afore mentioned interaction between the incident and reflected wavefields. As mentioned previously, $\mathcal{C}$ is zero for a perfectly elastic medium (in which case $\mathcal{R} + \mathcal{T} = 1$) but nonzero for a dissipative (viscoelastic) medium.

Hence, for perfectly elastic *SH* body waves, we have

$$\mathcal{R} = -\frac{\left\langle I_z^{(R)} \right\rangle_0}{\left\langle I_z^{(I)} \right\rangle_0} = -\frac{\left\langle \mathbf{I}^{(R)} \cdot \mathbf{n} \right\rangle_0}{\left\langle \mathbf{I}^{(I)} \cdot \mathbf{n} \right\rangle_0} = -\frac{-\frac{1}{2}\rho_1\beta_1\omega^2|B_R|^2\cos\phi_1}{\frac{1}{2}\rho_1\beta_1\omega^2|B_I|^2\cos\phi_1} = |R_S|^2, \qquad (3.48)$$

$$\mathcal{T} = \frac{\left\langle I_z^{(T)} \right\rangle_0}{\left\langle I_z^{(I)} \right\rangle_0} = \frac{\left\langle \mathbf{I}^{(T)} \cdot \mathbf{n} \right\rangle_0}{\left\langle \mathbf{I}^{(I)} \cdot \mathbf{n} \right\rangle_0} = \frac{\frac{1}{2}\rho_2\beta_2\omega^2|B_T|^2\cos\phi_2}{\frac{1}{2}\rho_1\beta_1\omega^2|B_I|^2\cos\phi_1} = \left(\frac{\rho_2\beta_2\cos\phi_2}{\rho_1\beta_1\cos\phi_1}\right)|T_S|^2, \quad (3.49)$$

where $\mathbf{n}$ is a unit vector normal to the interface ($\mathbf{n} = \mathbf{e}_z$ in this case) and where $R_S = B_R/B_I$ is the *SH* displacement reflection coefficient and $T_S = B_T/B_I$ is the *SH* displacement transmission coefficient.

Equivalently, as in (3.43), for perfectly elastic media, the *SH* energy coefficients could be defined in terms of absolute values as $\mathcal{R} = \left| \langle \mathbf{I}^{(R)} \cdot \mathbf{n} \rangle / \langle \mathbf{I}^{(I)} \cdot \mathbf{n} \rangle \right|_0$ and $\mathcal{T} = \left| \langle \mathbf{I}^{(T)} \cdot \mathbf{n} \rangle / \langle \mathbf{I}^{(I)} \cdot \mathbf{n} \rangle \right|_0$, which makes their positiveness manifest.

The results for the *P-SV* case in Equations (3.40) through (3.42) can be obtained in a similar way, including formulas for the interaction coefficients whose mean values are nonzero for anelastic media.

3.8 A Liquid–Liquid Interface

Suppose that media 1 and 2 in Figure 3.5 are both liquids. Since shear waves cannot propagate in a liquid, an incident *P* wave upon striking the interface generates only a reflected *P* wave and a transmitted *P* wave. Therefore, we need only two boundary conditions: the normal displacement u_z and the normal stress σ_{zz} must be continuous across the boundary. The tangential displacement u_x need not be continuous because the two liquid layers can slip or slide across each other at the interface. There is no condition for the tangential stress σ_{xz} because liquids do not support shearing stresses. The reflection and transmission coefficients can be derived in the usual way.

3.9 A Liquid–Solid Interface

In this case, an incident wave in either medium will generate only three reflected and transmitted waves, i.e., a *P* wave in the liquid and a *P* wave and *SV* wave in the solid. Therefore, we need only three boundary conditions: the normal displacement u_z and the normal stress σ_{zz} must be continuous across the interface, and the tangential stress σ_{xz} (in the solid) must be zero at the interface. The tangential displacement u_x need not be continuous because the liquid can slip or slide along the solid surface.

3.10 A Rigid Boundary

This is an idealized immovable interface. Waves striking it produce no motion of the boundary. An incident wave generates only reflected waves – there is no particle motion on the other side of the boundary (in the transmission medium). If the incidence medium is a solid, then there are two boundary conditions: the normal and tangential components of displacement, u_x and u_z, must be zero at the boundary.

3.11 Approximations

Approximations to the reflection and transmission coefficients in (3.28e), in particular the *PP* reflection coefficient, are often useful in amplitude versus offset (AVO) studies.

Suppose that the solids above and below the interface are similar, i.e., their medium parameters are close in numerical value. Let

$$\Delta\rho \equiv \rho_2 - \rho_1, \quad \rho \equiv (\rho_1 + \rho_2)/2, \quad \Delta\alpha \equiv \alpha_2 - \alpha_1, \quad \alpha \equiv (\alpha_1 + \alpha_2)/2, \quad (3.50)$$

$$\Delta\beta \equiv \beta_2 - \beta_1, \quad \beta \equiv (\beta_1 + \beta_2)/2, \quad \left|\frac{\Delta\rho}{\rho}\right| \ll 1, \quad \left|\frac{\Delta\alpha}{\alpha}\right| \ll 1, \quad \left|\frac{\Delta\beta}{\beta}\right| \ll 1 \tag{3.51}$$

$$\Delta\theta \equiv \theta_2 - \theta_1, \quad \theta \equiv (\theta_1 + \theta_2)/2, \quad \Delta\phi \equiv \phi_2 - \phi_1, \quad \phi \equiv (\phi_1 + \phi_2)/2. \quad (3.52)$$

Then it can be shown (Aki and Richards 1980, 2002) that to first order in the velocity jumps,

$$\Delta\theta = \frac{\Delta\alpha}{\alpha}\tan\theta, \quad \Delta\phi = \frac{\Delta\beta}{\beta}\tan\phi, \quad (3.53)$$

and

$$R_{PP} \approx \frac{1}{2}(1 - 4\beta^2 p^2)\frac{\Delta\rho}{\rho} + \frac{1}{2\cos^2\theta}\frac{\Delta\alpha}{\alpha} - 4\beta^2 p^2\frac{\Delta\beta}{\beta}, \quad (3.54)$$

where $p = (\sin\theta_l)/\alpha_l = (\sin\phi_l)/\beta_l$, $l = 1, 2$. It can also be shown that, to first order, $p = (\sin\theta)/\alpha = (\sin\phi)/\beta$. Note that the three inequalities in (3.51) imply that $|R_{PP}|$ will be small. Equation (3.54) holds only if all the angles are not near $90°$.

Approximations for the other coefficients can be found in Aki and Richards (1980, 2002).

Again, suppose that the parameters of the upper and lower media are close in numerical value. Then (Shuey, 1985)

$$R_{PP} \approx (R_{PP})_0 + B\sin^2\theta_1 + \frac{\Delta\alpha}{2\alpha}(\tan^2\theta_1 - \sin^2\theta_1), \quad (3.55)$$

where

$$(R_{PP})_0 = \frac{\rho_2\alpha_2 - \rho_1\alpha_1}{\rho_2\alpha_2 + \rho_1\alpha_1}, \quad B = \left[A_0\,(R_{PP})_0 + \frac{\Delta\sigma}{(1-\sigma)^2}\right], \quad (3.56)$$

$$\sigma = \frac{1}{2}(\sigma_1 + \sigma_2) = \frac{(\beta/\alpha)^2 - \frac{1}{2}}{(\beta/\alpha)^2 - 1}, \quad \Delta\sigma = \sigma_2 - \sigma_1, \quad (3.57)$$

$$A_0 = B_0 - 2(1 + B_0)\left(\frac{1 - 2\sigma}{1 - \sigma}\right), \quad B_0 = \frac{\Delta\alpha}{\alpha}\bigg/\left[\frac{\Delta\alpha}{\alpha} + \frac{\Delta\rho}{\rho}\right], \quad (3.58)$$

and where α and β are given by (3.50) and (3.51). Note that $(R_{PP})_0$ is the normal incidence reflection coefficient, and σ is the average Poisson ratio.

In (3.55), the first, second, and third terms dominate for small, intermediate, and near-critical angles, respectively. Consequently, if we are interested in small and intermediate angles only, we have

$$R_{PP} \approx (R_{PP})_0 + B\sin^2\theta_1, \quad (3.59)$$

where B is often called the **AVO gradient**.

Other approximations have been developed as well and can be found in the literature. See, e.g., Thomas, Ball, Blangy, and Tenorio (2016) for a list of other approximations. Some approximations have been extended to the anelastic case as well (see, e.g., Innanen, 2011, 2012a, 2012b; Moradi and Innanen, 2016).

Exercises

1. (a) Repeat the analysis in Section 3.1, except for the part on pressure coefficients, for a normally incident S wave pulse, i.e., analyze the problem of the reflection and transmission of an S wave pulse normally incident upon a boundary. The boundary conditions are that the shear displacement u_x and stress σ_{xz} are continuous across the boundary.

   (b) Carefully go through the analysis in Section 3.1, deriving formulas (3.4)–(3.15). Also, redraw Figure 3.1 for the case $\alpha_1 = 2\alpha_2$.

2. (a) Repeat the reflection/transmission analysis connected with Equations (3.17) to (3.20) with the following change: let $\mathbf{d} = +\mathbf{e}_z$ for all three waves; incident, reflected, and transmitted. Compare your results with those in the text.

   (b) Repeat the reflection/transmission analysis connected with Equations (3.17) to (3.20) with the following change: let the z axis point in the opposite direction. Compare your results with those in the text.

3. Repeat the analysis connected with Equations (3.17)–(3.20) for a normally incident SV wave. Compare your results with those for a normally incident P wave, and discuss the similarities and differences (mathematical form of the reflection and transmission coefficients, polarity reversals, etc.). Consider only the displacement coefficients, not the stress coefficients (i.e., not $\bar{R}$ or $\bar{T}$).

4. Compute R_{SH} and T_{SH} (both magnitude and phase) for incidence angles of $20°$ and $60°$ using the medium parameters in Figure 3.6.

5. An SH receiver is placed at a distance of 150 m from an SH source (see Figure 3.9). The source is activated at time $t = 0$, and the reflected signal arrives at the receiver at time $t = t_0$. However, the receiver trace is observed to have an amplitude of zero at $t = t_0$, where the reflection pulse was expected. Assuming there is nothing wrong with the equipment, calculate ϕ_1, t_0, and h.

6. (a) Apply the Zoeppritz equations (3.28) to derive the reflection and transmission coefficients for a normally incident P wave in medium 1. Do the same for a normally incident SV wave in medium 1, and compare the results with the P wave case.

   (b) Derive Equations (3.28a, b, and d).

7. Plot, as a function of incidence angle, the S_1S_1 and S_1S_2 reflection and transmission coefficients (amplitude and phase) for SH waves using the medium parameters of Figure 3.6.

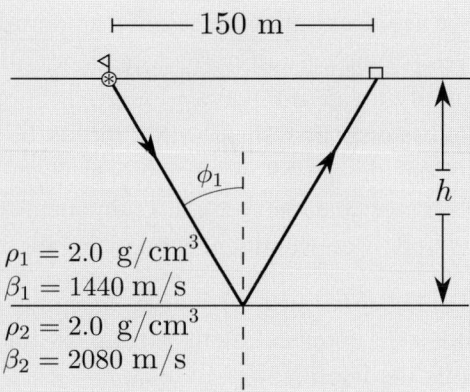

Figure 3.9 See exercise 5.

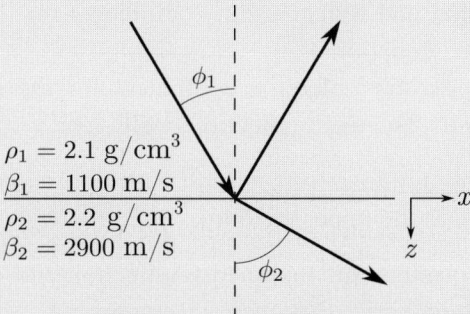

Figure 3.10 See exercise 9.

8. (a) Plot the reflection and transmission energy coefficients $\mathcal{R}$ and $\mathcal{T}$ for vertically incident P waves as functions of the impedance contrast η ($= Z_2/Z_1$). Superimpose your plots on one graph.
    (b) Calculate the values of η for which half of the incident wave energy is reflected and half is transmitted ($\mathcal{R} = \mathcal{T} = 1/2$).

9. Compute the value of the *SH* reflection coefficient (amplitude and phase) for the *SH* ray shown in Figure 3.10 for $\phi_1 = 20°$ and $30°$. The reflection coefficient is given by $(C_1 - C_2)/(C_1 + C_2)$, where $C_i = \rho_i \beta_i \cos \phi_i$, $i = 1, 2$. The parameter values are $\rho_1 = 2.1$ g/cm^3, $\beta_1 = 1.1$ km/s, $\rho_2 = 2.2$ g/cm^3, and $\beta_2 = 2.9$ km/s.

10. Consider a plane *SV* wave incident upon a free surface.

    (a) Are there any critical angles in this case? Derive the formulas for the particle displacement reflection coefficients by reducing and solving the Zoeppritz equations. Compare the formulas with those for the case of

an incident P wave. Determine the relative amplitudes and phases of the
reflected waves for the case of normal incidence, and compare with the
case of a normally incident P wave.

(b) Let the compressional and shear wave speeds be 4 km/s and 2 km/s
respectively. Consider an SV wave incident at an angle of 40°. What
is the phase change that the vertical component of displacement (the
quantity measured by conventional geophones) for the reflected SV wave
undergoes?

11. Consider a liquid–solid interface with a compressional wave incident at
some given angle in the liquid.

(a) List all the possible critical angles in this case.

(b) Derive the system of equations whose solutions are the particle displacement reflection and transmission coefficients for this case (reduce the
four Zoeppritz equations).

(c) Derive the pressure reflection coefficient for the reflected P wave in
terms of the displacement reflection coefficient.

12. Consider a liquid–liquid interface with a compressional wave incident at
some given angle in the upper medium.

(a) Derive the formulas for the energy reflection and energy transmission
coefficients, and verify that energy is conserved.

(b) Derive the formulas for the pressure reflection and pressure transmission
coefficients (derive expressions for the stress component σ_{zz} for all the
waves and form the appropriate ratios).

(c) Let the density and compressional wave speed be 1 g/cm^3 and 1.2 km/s
in the upper medium, and 0.8 g/cm^3 and 1.4 km/s in the lower medium.
What is the critical angle for the transmitted wave? What is the angle at
which there is no reflected wave, i.e., at which all the incident energy
passes into the transmitted wave? For incidence angles of 0°, 20°, and
70°, compute the amplitude and phase change that the pressure wave
undergoes upon reflection.

(d) The water–air interface is usually approximated as a free surface, rather
than a liquid–liquid interface. Compute the numerical values of the
energy reflection and energy transmission coefficients for a normally
incident compressional wave in the water to verify that the approximation is a good one. The density and compressional wave speed
are 1 g/cm^3 and 1.418 km/s for water, and 1.29×10^{-3} g/cm^3 and
0.331 km/s for air.

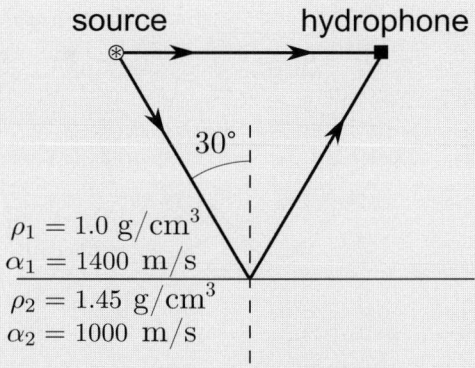

$\rho_1 = 1.0 \text{ g/cm}^3$
$\alpha_1 = 1400 \text{ m/s}$
$\rho_2 = 1.45 \text{ g/cm}^3$
$\alpha_2 = 1000 \text{ m/s}$

Figure 3.11 See exercise 13d.

13. Consider a liquid–liquid interface with a *P* wave incident in the upper medium.

(a) Derive the formulas for the displacement reflection and transmission coefficients by deriving and solving the equations for the boundary conditions (i.e., by carrying out a calculation analogous to that for the *SH* wave case in Section 3.4).

(b) Obtain the formulas of part (a) by applying the Zoeppritz equations.

(c) Derive the corresponding formulas for the pressure coefficients (use your expressions for σ_{zz} from part (a) to form the appropriate ratios).

(d) In Figure 3.11, do the direct wave and reflected wave recorded by the hydrophone have the same or opposite polarities?

14. Consider a *P* wave incident upon a rigid boundary.

(a) Calculate formulas for the two reflection coefficients. Write the formulas in their simplest form.

(b) Let the *P* and *S* wave speeds be 4 and 2 km/s respectively. Plot graphs of the two reflection coefficients as functions of the angle of incidence. Draw a graph of the relative phase of the normal component of displacement u_z of the reflected *P* wave at the boundary. For which angles of incidence does u_z show a polarity reversal? (u_z is the quantity measured by geophones – actually, velocity, not displacement, is measured, but this is not significant, as there is merely a constant 90° phase difference between velocity and displacement.) Do the same for u_x (for the reflected *P* wave).

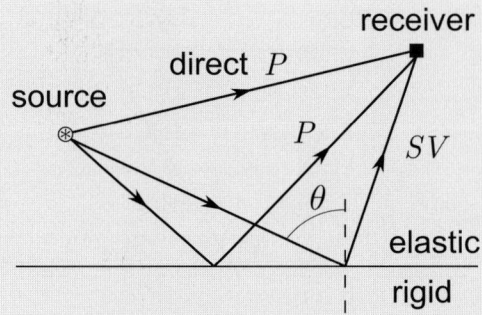

Figure 3.12 See exercise 15.

(c) At a certain angle of incidence, the reflection coefficient for the reflected *P* wave is zero, i.e., the incident *P* wave is entirely converted into a reflected *SV* wave. What is this angle?

15. Figure 3.12 shows a source of compressional waves in an elastic half-space overlying a rigid half-space. The geophone, which measures the vertical component of particle displacement, records three arrivals, namely the direct wave and the reflected *P* and *S* arrivals. Assuming the geophone records upward particle motion as an up-kicking waveform on the trace (the opposite of the SEG convention), what does the geophone trace look like? In other words, determine the phase relationship between the reflections and the direct wave. Assume $\theta < 45°$.

16. Repeat exercise 15 for a source of *SV* waves. θ is the angle for the reflected *P* wave.

17. Derive the formulas for the *SH* plane wave particle displacement reflection and transmission coefficients for a *linear slip interface* separating two rock layers (often called a *nonwelded contact interface*). The incident and scattered waves satisfy the following *displacement discontinuity boundary conditions* at the interface:

$$\left[u_x^{(2)} - u_x^{(1)}\right] = \left[c_x\sigma_{xz}\right], \quad \left[u_y^{(2)} - u_y^{(1)}\right] = \left[c_y\sigma_{yz}\right], \quad \left[u_z^{(2)} - u_z^{(1)}\right] = \left[c_z\sigma_{zz}\right],$$

$$\left[\sigma_{xz}^{(1)}\right] = \left[\sigma_{xz}^{(2)}\right], \quad \left[\sigma_{yz}^{(1)}\right] = \left[\sigma_{yz}^{(2)}\right], \quad \left[\sigma_{zz}^{(1)}\right] = \left[\sigma_{zz}^{(2)}\right],$$

where superscripts (1) and (2) denote media 1 and 2, and where c_x, c_y, and c_z are the *specific compliances* for the interface in the tangential and normal directions, respectively. In the conditions for u_x, u_y, and u_z, the stress components in either medium can be used for the right-hand sides (because the stress components are continuous across the boundary);

however, choosing those in medium 2 results in simpler equations to solve if the waves are incident from medium 1, and vice versa. Discuss your results. Describe how the coefficients differ from those for the familiar welded contact case. Describe what happens if there is no impedance contrast, i.e., if media 1 and 2 are identical (e.g., a crack in a single homogeneous medium). This type of interface is often used to describe the reflection and transmission of plane waves by joints, cracks, fractures, etc., for which the contact at the interface is nonwelded.

18. Do exercise 17 for normally incident plane *P* and *SV* waves.

4

Surface Waves, Head Waves, and Normal Modes

4.1 Surface Waves

A surface wave is a wave that travels along the surface of the Earth. Its amplitude of particle motion generally decreases with depth.

4.2 Surface Waves for a Homogeneous Half-Space

Consider a plane surface wave traveling in the positive x direction along the surface $z = 0$ of a solid homogeneous half-space. As usual, we let depth be in the positive z direction. How can we represent the surface wave mathematically in terms of plane waves? To answer this, we recall from Chapter 3 that evanescent waves have the physical properties of surface waves, i.e., they propagate along an interface, and their amplitude decreases with distance from the interface. This suggests that a surface wave could possibly be represented as a sum of evanescent waves, in general. The results of Chapter 3 also suggest that it may be possible to split up the problem into two cases: the P-SV case, and the SH case.

For the P-SV case, particle motion is in the xz plane. This suggests that the surface wave has both a P and an SV component of particle motion, in general. Consequently, we consider the surface wave to be a sum of a P evanescent wave and an SV evanescent wave. Let A be the displacement amplitude of the P wave, and let B be the displacement amplitude of the SV wave. The amplitudes A and B are related by the free-surface boundary conditions, i.e., $\sigma_{xz} = \sigma_{zz} = 0$ at $z = 0$. Referring to Figure 3.5, if we think of medium 1 as a vacuum and medium 2 as the half-space, and if we think of the two transmitted waves in medium 2 as the two evanescent parts of our surface wave (with θ_2 and ϕ_2 being complex angles of the form $90° + ia$), then the boundary conditions can be obtained from the last two Zoeppritz equations (Equations 3.28c and 3.28d): setting the amplitudes in medium 1 equal to zero, replacing A_T with A and B_T with B, and dropping the subscript "2" since we are dealing with only one solid, we obtain

$$2\rho\beta^2 pA \cos\theta \;+\; \rho\beta(1 - 2\beta^2 p^2)B \;=\; 0$$
$$\rho\alpha(1 - 2\beta^2 p^2)A \;-\; 2\rho\beta^2 pB \cos\phi \;=\; 0. \tag{4.1}$$

We know from Snell's law that $p = (\sin\theta)/\alpha = (\sin\phi)/\beta$. But since the P and SV components of our surface wave are both evanescent waves, we have $\sin\theta > 1$ and $\sin\phi > 1$, i.e., $p > (1/\beta) > (1/\alpha)$. Let c be the speed of the surface wave in the x direction. Then $p = 1/c$, and we then have $c < \beta < \alpha$. This states that the speed of propagation of the surface wave is less than the speeds of the shear and compressional body waves. A surface wave of this type is known as a **Rayleigh wave** on a homogeneous half-space. The events referred to as **ground roll** on a seismic data section are essentially Rayleigh waves on an *inhomogeneous* half-space, which are more complicated.

Since $\sin\theta > 1$ and $\sin\phi > 1$, we must use the imaginary forms for $\cos\theta$ and $\cos\phi$ (see Equation 3.35), i.e.,

$$
\begin{array}{lll}
\cos\theta = i\sqrt{\alpha^2 p^2 - 1} & \text{for} & \sin\theta = \alpha p = \alpha/c > 1 \\
\cos\phi = i\sqrt{\beta^2 p^2 - 1} & \text{for} & \sin\phi = \beta p = \beta/c > 1
\end{array}
\tag{4.2}
$$

in the boundary conditions above, which then become

$$2i\rho\beta^2 p\sqrt{\alpha^2 p^2 - 1}\,A \;+\; \rho\beta(1 - 2\beta^2 p^2)B \;=\; 0$$
$$\rho\alpha(1 - 2\beta^2 p^2)A \;-\; 2i\rho\beta^2 p\sqrt{\beta^2 p^2 - 1}\,B \;=\; 0. \tag{4.3}$$

A result from linear algebra states that the system of N linear equations in N unknowns expressed in matrix form as $\mathbf{Ax} = \mathbf{0}$ has only the trivial solution $\mathbf{x} = \mathbf{0}$ *unless* the determinant of $\mathbf{A}$ is zero. To see this for the case in which $N = 2$, consider the system

$$ax + by = 0 \qquad \text{and} \qquad cx + dy = 0, \tag{4.4}$$

where a, b, c, and d (all nonzero) are the coefficients of the unknowns x and y. If we solve the first equation for y and substitute it into the second, we obtain $(ad - bc)x = 0$ (assuming $b \neq 0$). Hence, $x = 0$ and consequently $y = 0$ *unless* $(ad - bc) = \det(\mathbf{A}) = 0$ (which also means that the two equations are linearly dependent, i.e, one is a constant times the other). To obtain the nontrivial solution to Equations (4.3), we must then set the determinant of the matrix of coefficients equal to zero, which gives

$$\alpha(1 - 2\beta^2 p^2)^2 = 4\beta^3 p^2 \sqrt{\alpha^2 p^2 - 1}\sqrt{\beta^2 p^2 - 1}. \tag{4.5}$$

Note that this equation can be obtained by setting the denominator in (3.30) equal to zero (with $\cos\theta$ and $\cos\phi$ given by Equation 4.2). This is **Rayleigh's equation**. Replacing p with $1/c$ and rationalizing the equation, we get, after some algebra,

$$\frac{c^2}{\beta^2}\left[\frac{c^6}{\beta^6} - 8\frac{c^4}{\beta^4} + \left(24 - 16\frac{\beta^2}{\alpha^2}\right)\frac{c^2}{\beta^2} - 16\left(1 - \frac{\beta^2}{\alpha^2}\right)\right] = 0. \tag{4.6}$$

The solution $c = 0$ is not of interest, which leaves us with the equation $[\cdots] = 0$. This is a cubic equation in c^2/β^2, and hence has three solutions. Two of the solutions are extraneous and unphysical – they result from the rationalizing process. The remaining root is real and positive and lies between 0 and β (note that if $c = \beta$, then $[\cdots]$ is positive, and if $c = 0$, then $[\cdots]$ is negative).

If one computes c/β for different values of Poissons's ratio σ, one finds that c is just a little less than β, and that c/β increases monotonically with σ. An approximate formula for c/β as a function of σ is (Achenbach, 1973, p. 192)

$$\frac{c}{\beta} \approx \frac{0.862 + 1.14\sigma}{1 + \sigma}, \tag{4.7}$$

which implies c/β lies between 0.862 and 0.955 for σ between 0 and $\frac{1}{2}$. For example, for $\sigma = \frac{1}{4}$, we obtain $c \approx 0.918\,\beta$.

Note also that c is not a function of frequency, meaning that Rayleigh waves on the surface of a homogeneous half-space are not **dispersive**. We will see later that Rayleigh waves on the surface of an inhomogeneous (e.g., layered) half-space *are* dispersive.

The particle motion due to Rayleigh waves is a combination of P and SV particle motion. To determine this motion, we compute the displacement vector $\mathbf{u}$, which is the sum of the P component of motion $\mathbf{u}_P$ and the SV component of motion $\mathbf{u}_S$. Referring again to the transmitted waves in Figure 3.5, the P component is given by

$$\mathbf{u}_P = A \exp\left[i\omega(\mathbf{s}_P \cdot \mathbf{x} - t)\right]\mathbf{d}_P, \qquad \text{where} \tag{4.8}$$

$$\mathbf{s}_P = \frac{1}{\alpha}(\sin\theta\,\mathbf{e}_x + \cos\theta\,\mathbf{e}_z), \qquad \mathbf{d}_P = \sin\theta\,\mathbf{e}_x + \cos\theta\,\mathbf{e}_z, \tag{4.9a}$$

$$\sin\theta = \alpha p = \alpha/c\,, \qquad \cos\theta = i\sqrt{\alpha^2 p^2 - 1}, \tag{4.9b}$$

where (4.2) has been used. Similarly, the SV component is given by

$$\mathbf{u}_S = B \exp\left[i\omega(\mathbf{s}_S \cdot \mathbf{x} - t)\right]\mathbf{d}_S, \qquad \text{where} \tag{4.10}$$

$$\mathbf{s}_S = \frac{1}{\beta}(\sin\phi\,\mathbf{e}_x + \cos\phi\,\mathbf{e}_z)\,, \qquad \mathbf{d}_S = \cos\phi\,\mathbf{e}_x - \sin\phi\,\mathbf{e}_z, \tag{4.11a}$$

$$\sin\phi = \beta p = \beta/c\,, \qquad \cos\phi = i\sqrt{\beta^2 p^2 - 1}, \tag{4.11b}$$

where (4.2) has been used. Using $p = 1/c$, this gives

$$\mathbf{u}_P = A \exp\left[-\frac{\omega}{\alpha}\sqrt{\frac{\alpha^2}{c^2} - 1}\,z\right]\exp\left[i\omega\left(\frac{x}{c} - t\right)\right]\mathbf{m}_P \tag{4.12a}$$

$$\text{where} \qquad \mathbf{m}_P \equiv \left(\frac{\alpha}{c}\mathbf{e}_x + i\sqrt{\frac{\alpha^2}{c^2} - 1}\,\mathbf{e}_z\right),$$

$$\mathbf{u}_S = B \exp\left[-\frac{\omega}{\beta}\sqrt{\frac{\beta^2}{c^2} - 1}\, z\right] \exp\left[i\omega\left(\frac{x}{c} - t\right)\right] \mathbf{m}_S \qquad (4.12b)$$

$$\text{where} \qquad \mathbf{m}_S \equiv \left(i\sqrt{\frac{\beta^2}{c^2} - 1}\,\mathbf{e}_x - \frac{\beta}{c}\mathbf{e}_z\right).$$

Note, from the form of $\mathbf{m}_P$, that the evanescent P wave, propagating in the x direction, has both an x and a z component of particle motion (not just an x component), and that the "i" in $\mathbf{m}_P$ indicates that the x and z components of particle motion are out of phase with each other by $90°$. A similar statement can be made for the evanescent SV wave. The displacement $\mathbf{u}$ of the surface wave is then $\mathbf{u} = \mathbf{u}_P + \mathbf{u}_S$.

For the next step, we obtain, from (4.3),

$$B = \frac{-2i\beta p\sqrt{\alpha^2 p^2 - 1}}{1 - 2\beta^2 p^2}A \qquad \text{and} \qquad B = \frac{-i\alpha(1 - 2\beta^2 p^2)}{2\beta^2 p\sqrt{\beta^2 p^2 - 1}}A, \qquad (4.13a,b)$$

where $p = 1/c$. If we work out $u_x = u_{Px} + u_{Sx}$ using (4.13b) to replace B, and then take the real part of the outcome (to obtain the actual physical displacement), and assume, without loss of generality, that A is positive real, we get

$$u_x = a(z)\cos\left[\omega\left(\frac{x}{c} - t\right)\right], \qquad a(z) \equiv \frac{\alpha}{c}\left[e^{-\zeta z} - \left(1 - \frac{c^2}{2\beta^2}\right)e^{-\xi z}\right]A, \qquad (4.14a)$$

$$\zeta \equiv \frac{\omega}{c}\sqrt{1 - \frac{c^2}{\alpha^2}}, \qquad \xi \equiv \frac{\omega}{c}\sqrt{1 - \frac{c^2}{\beta^2}}. \qquad (4.14b)$$

If we do the same for u_z, using (4.13a) to replace B, we get

$$u_z = b(z)\sin\left[\omega\left(\frac{x}{c} - t\right)\right], \qquad b(z) \equiv -\frac{\alpha}{c}\sqrt{1 - \frac{c^2}{\alpha^2}}\left[e^{-\zeta z} - \left(\frac{e^{-\xi z}}{1 - c^2/2\beta^2}\right)\right]A. \qquad (4.15)$$

The forms of u_x and u_z indicate that the surface wave is propagating in the x direction with speed c and decaying in the z direction, with the decay described quantitatively by $a(z)$ and $b(z)$. If we analyze (or plot) $a(z)$ and $b(z)$, we find that $a(z)$ starts off positive, i.e., $a(0) > 0$, then decreases and becomes negative and remains so, whereas $b(z)$ is positive for all values of z, and that both $a(z)$ and $b(z) \to 0$ as $z \to \infty$.

Using the fact that $\sin^2\gamma + \cos^2\gamma = 1$, we obtain from (4.14a) and (4.15)

$$\left(\frac{u_x}{a}\right)^2 + \left(\frac{u_z}{b}\right)^2 = 1. \qquad (4.16)$$

This is the equation of an ellipse in the $u_x u_z$ plane. Hence, particles displaced by Rayleigh waves move in an *elliptical* fashion. The term **ground roll** is often used to describe this kind of motion.

Consider the displacement experienced by a particle at position x on the surface at two times, t_1 and the later time t_2, where $t_1 = x/c$ and $t_2 = x/c + \pi/2\omega$ (note that t_2 occurs a quarter of a period or cycle after t_1). First of all, at the surface $z = 0$, we have

$$a(0) = \frac{\alpha c}{2\beta^2}A > 0, \qquad b(0) = \frac{\alpha c}{2\beta^2} \frac{\sqrt{1 - (c^2/\alpha^2)}}{1 - (c^2/2\beta^2)}A > 0. \qquad (4.17)$$

Hence, using (4.14a) and (4.15), one obtains

$$\text{At } t = t_1 = \frac{x}{c}: \qquad u_x = a(0) > 0, \quad u_z = 0, \qquad (4.18a)$$

$$\text{At } t = t_2 = \frac{x}{c} + \frac{\pi}{2\omega}: \qquad u_x = 0, \quad u_z = -b(0) < 0. \qquad (4.18b)$$

This shows that a particle on the surface that is disturbed by a Rayleigh wave experiences a counterclockwise or **retrograde** elliptical motion (see Figure 4.1a). More generally, one can simply use (4.14a) and (4.15) to plot graphs of u_x and u_z vs. time at the surface $z = 0$, from which the retrograde elliptical particle motion can be inferred.

Although $a(z)$ is positive at $z = 0$, it vanishes at a certain depth z_0 (i.e., $u_x = 0$ at $z = z_0$). $a(z)$ is negative below the depth z_0. This means that a particle at a depth $z > z_0$ experiences a clockwise or **prograde** elliptical motion. $z = z_0$ is called the **nodal plane**. As Poisson's ratio σ goes from 0 to $\frac{1}{2}$, z_0 goes from about $0.25w$ to $0.14w$, where w is the wavelength of the Rayleigh wave (see Figure 4.1b). For instance, if $c = 400$ m/s and the frequency $f = 10$ Hz, then $w = c/f = 40$ m and z_0 goes from about 10 m to 5.6 m.

Another way to derive the preceding results is to use displacement potentials (see, e.g., Achenbach, 1973, p. 189; Grant and West, 1965, p. 59). Since the potentials satisfy standard wave equations, trial solutions, such as $\phi = f(z)$ $\exp[i\omega(px - t)]$ for the P wave component, and a similar one for the SV wave component, can be substituted into the wave equations to determine $f(z)$, etc.; boundary conditions can be applied; and (2.69a), $\mathbf{u} = \nabla\phi + \nabla \times \boldsymbol{\psi}$, can be used to obtain $\mathbf{u}$.

The preceding discussion assumed plane waves. In actuality, wavefronts are curved. We have seen before that spherical waves in a homogeneous medium decay as $1/r$, where r is the distance traveled by the waves. Rayleigh waves, however, decay as $1/\sqrt{r}$, i.e., they decay more slowly than body waves. At great distances, they carry most of the energy in a signal.

For SH waves, the same type of analysis as preceding can be applied to show that surface SH waves *cannot* exist on the surface of a homogeneous half-space. We will see, though, that they *can* exist on the surface of an *inhomogeneous* half-space.

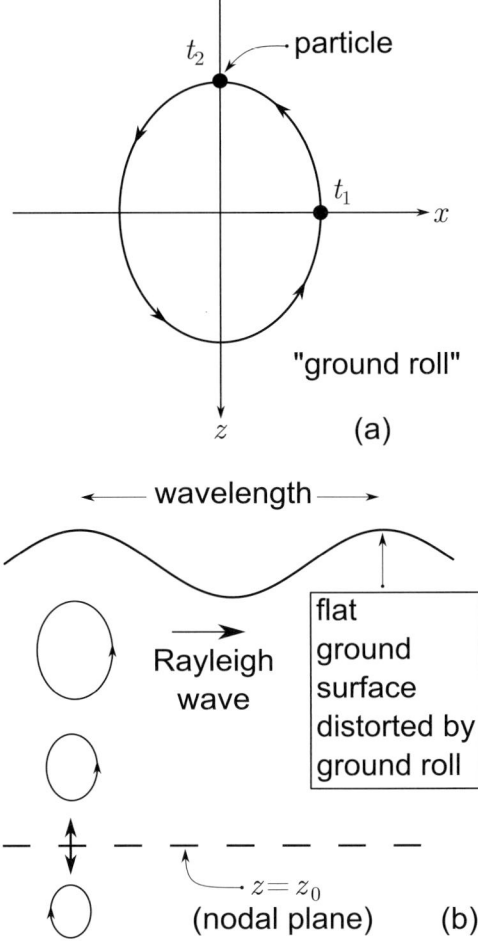

Figure 4.1 Schematic diagrams of (a) retrograde elliptical particle motion for a Rayleigh wave, and (b) decay and reversal of particle motion with depth.

4.3 Interface Waves

The preceding analysis can be generalized to surface waves traveling along the interface between two homogeneous half-spaces. An interface wave has *P* and *SV* components from both sides of the interface (four components altogether) and its amplitude decays with distance from the interface on both sides. Since we have a solid–solid interface in this case, we must apply all four boundary conditions (for u_x, u_z, σ_{xz}, and σ_{zz}), which makes sense because the interface wave has four components. It turns out that interface waves can exist only for certain ranges of values of the ratios ρ_1/ρ_2 and β_1/β_2. These interface waves are called **Stoneley**

waves. They can always exist at a liquid–solid interface. Like Rayleigh waves, they are nondispersive (for *homogeneous* half-spaces).

For the *SH* case, one finds that interface *SH* waves *cannot* exist for two homogeneous half-spaces.

More advanced mathematical methods can be used to show that Rayleigh and Stoneley waves are generated by the diffraction of curved wavefronts of body waves at a plane boundary.

4.4 Huygens' Principle and Applications

Huygens' principle states that every point on a wavefront can be regarded as a new source of waves. In Figures 4.2a and 4.2b, the wavefronts at time $t + \Delta t$ are the envelopes of the small circular arcs.

For a homogeneous isotropic medium, plane and spherical wave fronts do not change in shape as they propagate (plane fronts remain plane and spherical fronts remain spherical). However, generally curved wavefronts *do* change in shape as they propagate. For example, consider the triangular wavefront (generated by a complicated source, say) in Figure 4.2c. As the wavefront propagates, it becomes more and more spherical. At a great enough distance from the source, the wavefront is practically spherical, and so if we record the wavefront at a great distance, we cannot directly get detailed information about the source. This is also a problem in earthquake seismology, where the aim is to discover the source mechanism of earthquakes.

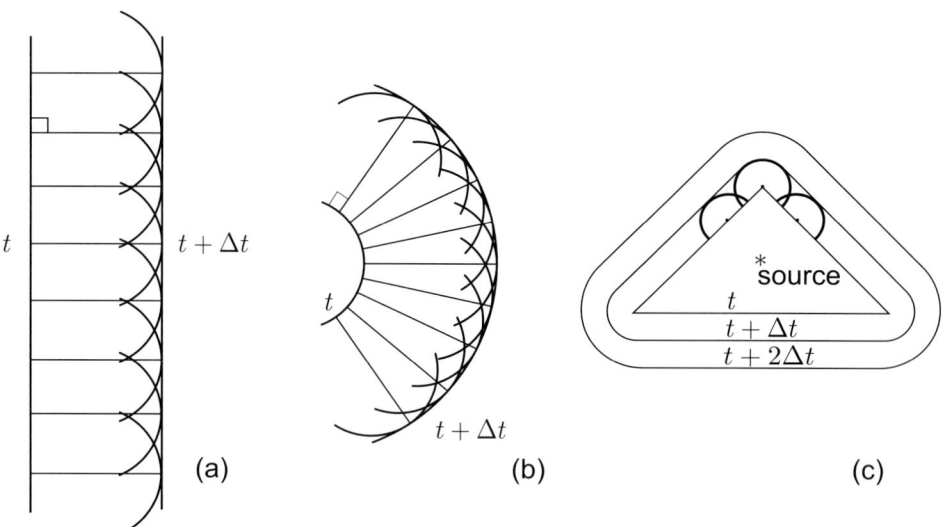

Figure 4.2 Illustration of Huygens' principle for (a) plane, (b) spherical, and (c) triangular wavefronts.

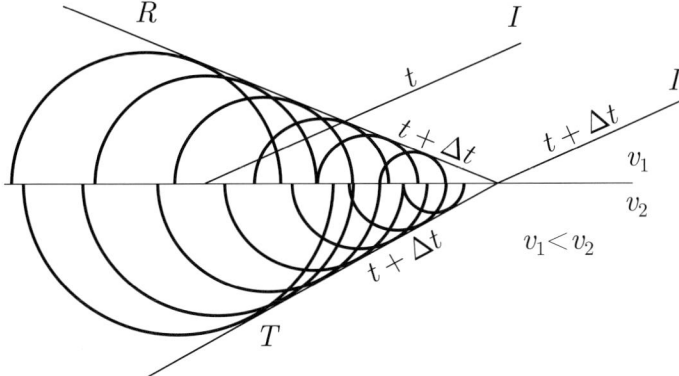

Figure 4.3 Huygens' principle applied to the reflection and transmission of plane waves. I, R, and T indicate the incident, reflected, and transmitted wavefronts.

Huygens' principle can be used to discuss reflected and transmitted waves at a boundary, head waves (refraction events), and diffraction events.

For reflected and transmitted plane waves, a careful consideration of the geometry in Figure 4.3 leads to Snell's law of reflection and transmission (see, e.g., Sheriff and Geldart, 1982).

Head waves are generated by the reflection and transmission of curved wave fronts. An advanced mathematical analysis of the problem using spherical wavefronts leads to the following description (see, e.g., Červený and Ravindra, 1971). Referring to Figure 4.4, before the critical angle is reached, point Q (the intersection of the three wavefronts) travels along the boundary with speed $v_Q = v_1/\sin\theta$, where θ is the angle of incidence of the ray from the source to the point Q (note that v_Q is just the plane wave apparent velocity in the horizontal direction – see Appendix 2A – and that for a small enough region around the point Q, the curved wavefronts can be approximated as being plane). Snell's law implies $v_Q > v_2$. At the critical angle θ_c, we have $v_Q = v_1/\sin\theta_c = v_1/(v_1/v_2) = v_2$. For $\theta > \theta_c$, we have $v_Q < v_2$. When $\theta = \theta_c$, the curved transmitted wavefront breaks away from point Q and travels at the speed v_2, and by Huygens' principle, generates a wave, i.e., the head wave, which travels upward at the angle θ_c (in plane wave theory, the plane transmitted wavefront would still be connected to the other two plane wavefronts at point Q and therefore would not generate an upgoing head wave). The head wave front is plane for a line source, or conical (the frustrum of a cone) for a point source.

An interpretation of the head wave in terms of geometrical ray paths is shown in Figure 4.5: a geophone at x will record the direct wave, the reflected body wave and the head wave.

Geometrical ray theory using plane waves cannot explain the existence of head waves, because the plane wave energy transmission coefficient indicates that the

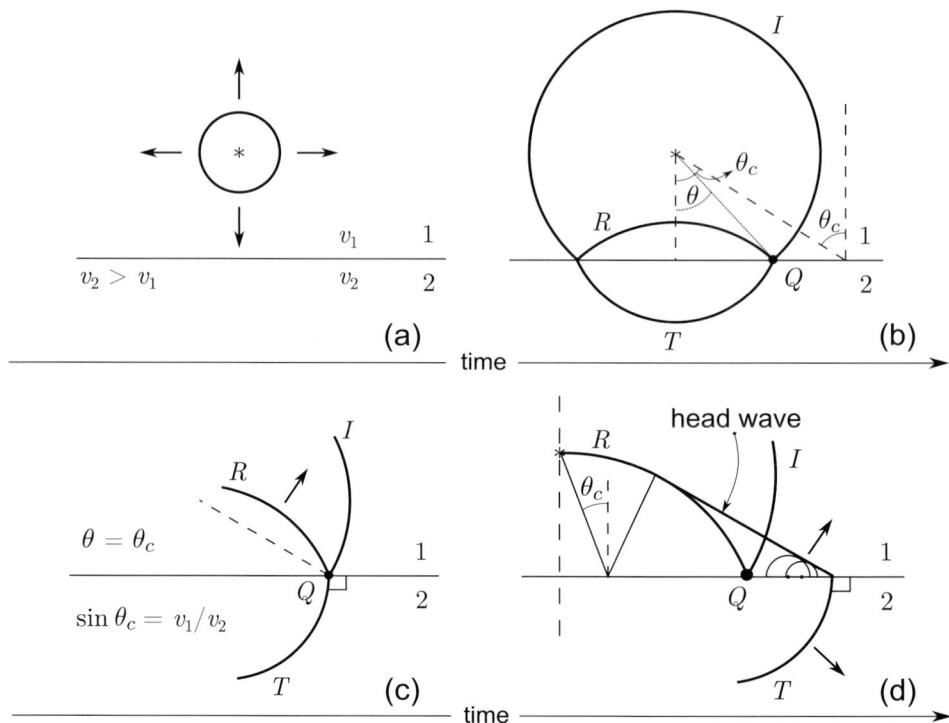

Figure 4.4 Head wave geometry. "I," "R," and "T" indicate the incident, reflected, and transmitted wavefronts. $v_2 > v_1$, and θ_c is the critical angle. Time t progresses from (a) to (d).

transmitted wave has no vertical component of energy flow beyond the critical angle (see the discussion on $\mathcal{T}_{PP}$ after Equation 3.43), and hence cannot send any energy back into the incidence medium. Also, as previously mentioned, the transmitted plane wavefront does *not* break away from the incident and reflected wavefronts in plane wave theory (see Chapter 3) as in Figure 4.4, meaning that plane waves alone cannot be used to explain head waves. A more accurate theory based on curved wavefronts must be used to explain head waves.

At large enough distances, such as x in Figure 4.5, the head wave arrives at the receiver *before* the direct wave. No head wave arrives for distances less than the **critical distance** x_c, where

$$x_c = 2h \tan \theta_c = 2h \frac{\sin \theta_c}{\sqrt{1 - \sin^2 \theta_c}} = \frac{2h}{\sqrt{(v_2/v_1)^2 - 1}}. \tag{4.19}$$

The travel time curve for the head wave is a straight line. The t-x equation for the curve can be worked out geometrically, and the result is (see, e.g., Telford, Geldart, Sheriff, and Keys, 1976),

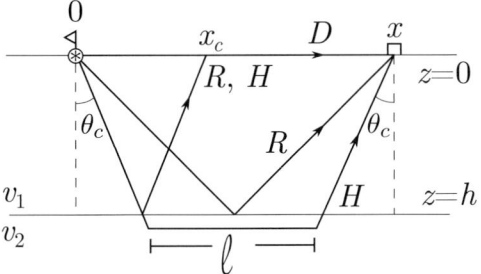

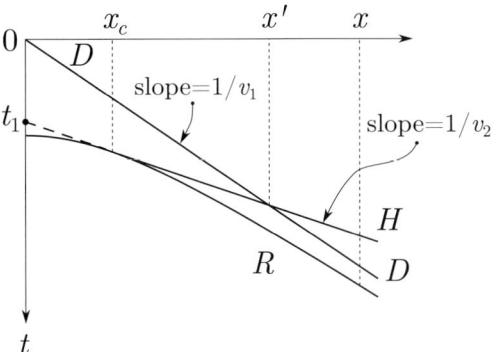

Figure 4.5 Ray paths and travel times for the direct wave D ($t = x/v_1$), the reflected wave R ($t^2 = t_0^2 + (x/v_1)^2$, $t_0 = 2h/v_1$), and the head wave H (t is given by Equation 4.20) for a one-layer model with $v_1 < v_2$.

$$t = \frac{x}{v_2} + t_1, \qquad \text{where} \qquad t_1 = \frac{2h\cos\theta_c}{v_1} = \frac{2h}{v_1}\sqrt{1 - \left(\frac{v_1}{v_2}\right)^2}. \qquad (4.20)$$

The t-x line for the head wave is tangent to the t-x curve for the reflection at $x = x_c$. The **crossover distance** x' (the distance beyond which the head wave arrives before the direct wave) is given by

$$x' = 2h\sqrt{\frac{v_2 + v_1}{v_2 - v_1}}. \qquad (4.21)$$

The more accurate and mathematically advanced theory used to describe head waves shows that the head wave amplitude decays as $(i/\omega)x^{-1/2}\ell^{-3/2}$ for a point source (see, e.g., Grant and West, 1965, p. 180), where ℓ is shown in Figure 4.5. For large distances, $\ell \approx x$, and the decay goes as $1/x^2$, i.e., head waves decay faster than body waves (which decay approximately as $1/x$). This also indicates that for smaller offsets, i.e., nearer the source, the head wave amplitude decay with offset is less rapid than for larger offsets. The "i/ω" indicates that the amplitude decreases as the frequency increases, and that the head waveform (a superposition of frequencies) is basically a time integral of the body waveform, making the head waveform smoother than the body waveform.

Since there can be two reflected and two transmitted waves for an incident P or SV wave, there can be more than one head wave, depending on the velocity relationships. Figure 4.6 shows the three possible head waves generated by an

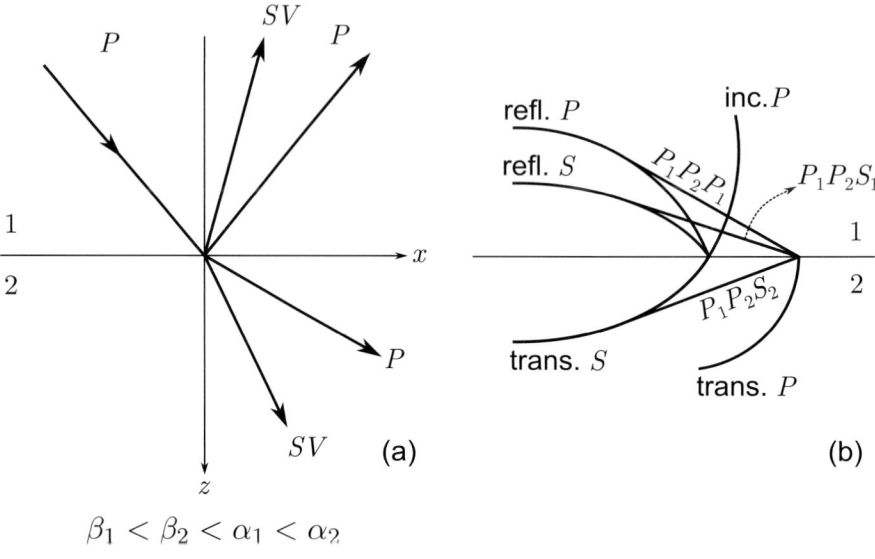

$$\beta_1 < \beta_2 < \alpha_1 < \alpha_2$$

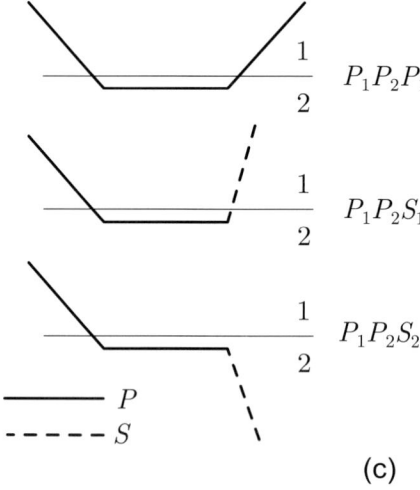

Figure 4.6 The head waves $P_1P_2P_1$, $P_1P_2S_1$, and $P_1P_2S_2$. (a) The ray paths for the incident, reflected, and transmitted waves, (b) the wavefront diagram showing the three possible head waves, and (c) the ray path diagrams for the head waves.

incident P wave when $\beta_1 < \beta_2 < \alpha_1 < \alpha_2$. Usually, only the $P_1P_2P_1$ head wave is of interest in exploration seismology.

To find the angle that the head wave makes with the vertical, one may think of the ray segment that is parallel to the interface (the segment of length ℓ in Figure 4.5) as an incident wave that "generates" the head wave and use Snell's law (see, e.g., Červený and Ravindra, 1971, p. 102). For example, if the head wave angle for the $P_1P_2S_1$ head wave in Figure 4.6 is ϕ_{1H} (the angle of the dashed line), and if the angle of the transmitted P wave (that ultimately travels parallel to the interface) is θ_2, then

$$\frac{\sin \phi_{1H}}{\beta_1} = \frac{\sin(\theta_2 = 90°)}{\alpha_2} = \frac{1}{\alpha_2} \quad \Rightarrow \quad \sin \phi_{1H} = \frac{\beta_1}{\alpha_2}.$$

In general, if the head wave angle is a_H, then $\sin a_H = v_H/v_c$, where v_H is the speed of the head wave and v_c is the speed of the wave traveling parallel to the interface that "generates" it.

An example of the use of Huygens' principle to illustrate diffraction is shown in Figure 4.7. By Huygens' principle, the receiver at x records both the reflection from the flat interface and the diffraction from the sharp point.

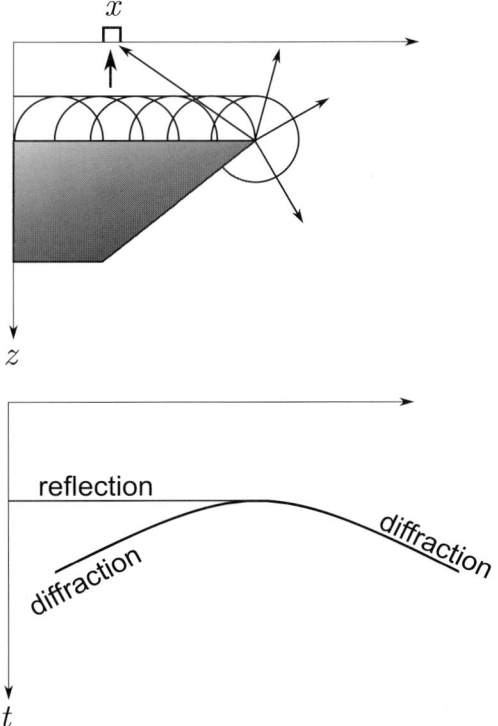

Figure 4.7 The use of Huygens' principle to explain a diffraction event.

4.5 Phase Velocity, Group Velocity, and Dispersion

Consider two harmonic waves traveling in the x direction. Suppose that they both have the same amplitude A, but different frequencies ω_1 and ω_2. Suppose also that they are traveling with different speeds c_1 and c_2 (dispersion is present). These speeds are called **phase velocities** since they are the velocities at which the planes of constant phase (the wavefronts) travel in the direction of propagation. In three dimensions, the phase velocity of a wave with slowness $\mathbf{s}$ is $(1/s)\mathbf{n}$, where $\mathbf{n}$ is a unit vector in the direction of $\mathbf{s}$. The total disturbance u due to the two harmonic waves is the sum of the individual disturbances, i.e.,

$$u = A\cos(\kappa_1 x - \omega_1 t) + A\cos(\kappa_2 x - \omega_2 t), \tag{4.22}$$

$$\text{where} \quad \kappa_1 = \frac{\omega_1}{c_1} \quad \text{and} \quad \kappa_2 = \frac{\omega_2}{c_2}. \tag{4.23}$$

Using the identity

$$\cos\theta + \cos\phi = 2\cos\left(\frac{\theta + \phi}{2}\right)\cos\left(\frac{\theta - \phi}{2}\right), \tag{4.24}$$

we can rewrite u as

$$u = 2A\cos\left[\left(\frac{\kappa_1 + \kappa_2}{2}\right)x - \left(\frac{\omega_1 + \omega_2}{2}\right)t\right]\cos\left[\left(\frac{\kappa_1 - \kappa_2}{2}\right)x - \left(\frac{\omega_1 - \omega_2}{2}\right)t\right]. \tag{4.25}$$

Or, with

$$\Delta\omega \equiv \frac{\omega_1 - \omega_2}{2}, \quad \Delta\kappa \equiv \frac{\kappa_1 - \kappa_2}{2}, \quad \omega_0 \equiv \frac{\omega_1 + \omega_2}{2}, \quad \kappa_0 \equiv \frac{\kappa_1 + \kappa_2}{2}, \tag{4.26}$$

we have

$$u = U\cos(\kappa_0 x - \omega_0 t), \quad \text{where} \quad U \equiv 2A\cos\left[(\Delta\kappa)x - (\Delta\omega)t\right]. \tag{4.27}$$

If ω_1 and ω_2 are fairly close in value, and if c_1 and c_2 are as well, then κ_1 and κ_2 are fairly close, i.e., $\Delta\omega$ and $\Delta\kappa$ are small compared to ω_0 and κ_0 (which are the *average* frequency and *average* wavenumber). Hence, the preceding equations describe a "carrier" wave with a high frequency, ω_0, and a short wavelength, $2\pi/\kappa_0$, modulated by an "envelope" with a low frequency, $\Delta\omega$, and a long wavelength, $2\pi/\Delta\kappa$ (see Figure 4.8a). The carrier wave moves with the phase velocity $c_0 = \omega_0/\kappa_0 = (\omega_1 + \omega_2)/(\kappa_1 + \kappa_2)$. However, modulation results in the formation of "groups" (defined by the "envelopes"), which move with the **group velocity** $c_g = \Delta\omega/\Delta\kappa = (\omega_1 - \omega_2)/(\kappa_1 - \kappa_2)$. The energy in the wave propagates at the group velocity, i.e., c_g is the speed of energy flow, because no energy can pass through the nodes.

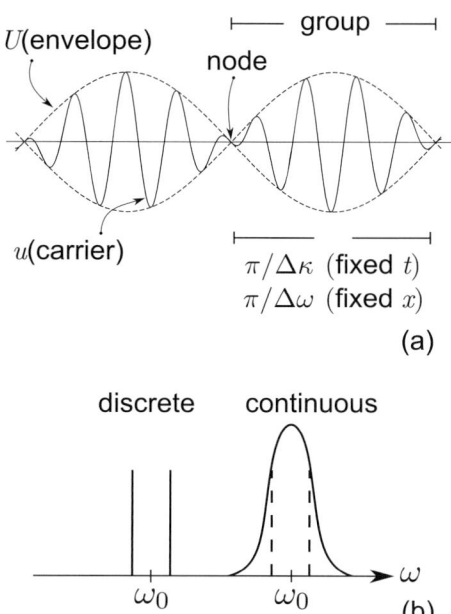

Figure 4.8 (a) A modulated carrier wave, and (b) the discrete and continuous spectra of wave groups.

Realistic waveforms, e.g., those appearing on seismograms, consist of a continuous spectrum of frequencies. So, instead of just two discrete frequencies that are close in value, consider a continuous narrow band of frequencies, with ω_0 being the average frequency in the band (see Figure 4.8b). Surface wave spectra are often of this type. We will see later that for certain kinds of waves, ω and κ are not linearly related, i.e., in general we have $\omega = \omega(\kappa)$ or $\kappa = \kappa(\omega)$. Note then that the formula given earlier for c_g, i.e., $c_g = \Delta\omega/\Delta\kappa = (\omega_1 - \omega_2)/(\kappa_1 - \kappa_2)$, gives approximately the slope, $d\omega/d\kappa$, of the ω versus κ curve at $\kappa = \kappa_0$. In fact, we will show later that for a continuous carrier wave with a sharply peaked spectrum whose average frequency is ω_0, the group velocity is given by

$$ c_g = \left(\frac{d\omega}{d\kappa}\right)_{\kappa=\kappa_0} = 1 \bigg/ \left(\frac{d\kappa}{d\omega}\right)_{\omega=\omega_0}, \tag{4.28} $$

where $\kappa_0 = \kappa(\omega_0)$. The phase velocity of the carrier (the speed at which the peaks and troughs of the carrier move) is $c_0 = \omega_0/\kappa_0$. The phase velocity of each individual Fourier component of the pulse is $c = c(\omega) = \omega/\kappa(\omega)$.

These results apply also to the case of a dispersive wave consisting of a discrete sum of N harmonic waves, with the nth wave having an amplitude A_n, a frequency ω_n, a wave speed c_n, and a wavenumber κ_n (a generalization of Equation 4.22).

The frequencies and wavenumbers are related through the wave's **dispersion relation**, $\omega = \omega(\kappa)$. As long as the spectrum is a narrowband one, i.e., as long as $|\omega_N - \omega_1|/\omega_0 < 1$ and $|\kappa_N - \kappa_1|/\kappa_0 < 1$ (rough rules of thumb), where ω_0 and κ_0 are the average frequency and wavenumber, the groups will propagate at the group velocity c_g given by (4.28).

Any arbitrary pulse on a seismogram for the distance x can be represented by a continuous superposition of harmonic waves with different frequencies as

$$\text{pulse}\,(x, t) = \int_{\Delta\omega} A(\omega)e^{i[\kappa(\omega)x - \omega t]}\, d\omega, \tag{4.29}$$

where $A(\omega)$ is the amplitude of the Fourier component (harmonic wave) of frequency ω. Also, $A(\omega)$ is a narrowband spectrum, meaning the integrand is nonzero only in a relatively small interval $\Delta\omega$, the range of integration. The exponential term represents an approximately sinusoidal oscillation. If the phase, $\kappa(\omega)x - \omega t$, changes rapidly with ω, the exponential term will change rapidly as well. There will be a negligible contribution to the integral from those frequency zones in $\Delta\omega$ in which the exponential changes rapidly with ω compared to $A(\omega)$, because in those zones, the integrand is a rapidly oscillating function of ω, producing a self-canceling effect (the area under the integrand in those zones will be very small because the many positive and negative areas of the oscillating integrand will tend to cancel each other out). The only significant contributions to the integral will come from those frequencies ω_0 at which the sinusoidal Fourier components $Ae^{i[\kappa x - \omega t]}$ are *in phase*, i.e., where the phase, $\kappa x - \omega t$, (and consequently $Ae^{i[\kappa x - \omega t]}$) does not change with frequency ω. Expressing this mathematically, we have

$$\left(\frac{d}{d\omega}[\kappa(\omega)x - \omega t]\right)_{\omega=\omega_0} = 0 \quad \Longrightarrow \quad t = \left(\frac{d\kappa}{d\omega}\right)_0 x = \frac{x}{c_g}, \tag{4.30a,b}$$

where the subscript "0" means the quantity is evaluated at $\omega = \omega_0$ (or $\kappa = \kappa_0$). The pulse then consists of a group of constructively interfering sinusoids, and the group travels at the group velocity $c_g = (d\kappa/d\omega)_0^{-1} = (d\omega/d\kappa)_0$, which is the average velocity of the group. A pulse on a seismogram with an average frequency around ω_0 has traveled a distance $x = c_g(\omega_0)t$ in a time t. Solving this equation for $\omega_0 = \omega_0(x, t)$ gives the average frequency ω_0, which is expected to dominate at time t and travel distance x.

To summarize, dropping the subscript "0" for convenience, if a pulse with a narrowband spectrum and an average frequency around ω has traveled a distance x in a time t, then its average velocity is the group velocity, $c_g(\omega) = x/t$.

The method used to obtain (4.30) is part of a more general method for doing integrals like (4.29), known as the **method of stationary phase**. It is described in more advanced texts.

The group velocity was discussed with reference to a waveform on a seismogram, i.e., a pulse at a fixed value of x, which is represented by (4.29). However, it is also possible to derive the group velocity formula by considering a "snapshot" of a waveform in a solid, i.e., a pulse at a fixed value of t, in which case the pulse is represented by a continuous superposition of wavenumbers, i.e.,

$$\text{pulse } (x, t) = \int_{\Delta\kappa} B(\kappa) e^{i[\kappa x - \omega(\kappa)t]} \, d\kappa. \tag{4.30c}$$

If the preceding method, which resulted in (4.30b), is applied to this, the same formula, (4.28), is obtained for the group velocity. Equation (4.30c) can be extended to 2D (or 3D) in a straightforward way to obtain the group velocity formulas. For example, for 2D, with $\omega = \omega(\kappa_x, \kappa_z)$, they are

$$c_{gx} = \frac{\partial\omega}{\partial\kappa_x}, \qquad c_{gz} = \frac{\partial\omega}{\partial\kappa_z}. \tag{4.30d}$$

Since c and c_g are different in value, a given wavefront of the carrier wave moves with the speed c_0 from one end of the wave group to the other as time passes. The wave group (the pulse, sometimes called a **wave packet**) moves with the speed c_g. The group velocity c_g can be related to the phase velocity c as follows:

$$\kappa = \frac{\omega}{c} \quad \Rightarrow \quad \frac{d\kappa}{d\omega} = \frac{1}{c} - \frac{\omega}{c^2}\frac{dc}{d\omega} = \frac{1}{c_g} \quad \Rightarrow \quad c_g = \frac{c}{1 - \dfrac{\omega}{c}\dfrac{dc}{d\omega}} = \frac{c}{1 + \dfrac{T}{c}\dfrac{dc}{dT}}, \tag{4.31}$$

where T is the period.

Both c and c_g vary with frequency. This is known as **dispersion**. If there is no dispersion ($dc/d\omega = 0$), then $c_g = c$. If $dc/d\omega$ is negative, then $c_g < c$ – this is called **normal dispersion** and it is the usual situation. If $dc/d\omega$ is positive, then $c_g > c$ – this is called **anomalous dispersion**. If dispersion is present, then the shape of the pulse changes as it propagates because the different sinusoidal Fourier components travel at different phase speeds. If there is no dispersion, the pulse shape stays the same.

It should also be mentioned that, in the strictest sense, the group velocity c_g is to be evaluated at the average frequency or wavenumber, ω_0 or κ_0, not the dominant frequency (see Equation 4.28). In practice, the average and dominant frequencies can be close in value, and for a simple symmetrically shaped spectrum, e.g., a Gaussian bell curve, they are identical. But in general, they are different.

4.6 Normal Modes in a Surface Layer over a Rigid Half-Space

Consider the case of surface *SH* waves propagating in a surface layer overlying a rigid half-space. Although the assumption of a perfectly rigid medium is not physically realistic, it does mean that the problem is relatively easy to treat mathematically. The results then provide us with some insight into the nature of surface waves. The surface layer acts as a **waveguide**, and as a consequence, the waves are often called **guided waves**. They can be thought of as a superposition of **normal modes**, just as the shape of a vibrating string can be expressed as a superposition of the string's normal modes of vibration.

In Cartesian coordinates, the displacement $\mathbf{u}$ for a plane *SH* wave has only a y component, so we can write $\mathbf{u} = u\mathbf{e}_y$. As usual, for simplicity we consider the 2D problem in which plane waves propagate in the xz plane ($s_y = 0$) and there is no variation in the y direction, i.e., $\partial(\cdots)/\partial y = 0$ (which comes from the fact that $s_y = 0$). Since $\nabla \cdot \mathbf{u} = 0$ for *SH* waves, the equation of motion (Equation 2.45) gives

$$\nabla^2 u = \frac{\partial^2 u}{\partial x^2} + \frac{\partial^2 u}{\partial z^2} = \frac{1}{\beta^2} \frac{\partial^2 u}{\partial t^2}. \tag{4.32}$$

Consider a disturbance propagating in the x direction in the layer with speed $c\,(= \omega/\kappa)$ for which the amplitude along a given vertical wavefront varies with depth z:

$$u = U(z)e^{i(\kappa x - \omega t)}. \tag{4.33}$$

This has the basic form of a surface wave. If we substitute this into Equation (4.32), we get

$$\frac{d^2 U}{dz^2} = -\gamma^2 U, \qquad \gamma^2 \equiv \frac{\omega^2}{\beta^2} - \kappa^2 = \frac{\omega^2}{\beta^2} - \frac{\omega^2}{c^2}. \tag{4.34}$$

Depending on the sign of γ^2, this equation has either a sinusoidal solution ($\gamma^2 > 0$) or an exponential solution ($\gamma^2 < 0$). Normal modes are associated with sinusoidal oscillations, hence we assume $\gamma^2 > 0$, which implies $c > \beta$. The solution to equation (4.34) is then

$$U(z) = Ae^{i\gamma z} + Be^{-i\gamma z}. \tag{4.35}$$

Equation (4.33) can then be written as

$$u = Ae^{i(\kappa x + \gamma z - \omega t)} + Be^{i(\kappa x - \gamma z - \omega t)}. \tag{4.36}$$

We will see that $\gamma > 0$. Hence, with z increasing downward, we see that u is a superposition of a downgoing wave with amplitude A and an upgoing wave with amplitude B. Next, we apply the boundary conditions at the two interfaces. We assume that the free surface is at depth $z = 0$ and that the interface between the surface layer and the rigid half-space is at depth $z = h$. At the free surface, σ_{yz} must vanish, and at $z = h$, we must have $\mathbf{u} = 0$ since there is no motion in the rigid medium. Using Equation (3.21), we then have

$$\sigma_{yz} = \mu \frac{\partial u}{\partial z} = 0 \quad \text{at} \quad z = 0 \,, \qquad \text{and} \qquad u = 0 \quad \text{at} \quad z = h. \qquad (4.37)$$

Upon substitution of Equation (4.36), these conditions become

$$A - B = 0 \qquad \text{and} \qquad A e^{i\gamma h} + B e^{-i\gamma h} = 0. \qquad (4.38)$$

Substituting $B = A$ (from the first equation) into the second equation gives

$$e^{-i\gamma h} + e^{i\gamma h} = 2\cos(\gamma h) = 0 \quad \Longrightarrow \quad \gamma h = \left(n + \tfrac{1}{2}\right)\pi \,, \quad n = 0, 1, 2, \ldots \qquad (4.39)$$

This could also have been obtained by setting the determinant of the coefficient matrix for (4.38) equal to zero (see the discussion around Equation 4.4). Substituting into (4.39) the last formula for γ from Equation (4.34), and solving for c, gives

$$c = c(\omega) = \frac{\beta}{\sqrt{1 - \left[\dfrac{\beta(n + \tfrac{1}{2})\pi}{\omega h} \right]^2}} \equiv c_n, \qquad (4.40)$$

for $n = 0, 1, 2, \ldots$ Hence, the phase speed in the x direction varies with frequency, indicating the presence of dispersion. The various values of the integer n denote the various normal modes. The **fundamental** mode has $n = 0$, and the first, second, etc., modes have $n = 1, 2$, etc. As $h \to \infty$, we have $c \to \beta$, as expected. Note that $c > \beta$ for all modes, as indicated in the discussion that precedes Equation (4.35). For any given mode n, only frequencies ω that satisfy

$$\frac{\beta(n + \tfrac{1}{2})\pi}{\omega h} < 1 \,, \qquad \text{i.e.} \qquad \omega > \frac{\beta(n + \tfrac{1}{2})\pi}{h} \equiv \omega_{cn}, \qquad (4.41)$$

will propagate as normal modes. ω_{cn} is called the **cutoff frequency** for mode n. If $\omega \leq \omega_{cn}$, then c is imaginary and hence the disturbance decays in the x direction (the normal mode does not propagate). To get an idea of the numerical values of the cutoff frequencies, let $h = 20$ m and $\beta = 500$ m/s. The cutoff frequencies $f_{cn} = \omega_{cn}/2\pi$ are then $f_{c0} = 6.25$ Hz, $f_{c1} = 18.75$ Hz, $f_{c2} = 31.25$ Hz, etc. If we calculate the group velocity from $1/c_g = d\kappa/d\omega$, we find that $c_g < \beta < c$ for all ω, i.e., the dispersion is of the "normal" type. From Equation (4.39), we can also obtain the **dispersion relation** $\omega = \omega(\kappa)$. It is given by

$$\omega = \beta \sqrt{\left[\frac{(n + \tfrac{1}{2})\pi}{h} \right]^2 + \kappa^2}, \qquad (4.42)$$

and is plotted in Figure 4.9a. Note that $\omega(\kappa)$ is a hyperbola. The velocities $c(\omega)$ and $c_g(\omega)$ are shown in Figure 4.9b.

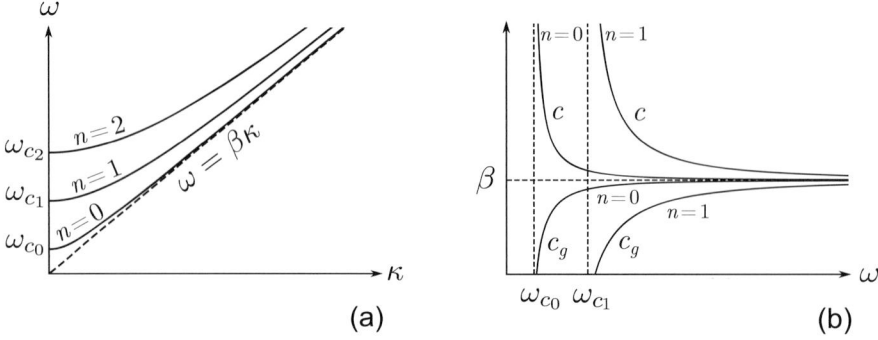

Figure 4.9 Plots of (a) ω versus κ, and (b) c and c_g versus ω.

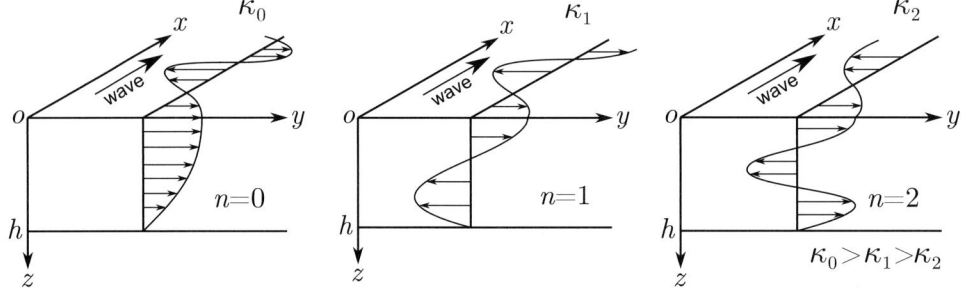

Figure 4.10 *SH* particle motion (Equation 4.43) for $t = 0$ and any value of ω ($> \omega_{cn}$).

Concerning the variation with depth, we have $A = B$ from Equations (4.38), meaning that Equation (4.35) can be written as $U(z) = 2A \cos(\gamma z)$. Inserting this into Equation (4.33) and taking the real part then gives, for the nth normal mode,

$$u = 2A \cos(\gamma z) \cos(\kappa x - \omega t) , \qquad \text{where} \qquad \gamma = \frac{(n + \frac{1}{2})\pi}{h} , \quad \kappa = \frac{\omega}{c_n}, \quad (4.43)$$

which shows how the particle motion varies with depth and offset. Note that the form of u indicates that it is a traveling wave in the x direction and a *standing wave* in the z direction. u is plotted in Figure 4.10 for the first three modes. A general displacement u is a superposition of the normal modes.

The guided waves just described are also often called **surface *SH* waves**, because they are confined to a near-surface zone, there being no motion for $z \geq h$. We have already seen that surface *SH* waves cannot exist at the free surface of a homogeneous half-space. The preceding analysis shows that they *can* exist, as guided waves, at the free surface of an *inhomogeneous* half-space (a layered one, in this case). The surface layer acts as a waveguide. The fact that surface *SH* waves are observed in nature has been used to infer that the Earth's crust is layered.

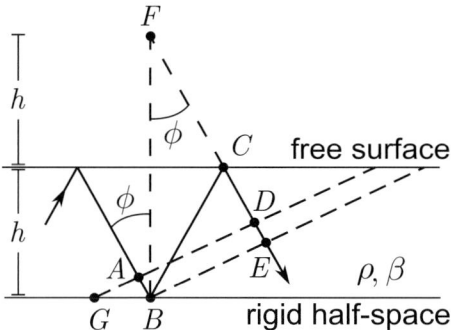

Figure 4.11 An internally reflected wave.

By looking at the problem from another viewpoint, we will see that these waves are due to the constructive interference of internally reflected upgoing and downgoing waves. Equation (4.36), which represents a downgoing wave (of amplitude A) coexistent with an upgoing wave (of amplitude B), supports this notion. Figure 4.11 shows an internally reflected wave. The wavefront associated with point A on the ray is coincident with the wavefront associated with point D.

For constructive interference, these two waves must be in phase with each other. The phase difference between points D and A is due to the extra travel path $\overline{ABCD}$ plus any phase changes that occur at the reflection points B and C. From Equation (3.31), we may conclude that SH waves do not experience a phase change upon reflection from a free surface. At the rigid boundary, the displacement must be zero. This boundary condition implies that SH waves experience a $180°$ phase change upon reflection from a rigid interface (set $B_T = 0$ in the first of Equations 3.25, or let $\beta_2 \to \infty$ (from $\mu_2 \to \infty$) in the first of Equations 3.26). This $180°$ phase change is a shift of half a wavelength. Hence, for constructive interference, we must have

$$\overline{ABCD} - \frac{1}{2}(\text{wavelength}) = \text{integral number of wavelengths.} \qquad (4.44)$$

Noticing that $\overline{ABCD} = \overline{BCDE} = \overline{FCDE} = 2h\cos\phi$, and letting λ be the wavelength, implies Equation (4.44) becomes

$$2h\cos\phi - \tfrac{1}{2}\lambda = n\lambda, \quad n = 0, 1, 2, \ldots \qquad (4.45)$$

The wavefront AD moves from point A to point B with speed β. Hence, the apparent speed c in the x direction, which is the speed at which point G moves to point B along the boundary, is $c = \beta/\sin\phi$. This can be seen as follows:

$$c = \frac{\overline{GB}}{t_{G \to B}} = \frac{\overline{GB}}{t_{A \to B}} = \frac{\overline{GB}}{\overline{AB}/\beta} = \frac{\beta}{\sin\phi},$$

where $\overline{GB} = \overline{AB}/\sin\phi$ has been used. Hence, using $\lambda = 2\pi\beta/\omega$ and also $\cos\phi = \sqrt{1 - \sin^2\phi} = \sqrt{1 - \beta^2/c^2}$ in Equation (4.45) and solving for c yields Equation (4.40).

The problem of P waves in a layer of liquid over a rigid half-space can be solved in the same way, but instead of u, we would start with the P wave displacement potential ϕ. The same dispersion equation for $c(\omega)$ (Equation 4.40) is obtained, with β replaced by α (see Grant and West, 1965). The results of the analysis of this type of problem are of importance in the interpretation of marine seismic data.

4.7 Love Waves

In the normal mode problem treated earlier, it is more physically realistic to let the half-space be elastic, rather than perfectly rigid. The analysis can be carried out in the same way, and as before, guided SH waves, i.e., surface SH waves, are found to exist. These waves are known as **Love waves** (after the mathematician A. E. H. Love, who worked out their properties). They exist only if the S wave speed in the surface layer is *less* than that in the half-space, as will be shown in this section. The surface layer then acts as a waveguide. General Love waves are superpositions of Love wave normal modes, just as the shape of a vibrating string can be expressed as a superposition of the string's normal modes of vibration.

Let the density and velocity be ρ_1 and β_1 in the surface layer, and ρ_2 and β_2 in the half-space. From Equation (4.36), the SH particle displacements in the surface layer, $u^{(1)}$, and in the half-space, $u^{(2)}$, are given by

$$u^{(1)} = A \exp\left[i(\kappa x + \gamma_1 z - \omega t)\right] + B \exp\left[i(\kappa x - \gamma_1 z - \omega t)\right] \qquad (4.46a)$$

$$u^{(2)} = C \exp\left[i(\kappa x + \gamma_2 z - \omega t)\right] + D \exp\left[i(\kappa x - \gamma_2 z - \omega t)\right], \qquad (4.46b)$$

where $\kappa = \omega/c$ and

$$\gamma_1^2 = \frac{\omega^2}{\beta_1^2} - \kappa^2, \qquad \gamma_2^2 = \frac{\omega^2}{\beta_2^2} - \kappa^2. \qquad (4.47)$$

Normal modes in the layer are associated with sinusoidal oscillations, so $\gamma_1^2 > 0$ (as for the previous rigid half-space problem). Since $\kappa = \omega/c$, this means that $c > \beta_1$, i.e., the phase speeds of the normal modes are greater than the SH body wave speed in the layer. Also, $\gamma_1^2 > 0$ means γ_1 is either a positive or negative real number. Without loss of generality, we can take γ_1 to be positive real, meaning that the wave in the layer, $u^{(1)}$, is a superposition of a downgoing harmonic wave with amplitude A and an upgoing harmonic wave with amplitude B (assuming that z increases downward).

In the half-space, either C or D must be zero, say, $D = 0$. To see this, note that $\gamma_2{}^2$ is either positive or negative. If $\gamma_2{}^2 > 0$, then γ_2 is positive or negative real – let's say it's positive real. This means that $C \exp[\cdots]$ is a downgoing wave – it is the transmitted wave produced by the downgoing incident wave in the surface layer – and $D \exp[\cdots]$ is an upgoing wave. But an upgoing wave in the half-space cannot exist because there is no reflector in the half-space to reflect the downgoing transmitted wave upward. Consequently, $D = 0$. On the other hand, if $\gamma_2{}^2 < 0$, then γ_2 is either positive imaginary or negative imaginary. Without loss of generality, we can take γ_2 to be positive imaginary. This means that $C \exp[\cdots]$ is a wave traveling in the x direction whose amplitude *decays* with depth (an evanescent wave), but that $D \exp[\cdots]$ is a wave traveling in the x direction whose amplitude *grows* with depth, which is unphysical. So again, we set $D = 0$.

Referring to the discussion in the preceding paragraph, we could have actually ignored the first case, $\gamma_2{}^2 > 0$, involving the downgoing transmitted wave, because if wave energy is propagating downward in the half-space, then we are no longer dealing with a "surface wave" problem (for which the waves are confined to zone near the surface), which is our interest here.

The boundary conditions are that $\sigma_{yz} = 0$ at the free surface $z = 0$, and that u_y and σ_{yz} are continuous across the interface $z = h$, i.e.,

$$\mu_1 \frac{\partial u^{(1)}}{\partial z} = 0 \qquad \text{at} \quad z = 0 \tag{4.48a}$$

$$u^{(1)} = u^{(2)} \qquad \text{at} \quad z = h \tag{4.48b}$$

$$\mu_1 \frac{\partial u^{(1)}}{\partial z} = \mu_2 \frac{\partial u^{(2)}}{\partial z} \qquad \text{at} \quad z = h. \tag{4.48c}$$

Substituting Equations (4.46) (with $D = 0$) into these conditions gives

$$A - B = 0 \tag{4.49a}$$

$$A \exp(i\gamma_1 h) + B \exp(-i\gamma_1 h) = C \exp(i\gamma_2 h) \tag{4.49b}$$

$$\mu_1 \gamma_1 A \exp(i\gamma_1 h) - \mu_1 \gamma_1 B \exp(-i\gamma_1 h) = \mu_2 \gamma_2 C \exp(i\gamma_2 h). \tag{4.49c}$$

The first equation implies $B = A$, and the last two then become

$$2A \cos(\gamma_1 h) = C \exp(i\gamma_2 h) \tag{4.50a}$$

$$2iA\mu_1 \gamma_1 \sin(\gamma_1 h) = \mu_2 \gamma_2 C \exp(i\gamma_2 h). \tag{4.50b}$$

The nontrivial solution ($A, C \neq 0$) can be obtained by setting the determinant of the coefficient matrix equal to zero (see the discussion on Equation 4.4). Or, since this is only a simple 2×2 system of equations, we can also just solve them directly quite easily. This gives

$$\tan(\gamma_1 h) = -\frac{i\mu_2 \gamma_2}{\mu_1 \gamma_1}. \tag{4.51}$$

Hence, since γ_1 is positive real, for solutions of Equation (4.51) to exist, γ_2 must be positive imaginary (e.g., $\gamma_2 = +i\nu$, where ν is a positive real number), confirming what we discovered in the paragraph in which we showed that $D = 0$. Substituting the formulas for γ_1 and γ_2 into Equation (4.51) gives

$$\tan\left[\omega\sqrt{\frac{1}{\beta_1^2} - \frac{1}{c^2}}\, h\right] = \frac{\mu_2\sqrt{\dfrac{1}{c^2} - \dfrac{1}{\beta_2^2}}}{\mu_1\sqrt{\dfrac{1}{\beta_1^2} - \dfrac{1}{c^2}}}. \tag{4.52a}$$

This equation is analogous to Equation (4.39) in the case of the rigid half-space. It is the **Love wave dispersion relation**. The quantities under the square roots must be positive. This means that Love waves exist for $\beta_1 < c < \beta_2$, i.e., β_1 must be less than β_2, and the Love wave speed c is between β_1 and β_2. Note also that γ_2 being positive imaginary means that $u^{(2)} = C\exp[\cdots]$ in Equation (4.46b) is an evanescent wave (it travels in the x direction along the interface $z = h$ and its amplitude decays with depth z in the half-space). This is consistent with the physical behavior expected of guided normal mode surface waves – they are essentially confined to a surface layer, or guided by it in the lateral direction.

Equation (4.52a) must be solved numerically for c (it is a transcendental equation). If, in (4.52a), we let

$$X = X(c) = \sqrt{\frac{1}{\beta_1^2} - \frac{1}{c^2}}\, h, \tag{4.52b}$$

then (4.52a) can be written as

$$\tan(\omega X) = \left(\frac{h\mu_2}{\mu_1 X}\right)\sqrt{\frac{1}{\beta_1^2} - \frac{1}{\beta_2^2} - \frac{X^2}{h^2}}. \tag{4.52c}$$

Consider Figure 4.12, which is a graphical representation of Equation (4.52c). Noting that the nth $\tan(\omega X)$ branch crosses the X axis at $X_n = n\pi/\omega$, we see that as $\omega \to 0$, the branches $n = 1, 2, \ldots$ move rightward out to infinity, i.e., outside the window defined by the vertical dashed lines, and the $n = 0$ branch flattens out and approaches the X axis, coinciding with it for $\omega = 0$. Hence, for $\omega = 0$, there is only one root, which is on the $n = 0$ branch. As ω gets larger, the $n = 1$ branch crosses the $c = \beta_2$ dashed line from the right. It crosses when $\pi/\omega = X(\beta_2)$, i.e., when $\pi/\omega = h\sqrt{(1/\beta_1^2) - (1/\beta_2^2)}$. Solving this for ω shows that the $n = 1$ mode

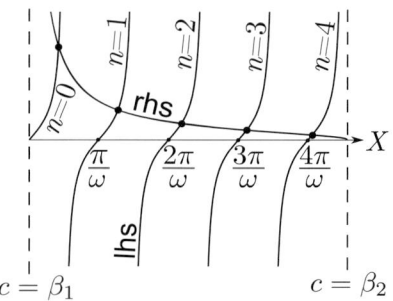

Figure 4.12 A graphical study of Equation (4.52c). "lhs" and "rhs" denote the left-hand side and right-hand side of (4.52c). The heavy dots on the graph are the roots of the equation, i.e., the values of c that solve (4.52a).

can exist if $\omega \geq \omega_{c1} = \pi\beta_1/h\sqrt{1 - (\beta_1/\beta_2)^2}$. As ω increases, more branches of $\tan(\omega X)$ will enter the picture. The cutoff frequency for the nth mode is

$$\omega_{cn} = \frac{n\pi\beta_1}{h\sqrt{1 - (\beta_1/\beta_2)^2}}. \tag{4.53}$$

This equation can also be derived by putting $c = \beta_2$ into Equation (4.52a), since that is the value of c when the nth branch crosses the $c = \beta_2$ dashed line. This gives $\tan[\omega X(\beta_2)] = 0$, meaning $\omega X(\beta_2) = n\pi$, $n = 0, 1, 2, \ldots$, which gives the cutoff frequencies in (4.53) when solved for ω.

Note also that as ω gets larger, the root for the nth branch moves leftward, meaning that c decreases as ω increases (as we'll see in Figure 4.13).

For an example of a typical numerical value of a cutoff frequency, let $\beta_1 = 500$ m/s, $\beta_2 = 800$ m/s, and $h = 20$ m. Then $f_{c1} = 16$ Hz. Note that the fundamental mode ($n = 0$) has *no* cutoff frequency. If $c_n = c_n(\omega)$ is the phase velocity for the nth mode, then $c_n(\omega_{cn}) = \beta_2$. As $\omega \to \infty$, $c_n \to \beta_1$ (for all n). The group velocity c_g can be obtained from Equation (4.31), or from a finite difference calculation – see Equation (4.60c) later in this chapter. Figure 4.13 shows the variation of c and c_g with ω for two modes.

The particle displacement can be determined from Equations (4.46). For the motion in the surface layer, we replace B with A in the equation for $u^{(1)}$, and take the real part to obtain

$$u^{(1)} = 2A \cos\left[\omega\sqrt{\frac{1}{\beta_1^2} - \frac{1}{c_n^2}} z\right] \cos(\kappa x - \omega t). \tag{4.54}$$

For the motion in the half-space, we note that Equation (4.50a) gives

$$C/A = 2\exp(-i\gamma_2 h)\cos(\gamma_1 h).$$

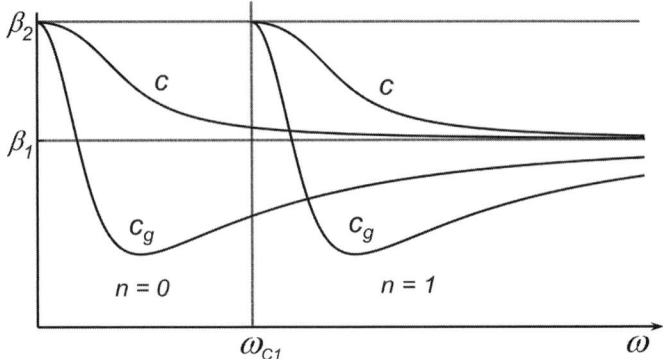

Figure 4.13   A plot of c and c_g versus ω for Love waves.

Using this (and $D = 0$) in the equation for $u^{(2)}$, and taking the real part, remembering that γ_2 is positive imaginary, gives

$$u^{(2)} = 2A \cos\left[\omega \sqrt{\frac{1}{\beta_1^2} - \frac{1}{c_n^2}}\, h\right] \exp\left[-\omega \sqrt{\frac{1}{c_n^2} - \frac{1}{\beta_2^2}}\, (z - h)\right] \cos(\kappa x - \omega t).$$

(4.55)

$u^{(1)}$ shows a sinusoidal depth variation in the particle motion in the layer, and $u^{(2)}$ shows an exponential decay with depth (meaning that it is an evanescent wave). The variation with depth of the particle motion depends on the frequency ω, unlike the case for the rigid half-space. Figure 4.14 shows some examples of Love wave particle motion. The motions in Figure 4.14 were determined from Equations (4.54) and (4.55), Figures 4.12 and 4.13, and Equation (4.53). Note that the form of (4.54) shows that the Love wave normal modes are traveling waves in the x direction and *standing waves* in the z direction. Equation (4.55) shows that $u^{(2)} \to 0$ as $\omega \to \infty$ (the exponential decay in the half-space vanishes at $\omega \to \infty$, resulting in an exact $\frac{1}{4}$-cosine in the layer for $n = 0$ and an exact $\frac{3}{4}$-cosine in the layer for $n = 1$). Note the similarities and differences between the particle motions for Love wave normal modes and those for the case of a rigid half-space in Figure 4.10.

     From a mathematical viewpoint, the determination of c_n, $u^{(1)}$ and $u^{(2)}$ for Love waves is an **eigenvalue** problem. The differential equation (4.34) can be written as $LU = \lambda U$, where L is the differential operator $d^2/dz^2 + (\omega^2/\beta^2)$, and $\lambda = \kappa_n^2 = \omega^2/c_n^2 = \lambda_n$. The numbers λ_n are the **eigenvalues** (or **characteristic values**), and $U = U_n$ are the **eigenfunctions** (or **characteristic functions**). In linear algebra, L may be a matrix representing a linear transformation (i.e., a rotation), λ_n are the eigenvalues that satisfy the equation $LU = \lambda U$, and U_n are the eigenvectors (column vectors). For our Love wave problem, the numbers κ_n^2 are the eigenvalues, and $u^{(1)}$ and $u^{(2)}$ in Equations (4.54) and (4.55) are the eigenfunctions.

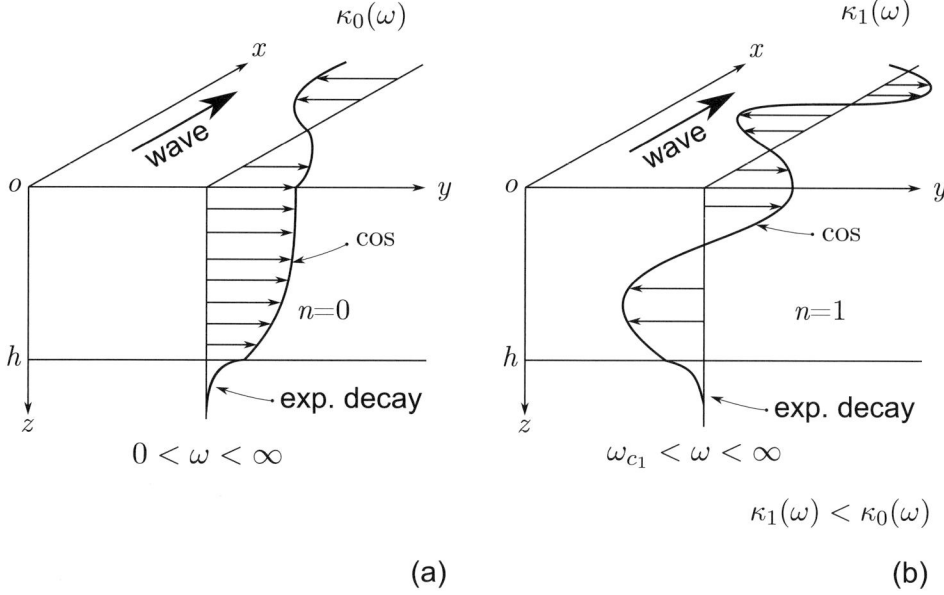

Figure 4.14 *SH* particle motion for (a) the fundamental Love wave mode $n = 0$, and (b) the Love wave mode $n = 1$. Both figures were drawn using Equations (4.54) and (4.55) for $t = 0$.

As mentioned previously, the surface layer acts as a waveguide. Love waves are produced by the constructive interference of internally reflected waves. Since the wave in the half-space is evanescent, the internal reflection in the layer is total, i.e., the waves in the layer propagate at supercritical incident angles ($\phi > \phi_c = \sin^{-1}(\beta_1/\beta_2)$ in Figure 4.11). This is also trivially true in the case of the rigid half-space normal mode problem: in that problem, $\beta_1 = \beta$ and $\mu_2 = \infty \Rightarrow \beta_2 = \infty$, hence $\phi_c = 0$, i.e., *all* incident angles are supercritical in that case. For Love waves, the dispersion equation (4.52a) can be derived from an argument based on the constructive interference of wavefronts, like the one connected with Figure 4.11 in the case of a rigid half-space. However, the phase change at the solid–solid interface (at point *B* in Figure 4.11) must be calculated from the *SH-SH* reflection coefficient for supercritical incidence angles (Equation 3.37).

As a point of interest, Love waves have been used in exploration seismology to estimate static corrections for shear wave profiling – see Mari (1984).

Examples of guided waves in real data can be found in Yilmaz (1987). See also the prominent dispersive events on Figures 7.12 and 7.13. On a shot record, guided waves tend to be linear events emanating from a source point, like the ground roll line in the shot record in Figure 6.2.

4.8 The Airy Phase

Figure 4.13 has an interesting feature, namely, the existence of a minimum for the group velocity. This has some interesting consequences. Imagine an *SH* source in the surface layer (or half-space) and a receiver at a large distance x. What kind of response can we expect at the receiver? First, remember that the group velocity at a given frequency ω_0 is the average speed of a pulse or wave group with average frequency ω_0. The first arrival at the receiver would be the head wave ($S_1 S_2 S_1$). Then there would come a dispersive signal from the left sides of the group velocity curves. If most of the energy is in the fundamental mode, then this would be a low-frequency signal, with frequency steadily increasing with time until the time $t = x/\beta_1$ is reached. At this time, a high-frequency signal (from the right end of the c_g curve) arrives, and we have a high-frequency "rider" superimposed on the low-frequency signal. As time passes, the frequency of the rider rapidly decreases, and the two waves merge into a single high-amplitude wave pattern at the group velocity minimum, which then abruptly terminates. This final single wave pattern is known as the **Airy phase** (see Figure 4.15).

The Airy phase has a high amplitude because a wave group is *strongly* coherent if its (narrow) frequency band is located at the frequency corresponding to the group velocity minimum, i.e., the Fourier components of the wave group are much more in phase at this frequency than at others. To see this, let the group velocity minimum be at $\omega = \omega_0$ (this is the average frequency of the Airy phase pulse). The phase factor $\eta = \kappa(\omega)x - \omega t$ can be expanded about ω_0 using a Taylor series:

$$\eta(\omega) = \eta(\omega_0) + \left(\frac{d\eta}{d\omega}\right)_0 (\omega - \omega_0) + \frac{1}{2}\left(\frac{d^2\eta}{d\omega^2}\right)_0 (\omega - \omega_0)^2 + \frac{1}{6}\left(\frac{d^3\eta}{d\omega^3}\right)_0 (\omega - \omega_0)^3 + \cdots$$

$$(4.56)$$

where the subscript "0" on the derivatives means that they are to be evaluated at $\omega = \omega_0$. The wave group consisting of frequencies near ω_0 will arrive last because

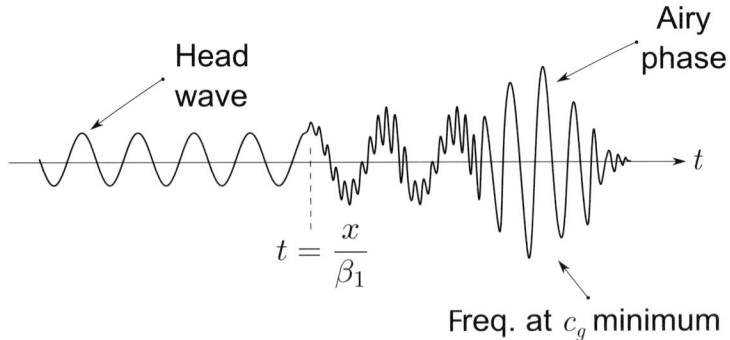

Figure 4.15 A schematic diagram of the Airy phase.

it has the lowest group velocity. From Equation (4.30a), we have $(d\eta/d\omega)_0 = 0$. Hence, the phase is constant *to first order* in $\omega - \omega_0$. However, at the group velocity minimum, we have $(d^2\eta/d\omega^2)_0 = 0$ as well, which means that the phase η is constant *to second order* in $\omega - \omega_0$, making the wave group much more coherent. To prove that $(d^2\eta/d\omega^2)_0 = 0$, note that

$$\frac{d}{d\omega}\left[\frac{1}{c_g}\right] = \frac{d}{d\omega}\left[\frac{d\kappa}{d\omega}\right] \implies -\frac{1}{c_g{}^2}\frac{dc_g}{d\omega} = \frac{d^2\kappa}{d\omega^2}.$$

Therefore, at $\omega = \omega_0$, where c_g is a minimum,

$$0 = \left(\frac{dc_g}{d\omega}\right)_0 = \left(-c_g^2\frac{d^2\kappa}{d\omega^2}\right)_0, \text{ i.e. } \left(\frac{d^2\kappa}{d\omega^2}\right)_0 = 0$$

$$\implies \left(\frac{d^2\eta}{d\omega^2}\right)_0 = \left(\frac{d^2\kappa}{d\omega^2}\right)_0 x = 0. \tag{4.57}$$

4.9 Other Cases of Normal Modes

Seismograms of the type shown in Figure 4.15 have been observed in marine seismic data. They were first interpreted by Pekeris (1948), who worked out the normal mode problem for a liquid layer over a liquid half-space, which is very similar to the Love wave problem. His theoretical seismograms matched the observed ones (Figure 4.15) very well. The fact that the results for a liquid half-space (rather than a solid one) matched the data well is consistent with the unconsolidated nature of the water bottom substratum. The problem of normal modes for a liquid layer over a solid half-space can also be worked out, although it is more complicated. Similar results are found.

The normal mode problem for *P-SV* waves in a solid surface layer over a solid half-space can also be worked out. The dispersion curves (c and c_g versus ω) are more complicated. The lowest mode is known as the **Rayleigh mode** because at the low frequency limit the phase velocity approaches that of Rayleigh waves in the half-space, and at the high frequency limit the phase velocity approaches that of Rayleigh waves in the surface layer. Rayleigh waves for a vertically inhomogeneous medium (such as a layer over a half-space) are dispersive, whereas for a homogeneous medium (such as a homogeneous half-space) they are not dispersive. The ground roll observed on a seismic shot record is identified with the Rayleigh mode.

Another feature of the *P-SV* normal mode problem is that, for the higher modes, the group velocity curves can have *several* minima, and several *maxima* as well, resulting in several Airy phases.

4.10 Measurement of Phase and Group Velocities

To measure the phase velocity, we assume that a wave group (or wavelet or pulse) w traveling in the x direction can be expressed as a superposition of sinusoids (harmonic waves) with different frequencies, amplitudes, and phases:

$$w(x,t) \;=\; \frac{1}{2\pi} \int_{-\infty}^{\infty} W(\omega) \exp\!\left[i\omega\!\left(\frac{x}{c(\omega)} - t \right) \right] d\omega. \qquad (4.58)$$

This could represent a horizontally traveling wave group, such as a surface wave, observed as a wavelet at the source–receiver offset x and time t on a seismic record, with $w(x,t)$ being the mathematical form of the wavelet. $c(\omega)$ is the phase velocity of the harmonic wave with frequency ω in the wavelet.

Note that Equation (4.58) has the form of an inverse Fourier transform. This means that the quantity $W(\omega)\exp[i\omega x/c(\omega)]$ in the integrand is the frequency spectrum of the wavelet $w(x,t)$ (i.e., its Fourier time transform). $W(\omega)$ is then the frequency spectrum of the wavelet $w(0,t)$, the source pulse. Or, more generally, $W(\omega)$ is the frequency spectrum of the wavelet at $x = 0$. $W(\omega)$ is generally a complex number. If we write $W(\omega) = |W(\omega)| \exp[i\theta(\omega)]$, then $|W(\omega)|$ is the amplitude of the harmonic wave with frequency ω in the source pulse, and $\theta(\omega)$ is its phase.

Note that the frequency spectrum of the pulse at x can be obtained by simply multiplying the frequency spectrum of the pulse at $x = 0$ (the source pulse) by the phase factor $\exp[i\omega x/c(\omega)]$. Also, the frequency spectrum of the pulse at x can be written as

$$\text{spectrum of } w(x,t) \;=\; |W(\omega)| \exp\!\big[i\phi(\omega) \big] \quad \text{where} \quad \phi(\omega) = \frac{\omega x}{c(\omega)} + \theta(\omega).$$
$$(4.59)$$

$|W(\omega)|$ is the frequency-amplitude spectrum of $w(x,t)$ and $\phi(\omega)$ is its frequency-phase spectrum. Note that the phase at x is just the phase at $x = 0$ plus $\omega x/c(\omega)$.

Suppose the traveling wave group is recorded at two different distances, x_1 and x_2. The phase velocity $c(\omega)$ can then be determined by measuring, at every frequency, the difference between the phase spectra, ϕ_1 and ϕ_2, of the two recorded waveforms. From (4.59), this difference is

$$\phi_2(\omega) - \phi_1(\omega) \;=\; \frac{\omega}{c(\omega)}\left(x_2 - x_1 \right) \pm 2n\pi, \qquad (4.60a)$$

where n is an integer. It is necessary to add the term $\pm 2n\pi$ used in Equation (4.60a) because the phase is periodic, with a period of 2π radians (or $360°$). In other words, a measured phase difference of ϵ could actually be a physical phase difference of $\epsilon \pm 2n\pi$. If the two receivers at x_1 and x_2 are close enough to each other, then $n = 0$.

But if $n \neq 0$, and n is not known from a prior general knowledge of $c(\omega)$, then n can be obtained by using more than two distances. Solving for $c(\omega)$ gives

$$c(\omega) = \frac{\omega(x_2 - x_1)}{\phi_2(\omega) - \phi_1(\omega) \pm 2n\pi}. \tag{4.60b}$$

This method assumes that the wave group represents a single wave type (e.g., the fundamental mode of a surface wave).

Once the phase velocity is known, the group velocity c_g can be determined from Equation (4.31). Or, the group velocity could presumably also be determined in the following way. Once $c(\omega)$ is known, the wavenumber $\kappa(\omega) = \omega/c(\omega)$ can be computed. c_g could then be numerically computed using a finite difference formula. For example, if $\omega \pm \frac{1}{2}\Delta\omega$ are two closely spaced frequency values, then c_g at ω (halfway between them) could be computed from the central-difference formula

$$c_g(\omega) = \frac{\Delta\omega}{\Delta\kappa} = \frac{\Delta\omega}{\kappa(\omega + \frac{1}{2}\Delta\omega) - \kappa(\omega - \frac{1}{2}\Delta\omega)} \implies$$
$$= \Delta\omega \Big/ \left[\frac{\omega + \frac{1}{2}\Delta\omega}{c(\omega + \frac{1}{2}\Delta\omega)} - \frac{\omega - \frac{1}{2}\Delta\omega}{c(\omega - \frac{1}{2}\Delta\omega)} \right]. \tag{4.60c}$$

The presence of noise reduces the accuracy of the $c(\omega)$ determination. However, the accuracy can be improved by decreasing the value of the parameter $\lambda/(x_2 - x_1)$, where λ is the wavelength, i.e., by increasing the distance $x_2 - x_1$ between the two receivers so that it contains many wavelengths. This is like increasing the resolving power of an optical device by increasing the aperture width so that it is much larger than a wavelength.

Sometimes the group and phase velocities can be estimated directly from the recorded data without using Fourier analysis. This has often been done for surface wave trains. The basic ideas are described in the following paragraphs.

For the group velocity, we have $x = c_g(\omega)t$, where ω is the average frequency of the surface wave group that has arrived at x in a time t. On a shot record of the surface wave train, with the x axis pointing upward and the t axis pointing to the right, $x = c_g(\omega)t$ is a straight line with slope $c_g(\omega)$ emanating from the origin, and at every point where this line crosses a trace of the surface wave signal, the average frequency is ω. So, if we simply draw a straight line from the origin through the surface wave signal, we can obtain c_g by measuring the slope of the line and ω by measuring the frequency of the signal at any point where the line crosses a trace (or we could take an average at several crossing points). This gives us a single point on the c_g vs. ω curve. To get another point on the curve, we draw another line with a different slope, and repeat the process, and so on. We also need to take care that

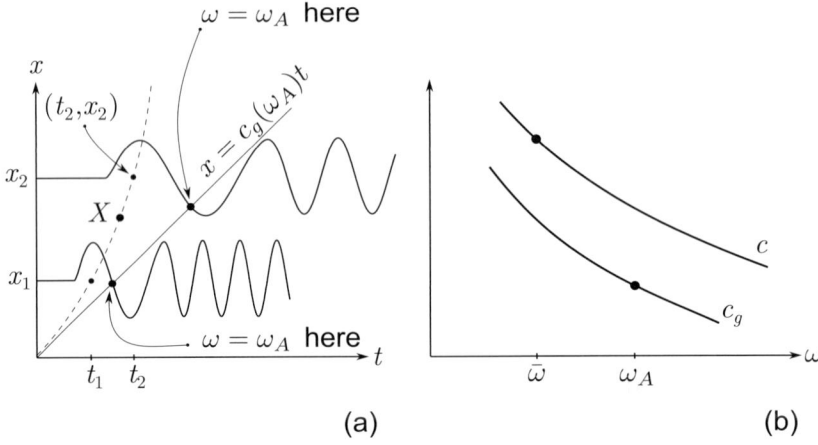

Figure 4.16 Measuring the group and phase velocity for a surface wave train. Schematic diagrams of (a) two surface wave traces from the shot record, and (b) the c_g-ω and c-ω curves obtained from the measurements. Note that c and c_g decrease as ω increases, meaning that the lower frequencies arrive earlier, as seen in part (a) for each trace.

the straight lines pass through the same wave type (e.g., the same mode of a surface wave). Figure 4.16 illustrates the procedure for two adjacent traces.

For the phase velocity, we know that a harmonic wave's phase must remain constant on a given wavefront as it propagates, i.e., for a harmonic wave of frequency $\bar{\omega}$, we have from (4.58)

$$\frac{x_1}{c(\bar{\omega})} - t_1 = \frac{x_2}{c(\bar{\omega})} - t_2 \implies c(\bar{\omega}) = \frac{x_2 - x_1}{t_2 - t_1}. \tag{4.61}$$

As an example of how to apply this, refer again to Figure 4.16. On the trace at x_1, we pick a wavefront, e.g., the peak of one of the cycles (the first peak in Figure 4.16), follow it to the adjacent trace at x_2, measure the times t_1 and t_2, and compute $c(\bar{\omega})$ from (4.61). Then we measure the frequencies ω_1 and ω_2 at the points (t_1, x_1) and (t_2, x_2). For adjacent traces, x_1 and x_2 are close in value, meaning ω_1 and ω_2 would also be close in value, and one could let $\bar{\omega}$ be either ω_1 or ω_2, or perhaps their average. This gives us a single point, $[\bar{\omega}, c(\bar{\omega})]$, on the c vs. ω curve. Note that $c(\bar{\omega})$ in (4.61) is basically the slope dx/dt of the dashed line in Figure 4.16 at the point X (i.e., at the time $(t_1 + t_2)/2$). One could measure this slope at other points along the dashed line as well, to obtain more points on the c vs. ω curve. Also, instead of following a peak or a trough in a signal from trace to trace, it may be easier to follow a small distortion, or "kink" or "glitch" in the signal (if one exists) from trace to trace.

These direct methods for determining c and c_g are not always applicable, and they are generally not as accurate as Fourier analysis. For other discussions of group and phase velocity measurements, see, e.g., Officer (1974, p. 245), Lay and Wallace (1995, pp. 140–147), and Stein and Wysession (2003, pp. 96–99).

When making phase and group velocity measurements on Rayleigh waves (ground roll), it is sometimes necessary to use geophones that measure particle motion in the x direction as well as geophones that measure z-motion. Then, a given sinusoidal-type oscillation can be positively identified as a Rayleigh wave if the oscillation on the x-trace lags the corresponding oscillation on the z-trace by 90°, showing retrograde elliptical motion. Dobrin, Simon, and Lawrence (1951) and Dobrin, Lawrence, and Sengbush (1954) measured c and c_g for Rayleigh waves from seismic data obtained from explosive charges. Their results are in fairly good agreement with theory. They also measured motion in the y direction (they used three-component detectors) and found that strong *SH* motion was generated, even though none is expected if the disturbance is purely compressional and the subsurface is perfectly horizontally stratified. The existence of *SH* motion was likely due to inhomogeneous conditions near the shot.

4.11 Air-Coupled Ground Roll

The ground roll observed on seismic records generated by a buried explosive charge usually consists of a dispersed train of Rayleigh waves. However, if the explosive charge is in the air, the ground roll on the air shot records consists essentially of a *constant* frequency train of waves (Ewing, Jardetzky, and Press, 1957, p. 236). The constant frequency is identical to the frequency at which the phase velocity of the dispersed Rayleigh waves (from the buried charge) is equal to the speed of sound in

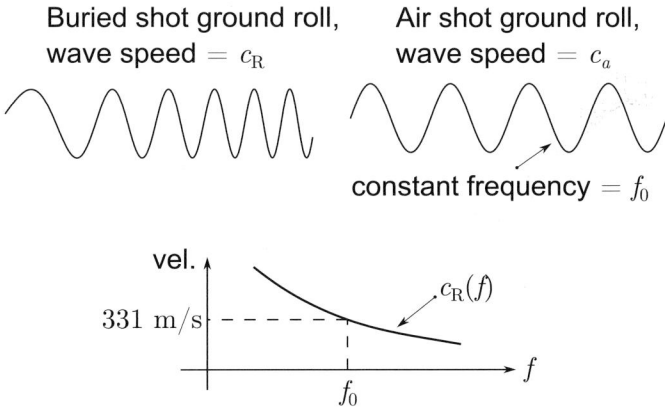

Figure 4.17 Air-coupled ground roll.

air (331 m/s). The phase velocity of the constant frequency wave train remains close to the speed of sound in air for several cycles – see Figure 4.17. The experimental results agree with theory.

Exercises

1. Consider a 10 Hz Rayleigh wave propagating along the surface of a homogeneous half-space. The compressional and shear wave speeds in the half-space are 2000 m/s and 1070 m/s respectively. Calculate the depth z_0 of the nodal plane. What fraction of the Rayleigh wavelength is it?

2. For the incident P wave shown in Figure 4.18, calculate all the critical angles, and sketch a wavefront diagram showing all the head waves generated.

3. Calculate the formula for the group velocity c_g, which is plotted in Figure 4.9b. Use $1/c_g = d\kappa/d\omega$. Use your formula to verify that the plot is correct.

4. Repeat the SH wave normal mode analysis in Section 4.6 for the case in which *both* interfaces ($z = 0$ and $z = h$) are free surfaces (e.g., a slab floating in a vacuum). This case is not very geophysically realistic, but it is instructive.

5. A waveform w at offset x and time t on a seismic section can be represented mathematically as a continuous superposition of sinusoids, i.e.,

$$w(x, t) = \frac{1}{2\pi} \int_{-\infty}^{\infty} W(\omega)\, e^{-i\omega\left[t - x/c(\omega)\right]}\, d\omega,$$

where $W(\omega)$ is the frequency spectrum of the source pulse $w(0, t)$ and $\omega = 2\pi f$, where f is the frequency. The two waveforms, 1 and 2, shown on the seismic section in Figure 4.19, represent the same type of wave. The waveforms are band-limited – they contain no frequencies lower than 5 Hz or higher than 15 Hz. The phase spectra of the two waveforms are shown in

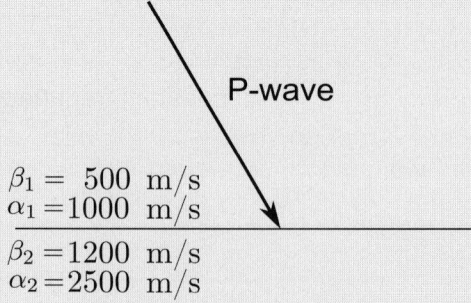

Figure 4.18 See exercise 2.

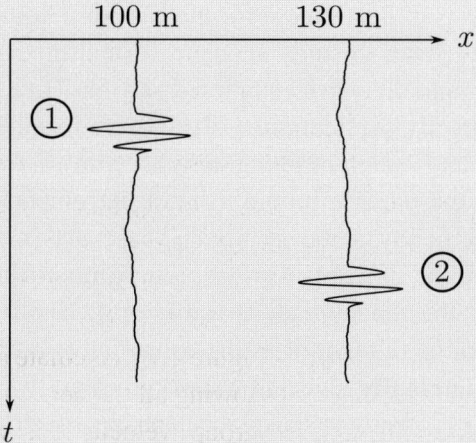

Figure 4.19 See exercise 5.

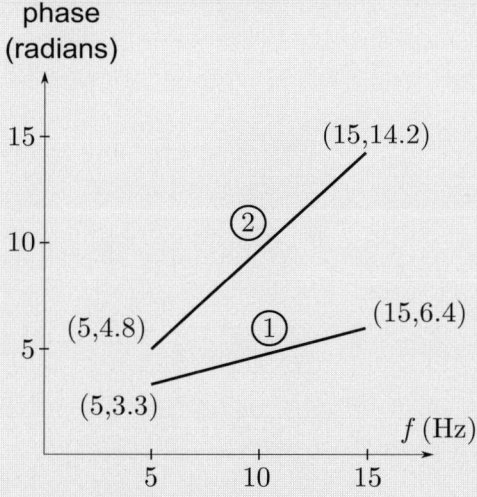

Figure 4.20 See exercise 5.

Figure 4.20. These phase spectra are "unwrapped" (for example, a phase of 370° would be plotted as 370°, not as 10°). Compute and sketch a graph of the phase velocity c and group velocity c_g as functions of the frequency f. The group velocity is given by

$$c_g(\omega) = c(\omega) \bigg/ \left(1 - \frac{\omega}{c}\frac{dc}{d\omega}\right).$$

6. Repeat the *SH* wave normal mode analysis in Section 4.6 for the case of a solid elastic layer sandwiched between two rigid half-spaces.

7. Repeat the analysis in Section 4.6 for the case of a liquid layer (e.g., water) over a rigid half-space, to study acoustic normal modes. Begin with the *P* wave potential ϕ and the wave equation it satisfies (like Equation 4.32), and remember that the displacement $\mathbf{u} = \nabla\phi$. Derive formulas for the phase and group velocities and sketch or plot graphs of them vs. frequency ω. Deduce, mathematically, the particle motion (the displacement) at all points in the liquid layer (specifically, at the surface and base, and inside the liquid layer). Sketch or plot graphs showing how the particle motion and the pressure vary with depth z for the modes $n = 0$ and $n = 1$. Show how these acoustic guided waves can be thought of as the constructive interference of upgoing and downgoing internally reflected waves.

8. Repeat exercise 7 using a displacement approach, rather than using potentials.

5

Waves in Heterogeneous Media

5.1 Waves in a Vertically Inhomogeneous Medium

A vertically inhomogeneous (or heterogeneous) medium is one for which the velocity v of a wave, the density ρ, etc., vary with depth z, but not laterally (i.e., with x). The depth variation can be discrete, such as in a medium consisting of a sequence of homogeneous flat horizontal layers, or continuous, such as in a medium in which v, ρ, are continuous functions of z, or both, such as in a horizontally layered medium with vertically inhomogeneous layers.

Consider a subsurface model consisting of a sequence of flat horizontal homogeneous layers. Suppose there is a source and receiver on the surface separated by the offset x, and consider a primary ray going from source to receiver that reflects off the base of layer n (see Figure 5.1a). Suppose the ray segments are all of the same type, i.e., either all P or all S. Let v_j be the velocity in layer j (i.e., the interval velocity), h_j be the thickness of layer j, and $\tilde{t}_j \equiv 2h_j/v_j$ be the two-way vertical travel time in layer j. Let $t = t_n(x)$ be the total two-way travel time of the primary ray. Note that $t_n(0) = \tilde{t}_1 + \cdots + \tilde{t}_n$.

The **average velocity** v_a to the base of layer n is the total vertical distance divided by the total one-way vertical travel time, i.e.,

$$v_a = v_{an} = \frac{h_1 + \cdots + h_n}{\frac{1}{2}(\tilde{t}_1 + \cdots + \tilde{t}_n)} = \frac{v_1\tilde{t}_1 + \cdots + v_n\tilde{t}_n}{\tilde{t}_1 + \cdots + \tilde{t}_n}. \tag{5.1}$$

The average velocity is a weighted average of the numbers $v_1, \cdots, v_n$, with the weights being $\tilde{t}_1, \cdots, \tilde{t}_n$. The **root-mean-square** (RMS) **velocity** $\bar{v}$ to the base of layer n is

$$\bar{v} = \bar{v}_n = \sqrt{\frac{v_1^2\tilde{t}_1 + \cdots + v_n^2\tilde{t}_n}{\tilde{t}_1 + \cdots + \tilde{t}_n}}. \tag{5.2}$$

Let us determine the travel time t and the relationship between t and x. First, for a single layer of thickness h and velocity v, we have for the primary ray

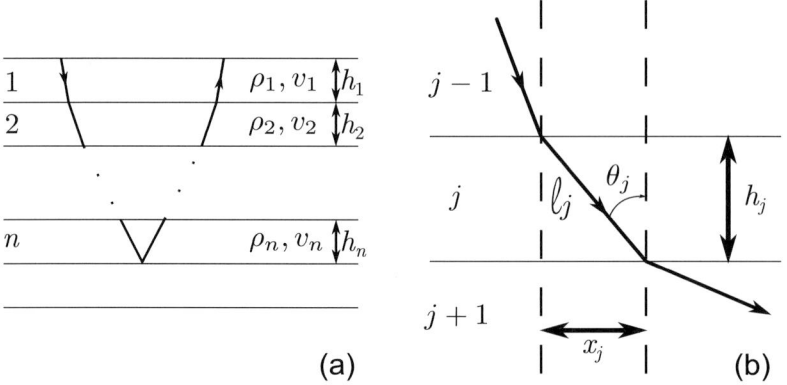

Figure 5.1 (a) The subsurface model, and (b) the jth layer.

$$t^2 = t(0)^2 + \frac{x^2}{v^2}, \qquad t(0) = \frac{2h}{v}. \tag{5.3}$$

We also have $v_a = \bar{v} = v$, and the normal moveout (NMO) is

$$t(x) - t(0) = \sqrt{t(0)^2 + \frac{x^2}{v^2}} - t(0) \approx \frac{x^2}{2v^2 t(0)} \quad \text{if} \quad x \ll vt(0), \tag{5.4}$$

where we have used the approximation

$$\sqrt{1+\epsilon} \approx 1 + \tfrac{1}{2}\epsilon \quad \text{for} \quad |\epsilon| \ll 1. \tag{5.5}$$

For n layers, we have for the primary ray (refer to Figure 5.1b)

$$t = 2\sum_{j=1}^{n} \frac{l_j}{v_j} = 2\sum_{j=1}^{n} \frac{h_j}{v_j \cos\theta_j} = 2\sum_{j=1}^{n} \frac{h_j}{v_j\sqrt{1-\sin^2\theta_j}} = 2\sum_{j=1}^{n} \frac{h_j}{v_j\sqrt{1-p^2 v_j^2}}, \tag{5.6a}$$

where Snell's law (3.27) has been used, i.e., $p = (\sin\theta_j)/v_j$, $j = 1,\ldots,n$. The two-way time t can also be written as

$$t = \tau + px, \qquad \text{where} \qquad \tau = \sum_{j=1}^{n} \tau_j = \sum_{j=1}^{n} \frac{2h_j}{v_j}\sqrt{1-p^2 v_j^2}. \tag{5.6b}$$

To prove this, substitute (5.7) for x into $t = \tau + px$ and do the math to get (5.6a).

Similarly, for the offset x, we have

$$x = 2\sum_{j=1}^{n} x_j = 2\sum_{j=1}^{n} h_j \tan\theta_j = 2\sum_{j=1}^{n} h_j \frac{\sin\theta_j}{\cos\theta_j}$$

$$= 2\sum_{j=1}^{n} h_j \frac{\sin\theta_j}{\sqrt{1-\sin^2\theta_j}} = 2\sum_{j=1}^{n} \frac{h_j v_j p}{\sqrt{1-p^2 v_j^2}}. \tag{5.7}$$

Equations (5.6) and (5.7) represent the equation $t = t(x)$ expressed in parametric form, with p being the parameter, i.e., $t = t(p)$ and $x = x(p)$. Note that these equations can be used to trace rays for the subsurface model in Figure 5.1a, assuming the v_j and h_j are known, without doing any "trial-and-error ray shooting." For a given offset x, Equation (5.7) is first solved for the ray parameter p (by using, say, the Newton–Raphson method). Then, this value of p is used in (5.6) to compute the travel time for the ray. As a practical point, when solving (5.7), it would be better, from a numerical viewpoint, to replace p with $(\sin\theta_1)/v_1$ in (5.7) and solve for $\sin\theta_1$ (from which p can then be computed) because $\sin\theta_1$ lies between 0 and 1.

As an example, suppose $n = 4$, $\{h_1, h_2, h_3, h_4\} = \{0.1, 0.2, 0.3, 0.4\}$ km, $\{v_1, v_2, v_3, v_4\} = \{1, 2, 3, 4\}$ km/s, and $x = 1$ km. Solving (5.7) gives $p = 0.143946$, and substituting this p value into (5.6a) gives $t = 0.8773$ s. In addition, if desired, the ray angles can be calculated from Snell's law $p = (\sin\theta_j)/v_j$ – one obtains $\{\theta_1, \theta_2, \theta_3, \theta_4\} = \{8.276, 16.732, 25.584, 35.155\}°$.

An important relationship is

$$p = \frac{dt}{dx}, \tag{5.8}$$

i.e., on a shot record, p is the value of the slope at x of the primary reflection curve $t = t(x)$. This can be seen from Figure 5.2. As $\Delta x \to 0$, ray segments l and l' in Figure 5.2b become more and more parallel. We have, referring to Figure 5.2b,

$$p = \frac{\sin\theta_1}{v_1} = \frac{L/\Delta x}{v_1} = \frac{v_1\Delta t}{v_1\Delta x} = \frac{\Delta t}{\Delta x} \longrightarrow \frac{dt}{dx} \quad \text{as} \quad \Delta x \to 0. \tag{5.9}$$

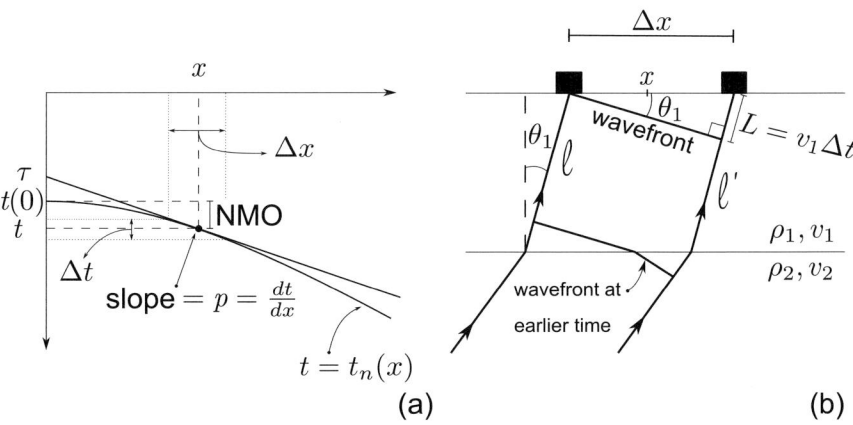

Figure 5.2 (a) The slope of a primary reflection t-x curve on a shot record, and (b) arriving rays for small Δx.

Equation (5.8) can also be derived more formally by writing

$$\frac{dt}{dx} = \frac{dt}{dp}\frac{dp}{dx} = \frac{dt}{dp} \bigg/ \frac{dx}{dp}, \tag{5.10}$$

substituting the p-derivatives of Equations (5.6a) and (5.7), and doing the math.

Note that p increases with x (because θ_1 increases with x), and that $p = 0$ for $x = 0$ (normal incidence, i.e., a vertically traveling wave in Figure 5.1a).

The **apparent velocity** c is the speed at which the wave *appears* to travel along the surface from the left geophone in Figure 5.2b to the right geophone, and is given by

$$c = \frac{\Delta x}{\Delta t} = \frac{v_1}{\sin\theta_1} = \frac{1}{p}. \tag{5.11}$$

See Appendix 2A of Chapter 2 for a more detailed discussion of apparent velocities. The ray parameter p is sometimes called the **horizontal slowness** (short for "the horizontal component of the slowness vector"). Note that if the wave front is normally incident upon the surface, then $c = \infty$ (since $\theta_1 = 0$).

Equation (5.3), which applies to the one-layer model, is an exact hyperbola. For n layers though, the t-x equation is no longer an exact hyperbola. It can be written as

$$t^2 = t(0)^2 + b_1 x + b_2 x^2 + b_3 x^3 + b_4 x^4 + \cdots. \tag{5.12}$$

For horizontal layers (Figure 5.1a), we have $b_1 = b_3 = b_5 = \cdots = 0$, since t must then be an even function of x (the ray paths are symmetric about $x = 0$). A typical seismic record shows, however, that the shapes of the t-x curves of the various reflection events can be closely approximated by hyperbolas. Hence, for any given reflection event, we can assume that $t^2 \approx t(0)^2 + b_2 x^2$. But what is b_2? For one layer, $b_2 = 1/v^2$, and the equation for t^2 is exact, not approximate. For n horizontal layers, it turns out that $b_2 = 1/\bar{v}^2$. This can be shown as follows. Expand $t = t(x)$ in a Taylor series about $x = 0$:

$$t = t(x) = t(0) + \left(\frac{dt}{dx}\right)_0 x + \frac{1}{2}\left(\frac{d^2t}{dx^2}\right)_0 x^2 + \cdots, \tag{5.13}$$

where the subscript "0" indicates that the associated quantity is to be evaluated at $x = 0$. Now, $(dt/dx)_0 = (p)_0 = 0$. That this should be zero can also be inferred from the fact that all coefficients of odd powers of x in (5.13) should be zero, because $t(-x) = t(x)$. Also, calculating dt/dp from Equation (5.6), we obtain

$$p = \frac{dt}{dx} = \frac{dt}{dp}\frac{dp}{dx} = \left[2\sum_{j=1}^{n} h_j v_j \left(1 - p^2 v_j^2\right)^{-3/2}\frac{dp}{dx}\right]p. \tag{5.14}$$

This implies $[\cdots] = 1$, which can be solved for dp/dx and evaluated at $x = 0$:

$$\left(\frac{dp}{dx}\right)_0 = \left(\frac{d^2t}{dx^2}\right)_0 = 1 \Big/ \left[2\sum_{j=1}^{n} h_j v_j \left(1 - p^2 v_j^2\right)^{-3/2}\right]_0$$

$$= \frac{1}{2\sum_j h_j v_j} = \frac{1}{\sum_j v_j^2 \tilde{t}_j} = \frac{1}{\bar{v}^2 t(0)}. \tag{5.15}$$

Another way to obtain this would be from $dp/dx = 1/(dx/dp)$, where dx/dp would be obtained by differentiating (5.7). Substitution of (5.15) into Equation (5.13) gives

$$t = t(0) + \frac{1}{2}\left(\frac{1}{\bar{v}^2 t(0)}\right)x^2 + \cdots . \tag{5.16}$$

Squaring t and neglecting terms in x of order 4 and higher, and inserting the subscript n, gives

$$t_n(x)^2 \approx t_n(0)^2 + \frac{x^2}{\bar{v}_n^2}. \tag{5.17}$$

Hence, the RMS velocity $\bar{v}_n$ can be used in fitting a hyperbola to a reflection event from layer n (although the error in the fit increases with x). This means that the RMS velocity is a good approximation to the stacking velocity (the velocity that best stacks a reflection event, i.e., the velocity $\hat{v}_n$ for which the curve $\hat{t}_n(x)^2 = \hat{t}_n(0)^2 + x^2/\hat{v}_n^2$ best fits the nth reflection event).

Equation (5.17) can be derived algebraically as well (see, e.g., Taner and Koehler, 1969).

The so-called x^2, t^2 method can be used to determine interval velocities and thicknesses from the reflection records. Referring to Equation (5.17), if one plots t_n^2 vs. x^2 for the nth reflection curve on a shot record, one obtains a straight line whose slope is approximately $1/\bar{v}_n^2$, from which $\bar{v}_n$ can be determined (for $n = 1, 2, \ldots$). Nowadays, $\bar{v}_n$ is usually determined from various velocity analysis methods such as the analysis of velocity spectra and constant velocity stacks. The interval velocities v_n can then be obtained from the Dix equation (Dix, 1955),

$$v_n^2 = \frac{\bar{v}_n^2 t_n(0) - \bar{v}_{n-1}^2 t_{n-1}(0)}{t_n(0) - t_{n-1}(0)}, \quad n = 1, 2, 3, \ldots \tag{5.18}$$

where $\bar{v}_1 = v_1$, and where $t_n(0)$ and $t_{n-1}(0)$ are obtained from the data. The Dix equation can be easily derived: from Equation (5.2), we have

$$\bar{v}_n^2 t_n(0) - \bar{v}_{n-1}^2 t_{n-1}(0) = (v_1^2 \tilde{t}_1 + \cdots + v_n^2 \tilde{t}_n) - (v_1^2 \tilde{t}_1 + \cdots + v_{n-1}^2 \tilde{t}_{n-1})$$

$$= v_n^2 \tilde{t}_n = v_n^2 [t_n(0) - t_{n-1}(0)]. \tag{5.19}$$

Solving for v_n^2 gives the Dix equation. Large errors in the computed values of v_n can arise for closely spaced reflection events. In other words, if $t_n(0) - t_{n-1}(0)$ is a relatively small number, then a small error in $t_n(0) - t_{n-1}(0)$ results in a large error in v_n (e.g., if $y = 1/x$, then the error dy in y resulting from an error dx in x is $dy = -(1/x^2)dx$, and if x is small, then dy can be large).

The thickness h_n can then be obtained from

$$h_n = \tfrac{1}{2}v_n\tilde{t}_n = \tfrac{1}{2}v_n\big[t_n(0) - t_{n-1}(0)\big]. \tag{5.20}$$

The interval velocities v_n can also be determined from NMO measurements. For the nth reflector, the NMO at offset x, denoted here by T_n^x, is given by

$$T_n^x = \sqrt{t_n(0)^2 + \frac{x^2}{v_n^2}} - t_n(0) \approx \frac{x^2}{2v_n^2 t_n(0)} \tag{5.21}$$

(where the error of the approximation increases with x). This implies

$$\bar{v}_n^2 t_n(0) - \bar{v}_{n-1}^2 t_{n-1}(0) \approx \frac{x^2}{2}\left[\frac{1}{T_n^x} - \frac{1}{T_{n-1}^x}\right] = v_n^2\big[t_n(0) - t_{n-1}(0)\big], \tag{5.22}$$

where Equation (5.19) has been used. Solving for v_n^2 gives

$$v_n^2 = \frac{x^2}{2\big[t_n(0) - t_{n-1}(0)\big]}\left[\frac{1}{T_n^x} - \frac{1}{T_{n-1}^x}\right], \tag{5.23}$$

where the NMO is measured at x. Once again, large errors can occur in computing v_n in situations where T_n^x or T_{n-1}^x are small (e.g., deep reflection events, short spread lengths, etc.).

Let us now consider the case of a vertically inhomogeneous half-space in which the velocity varies continuously with depth z, i.e., $v = v(z)$. In this case, the ray paths are curved. The equations for x and t can be obtained from Equations (5.6a) and (5.7) by replacing h_j with dz, v_j with $v(z)$, and the sum with an integral. For a ray reaching a depth of z_{max} and returning to the surface at x (see Figure 5.3a), we have

$$t = 2\int_0^{z_{max}} \frac{1}{v(z)\sqrt{1 - p^2 v(z)^2}}\, dz, \qquad x = 2\int_0^{z_{max}} \frac{pv(z)}{\sqrt{1 - p^2 v(z)^2}}\, dz. \tag{5.24a,b}$$

Equation (5.6b) can also be modified in the same way. The differential versions of these are also useful. They are as follows:

$$\frac{dt}{dz} = \pm\frac{1}{v(z)\sqrt{1 - p^2 v(z)^2}}, \qquad \frac{dx}{dz} = \pm\frac{pv(z)}{\sqrt{1 - p^2 v(z)^2}}. \tag{5.24c,d}$$

The $+ (-)$ sign applies to the downgoing (upgoing) part of the ray.

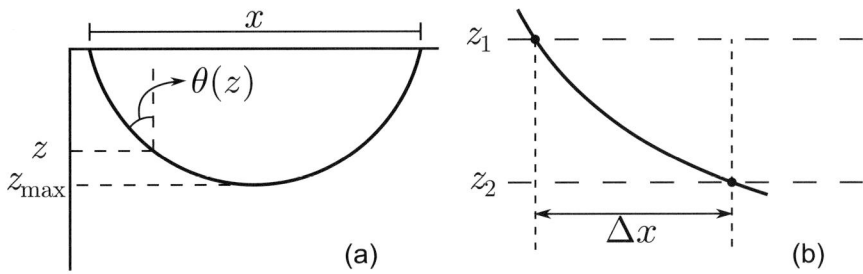

Figure 5.3 (a) A ray path in a medium in which v varies continuously with depth z, with $dv/dz > 0$ (v increases with depth). (b) An arc of the ray path between depths z_1 and z_2.

The ray parameter p for the ray is given by

$$p = \frac{\sin \theta(z)}{v(z)} = \frac{dt}{dx}. \tag{5.25a}$$

p is independent of z for the ray path. In other words, as one moves along the ray, $\theta(z)$ and $v(z)$ change by amounts that keep p constant. Another way to express this is as follows:

$$p = \frac{\sin \theta(z_1)}{v(z_1)} = \frac{\sin \theta(z_2)}{v(z_2)}, \tag{5.25b}$$

which relates the ray angles and velocities at two depths.

However, p *does* vary with x (different values of x correspond to different ray paths, and hence different ray parameters). Referring to Figure 5.3a, note that at $z = z_{max}$, θ is 90° and hence $p = 1/v(z_{max})$. Note also that z_{max} varies with p. Applying (5.25b) at the surface $z_1 = 0$ and at $z_2 = z_{max}$ then gives

$$p = \frac{\sin \theta(0)}{v(0)} = \frac{1}{v(z_{max})}. \tag{5.25c}$$

For the subsurface model studied here, i.e., a 1D vertically heterogeneous medium, the ray parameter p is a useful quantity because it is constant along the ray path (even though z, θ, and v vary along the ray path). However, in a 2D heterogeneous medium, where the wave velocity varies with both z and x (e.g., a stack of layers with flat interfaces having different dips), the ray parameter p is no longer constant along a ray path, and hence is no longer a useful quantity.

For a spherical Earth model with 1D heterogeneity, i.e., $v = v(r)$, where $v(r)$ is the wave speed at the radial distance r from the center of the Earth, the corresponding equation for the ray parameter is

$$p = \frac{r \sin \theta(r)}{v(r)} = \frac{dt}{d\Delta}, \tag{5.25d}$$

where r is the radial distance to a point on the ray path, $\theta(r)$ is the angle that the ray makes at the point, and Δ is the epicentral angle for the ray.

To obtain the travel time Δt along the arc of the ray between the depths z_1 and z_2 (see Figure 5.3b), and the corresponding distance Δx traveled in the horizontal direction, replace "$2\int_0^{z_{max}}$" with "$\int_{z_1}^{z_2}$", and "x" with "Δx" in Equations (5.24a,b).

As an example, consider a medium in which the velocity increases linearly with the depth, i.e., $v = v_0 + az$, where $a = dv/dz$ is a positive real constant. To determine the shape of the ray path, we use the x-equation, (5.24b), for a ray path between depths 0 and z on the downgoing part of the ray to obtain

$$x = \int_0^z \frac{pv(z')}{\sqrt{1 - p^2v(z')^2}}\, dz' = \frac{1}{pa}\int_{pv_0}^{pv(z)} \frac{w}{\sqrt{1 - w^2}}\, dw$$

$$= -\frac{1}{pa}\left[\sqrt{1 - w^2}\right]\Big|_{pv_0}^{pv(z)} = \frac{1}{pa}\left[\sqrt{1 - p^2v_0^2} - \sqrt{1 - p^2v(z)^2}\right], \qquad (5.26)$$

where the substitution $w = pv(z') = p(v_0 + az')$ was used in the integral. Equivalently, we could have started with (5.24d), with a + sign, and integrated from 0 to z to obtain (5.26).

Equation (5.26) gives the x coordinate of a point on the left (downgoing) half of the ray path at depth z. For a point on the right half at depth z, the equation would read $x = (1/pa)\left[\sqrt{\cdots} + \sqrt{\cdots}\right]$. To obtain this, one could integrate (5.24d) with a + sign from 0 to z_{max}, then add to it the result of integrating (5.24d) with a − sign from z_{max} to z.

If we substitute $v(z) = v_0 + az$ into Equation (5.26) and rearrange it, we obtain

$$(x - \bar{x})^2 + (z - \bar{z})^2 = R^2, \qquad (5.27a)$$

where

$$\bar{x} = \frac{1}{pa}\sqrt{1 - p^2v_0^2}, \qquad \bar{z} = -\frac{v_0}{a}, \qquad R = \frac{1}{pa} = \frac{v_0}{a\sin\theta_0}. \qquad (5.27b)$$

This is the equation of a circle with radius R, whose center is located at $(\bar{x}, \bar{z})$. In other words, the ray paths are circular arcs (see Figure 5.4). If we know the values of a, v_0, and the **take-off angle** θ_0, we can obtain the maximum depth of penetration of the ray, i.e.,

$$z_{max} = R - \frac{v_0}{a} = \frac{v_0}{a}\left(\frac{1}{\sin\theta_0} - 1\right). \qquad (5.28)$$

Also, from Equation (5.27) for $\bar{x}$, we see that x_r in Figure 5.4a is given by

$$x_r = \frac{2}{pa}\sqrt{1 - p^2v_0^2} = \frac{2v_0}{a}\cot\theta_0. \qquad (5.29)$$

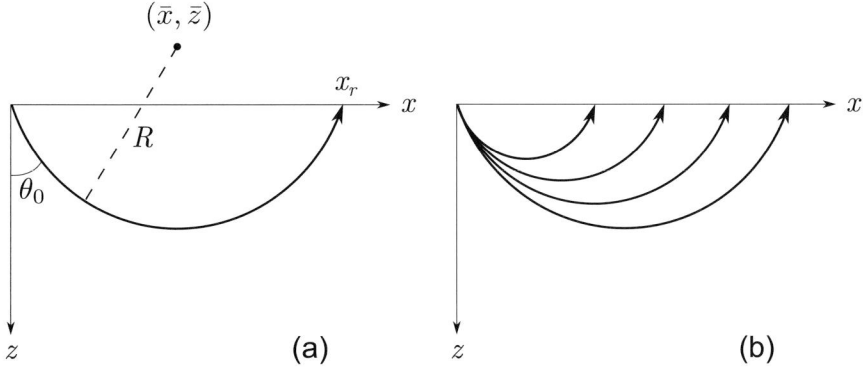

Figure 5.4 Circular ray paths (for $v = v_0 + az$, $a > 0$): (a) geometry, and (b) variation with offset.

The fact that ray paths in a medium with a constant velocity gradient ($dv/dz = a$) are circular arcs can also be shown by working out the **radius of curvature** R_c (see any first-year calculus text), where

$$R_c \equiv \left| \left[1 + \left(\frac{dx}{dz} \right)^2 \right]^{\frac{3}{2}} \bigg/ \frac{d^2x}{dz^2} \right|. \tag{5.30}$$

Substituting (5.24d), together with $v = v_0 + az$, into the formula for R_c gives $R_c = (1/pa)$, which is a constant, implying that the ray paths are circular arcs.

An equation for $t = t(z)$ could also be derived for this case from Equation 5.24a, by dropping the "2" and replacing $z_{\max}$ with z, or by integrating Equation 5.24c with a + sign from 0 to z. It would show how t varies with z as one moves along the ray.

Similarly, for this case of $v = v_0 + az$, (5.24a,b) can be integrated as they stand to obtain the travel time $t = t(p)$ and the source–receiver offset $x = x(p)$ for the whole ray (Figure 5.3a) with ray parameter p, and combined (by eliminating p) to obtain $t = t(x)$ giving the travel times for whole rays with different x (and p) values. Or this could be obtained by solving $x = x(p)$ for $p = p(x)$ and integrating $dt/dx = p(x)$.

5.2 Applications to Well Surveys and Logs

Consider Figure 5.5a, which illustrates a **velocity survey**. At each depth z of the receiver, the travel time from source to receiver is recorded, from which a t-z plot could be generated. The *vertical* traveltime t_v down to depth z is obtained by applying a geometric correction to the recorded times t to account for the source–borehole offset x (and the depth of the source if it is buried). For example, if the ray path from source to receiver is a straight line, and if θ is the angle that the ray path

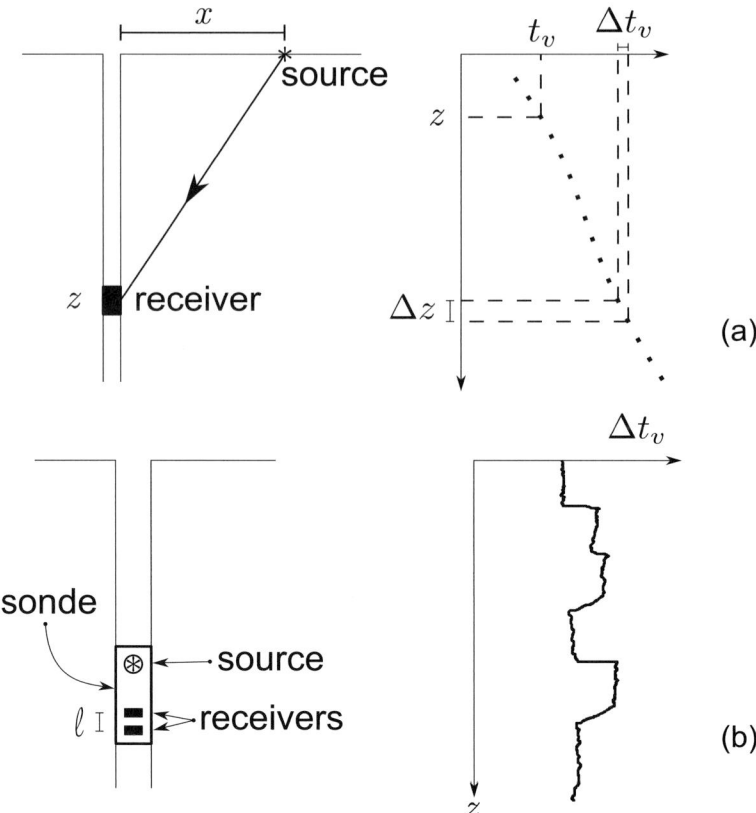

Figure 5.5 The source–receiver geometry and plotted data for (a) a velocity survey, and (b) a continuous velocity log. t_v is the vertical travel time down to depth z.

makes with the vertical direction, then $t_v = t \cos \theta = tz/\sqrt{x^2 + z^2}$. The vertical travel time t_v is plotted as a function of depth z (see Figure 5.5a). From the plot, one obtains, as a function of depth z, the average velocity z/t_v and the interval velocity $\Delta z/\Delta t_v \approx v(z)$.

It is usually assumed that the ray path from source to receiver is a straight line. However, in some areas this may be a poor assumption, and a curved ray path must be considered to avoid significant errors in the computed velocity and vertical travel time functions. A smooth line can be drawn through the recorded t-z data to simulate a continuous function (or interpolation techniques could be used), and then the velocity $v(z)$ can be obtained by numerically solving Equations (5.24) (see, e.g., Grant and West, 1965, pp. 134–137). If x is small enough, meaning $t \approx t_v$ and $p \approx 0$ for all the different ray paths going from the source point to the various receivers, then the **instantaneous velocity** dz/dt (obtained from the slope of the recorded t-z

data) may be a good enough estimate of $v(z)$. This can be seen by setting $p \approx 0$ in (5.24c), which gives $dt/dz \approx dt_v/dz = 1/v(z)$. The vertical travel time to depth z, $t_v(z)$, can then be obtained by integration, i.e., from $t_v = \int_0^z dz/v(z)$.

Consider Figure 5.5b, which illustrates a **continuous velocity log** or a **sonic log**. In the sonic log technique, the ray path curvature problem is eliminated because the source is part of the **sonde**, i.e., the logging tool lowered into the borehole.

Since the rock surrounding the borehole has a higher velocity than the drilling mud in the borehole (except perhaps near the surface), the first arrivals at the receivers are the waves that travel through the rock. The transit time Δt_v between the two receivers is plotted as a function of depth, and since l is very small (on the order of a foot), the velocity as a function of depth, $v(z)$, is effectively given by $l/\Delta t_v$ (which is $l \times$ the reciprocal of the log curve). The vertical travel time to depth z, $t_v(z)$, is obtained by integrating the recorded log curve Δt_v versus z, i.e., $t_v = \int_0^z dz/v(z) = (1/l) \int_0^z \Delta t_v(z)\, dz$ $(= (1/l) \times$ the area under the log curve).

In practice, more sophisticated sondes are used, having more sources and receivers, and transit times are averaged to give a single value of Δt_v at depth z.

5.3 The Wiechert–Herglotz Method

This is an **inverse** method, due to E. Wiechert and G. Herglotz (see Aki and Richards, 1980, 2002), for determining $v(z)$ from the t-x travel time data, e.g., a seismic record (**forward** methods estimate the data from a velocity model). For the relatively simple case in which $v(z)$ increases with depth z (see Figure 5.6), the equation for the maximum depth z_1 reached by the ray with offset x_1 is (Grant and West, 1965, p. 139)

$$z_1 = -\frac{1}{\pi} \int_{p_0}^{p_1} \frac{x(p)}{\sqrt{p^2 - p_1{}^2}}\, dp. \tag{5.31}$$

Apply to this the integration-by-parts rule

$$\int f\, dg = fg - \int g\, df \quad \text{with} \quad f = x(p) \quad \text{and} \quad dg = dp/\sqrt{p^2 - p_1{}^2}.$$

Then

$$g = \ln[(p/p_1) + \sqrt{(p/p_1)^2 - 1}] = \cosh^{-1}(p/p_1)$$

and Equation (5.31) becomes

$$z_1 = \frac{1}{\pi} \int_0^{x_1} \cosh^{-1}\left[\frac{p(x)}{p(x_1)}\right] dx. \tag{5.32}$$

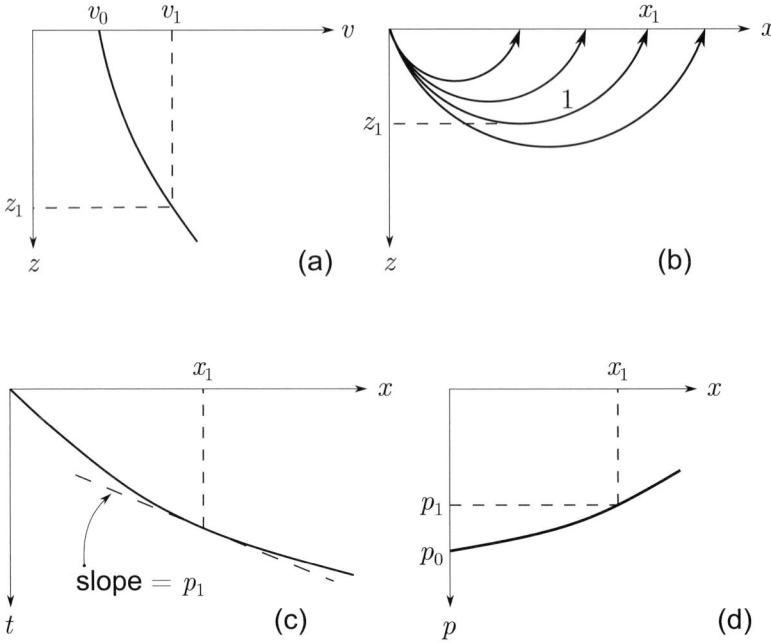

Figure 5.6 Schematic diagrams of (a) $v = v(z)$, (b) the ray paths, and the curves (c) $t = t(x)$ and (d) $p = p(x) = dt/dx$ used to illustrate the Wiechert–Herglotz inverse method.

The ray parameter $p_1 = p(x_1)$ for the ray with offset x_1 can be obtained by evaluating $p = [\sin\theta(z)]/v(z)$ (Equation 5.25a) at any depth z. In particular, if we choose the maximum depth z_1 of the ray, we have $\theta(z_1) = 90°$. Hence,

$$p_1 = p(x_1) = \frac{1}{v(z_1)} = \frac{1}{v_1}. \tag{5.33}$$

This equation also shows that $p(0) = 1/v(0)$, i.e., $p_0 = 1/v_0$, since $z_1 \to 0$ as $x_1 \to 0$.

Equations (5.31) and (5.32) are the **Wiechert–Herglotz formulas**. They can be used to determine the velocity–depth relation $v = v(z)$ in the following way. First, from the slope of the t-x data curve, obtain $p = p(x) = dt/dx$. Then, choose a value of x_1 and obtain $p_1 = p(x_1)$ from the p-x curve. Evaluate either Wiechert–Herglotz integral (numerically) to obtain z_1. The velocity at depth z_1 is then given by $v_1 = 1/p_1$. The procedure is repeated for many choices of x_1 to obtain a table of values of $v(z)$.

Another way to obtain $v = v(z)$ goes as follows. If the range of values assumed by the velocity is known, then Equation (5.31) can be used to determine the depth at which the velocity has a certain value. To do this, replace p_0 with $1/v_0$ and

p_1 with $1/v_1$ in Equation (5.31). Evaluating the integral then gives the depth z_1 corresponding to the velocity v_1. This can be repeated for other values of v_1 to obtain a table of values of $v(z)$.

The Wiechert–Herglotz method can also be applied to the case in which a zone of rapid velocity increase produces a **triplication** in the *t-x* curve (see Figure 5.7a). Although $p(x)$ is a multiple-valued function of x for the triplication, $x(p)$ is a *single-valued* function of p. Hence, Equation (5.31) is the appropriate formula to apply in this case. Each branch of the triplication must be used in the integral. In practice, it is difficult to identify the later branches because of interference from the earlier ones. Note that the *p-x* curve is "simpler" than the *t-x* curve (the *p-x* curve does not cross itself). Seismic records are sometimes easier to analyze in the *p-x* domain than in the *t-x* domain.

The Wiechert–Herglotz formulas cannot be applied to the case of a **shadow zone** (see Figure 5.7b), which is created by a subsurface region in which the velocity decreases with depth, or by a low-velocity region. However, the method can be extended to include such cases (see, e.g., Aki and Richards, 1980, p. 408, 2002, p. 414).

The Wiechert–Herglotz method can be applied in the case of a 1D spherical Earth model as well, to determine the velocity profile $v = v(r)$, but the formulas are different from (5.31) and (5.32). For the relatively simple case in which the velocity $v(r)$ decreases slowly with the radial distance r measured from the center of the Earth (or increases slowly with depth), the formula analogous to Equation (5.32) is

$$\pi \ln\left(\frac{r_0}{r_1}\right) = \int_0^{\Delta_1} \cosh^{-1}\left[\frac{p(\Delta)}{p(\Delta_1)}\right] d\Delta,$$

where Δ, which is called the epicentral angle, or more often, the epicentral distance, is the angle (as measured at the center of the Earth) spanned by a ray going through the Earth from one point on the spherical surface to another; r_0 is the radius of the Earth (about 6371 km); r_1 is the value of r at the midpoint of the ray whose epicentral distance is Δ_1, i.e., the value of r at the deepest point or turning point of the ray; $p(\Delta)$ is the ray parameter defined in (5.25d); $p(\Delta_1) = r_1/v_1$ is the value of p at $\Delta = \Delta_1$ (evaluated on the ray where $\theta(r)$ in Equation 5.25d is $90°$); and $v_1 = v(r_1)$. The velocity profile $v(r)$ can be obtained in a manner very similar to that described in the first paragraph following (5.33). If I_1 is the value of the integral on the right side of the preceding equation, then $r_1 = r_0 \exp(-I_1/\pi)$ and $v_1 = r_1/p(\Delta_1)$, giving the point (r_1, v_1) on the velocity profile. For more details, see, e.g., Aki and Richards, 1980, p. 408, 2002, p. 414.

In refraction seismology for a layered subsurface, the times of the first breaks, i.e., first arrivals, are measured. If a low-velocity layer is present, the first breaks data cannot be unambiguously interpreted. For instance, referring to Figure 5.8, the

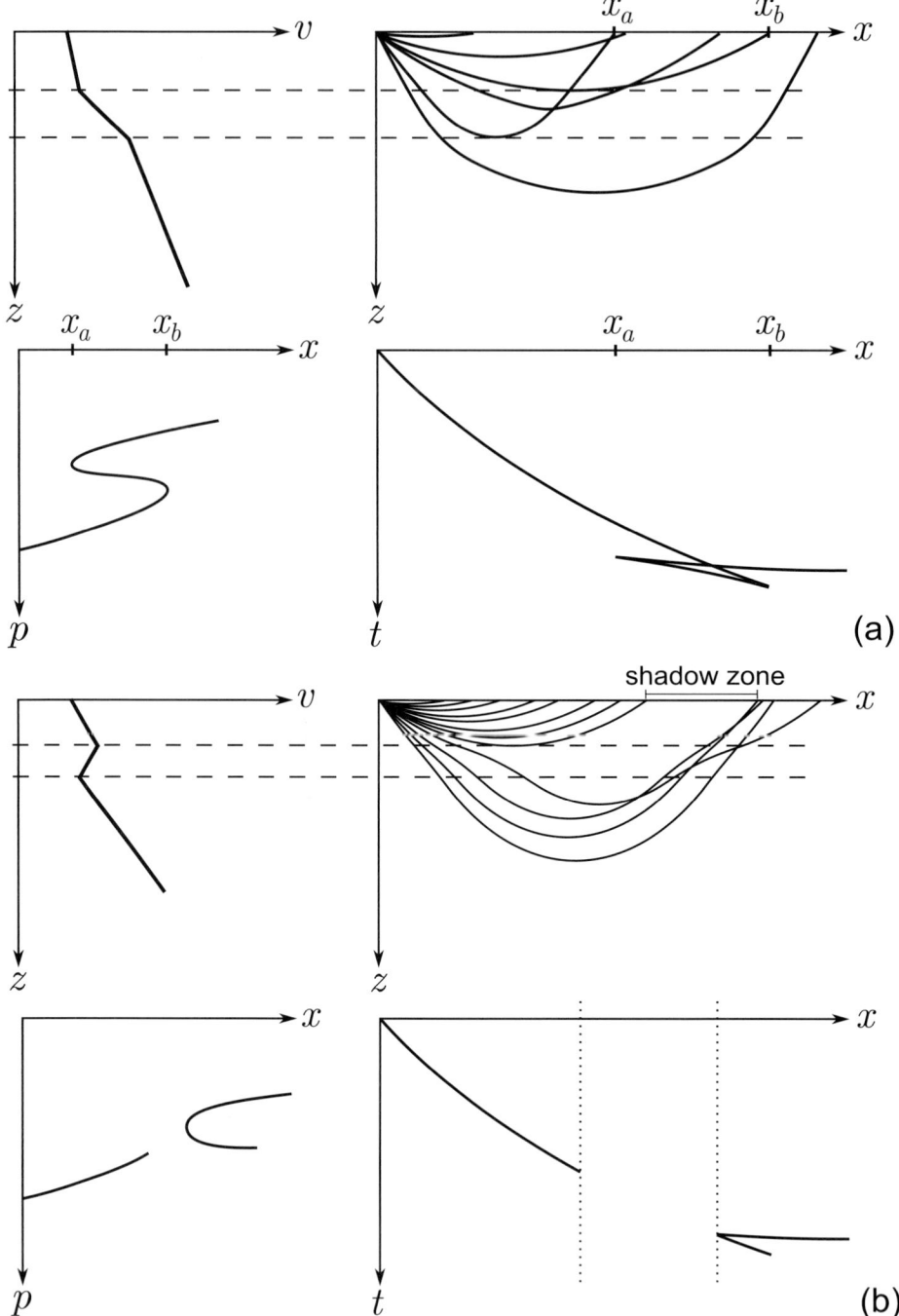

Figure 5.7 Schematic diagrams illustrating (a) a triplication in the *t-x* curve (see also Figure 4.5), and (b) a shadow zone.

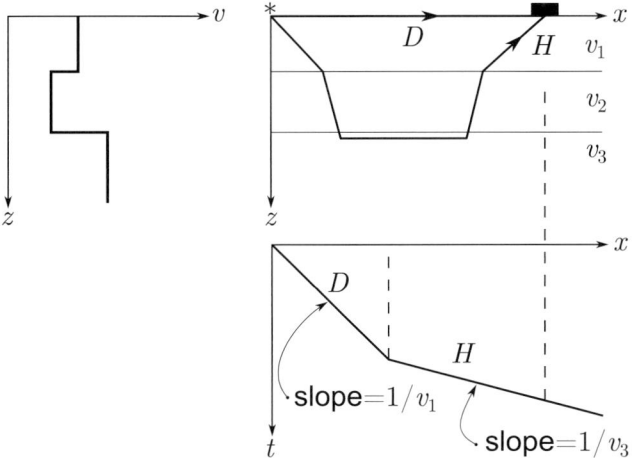

Figure 5.8 A low-velocity layer. D and H indicate the direct wave and the head wave, respectively.

first breaks data look like those for a single layer of velocity v_1 over a half-space of velocity v_3 (see, e.g., Figure 4.5) — the low-velocity layer is not detected.

5.4 Wave Propagation in a Slowly Varying Vertically Inhomogeneous Medium

If the velocity v varies with depth z (i.e., if the Lamé parameters λ and μ vary with z), then the equation of motion for the medium is more complicated than it is for a homogeneous medium (Equation 2.41). Typically, pure P and SV waves no longer exist independent of each other – P and SV particle motions are coupled. However, if $v(z)$ varies slowly enough with z, then the equation of motion can be somewhat simplified, and approximate analytical solutions involving pure or nearly pure P, SV, and SH waves are possible in some cases (see, e.g., Grant and West, 1965, pp. 45, 128).

If $v(z)$ varies slowly enough with z, and if we assume no variation in the y direction, then it can be shown that pure SH waves can propagate in an arbitrary direction in the x-z plane, and that nearly pure P and SV waves can propagate in near-vertical (z) directions in the x-z plane (they are pure if they travel strictly in the vertical direction).

The displacement u_y of the SH waves is given by

$$u_y = \frac{\chi}{\sqrt{\mu(z)}}, \quad \text{where} \quad \frac{\partial^2 \chi}{\partial x^2} + \frac{\partial^2 \chi}{\partial z^2} \approx \frac{1}{v^2}\frac{\partial^2 \chi}{\partial t^2}, \qquad v = v(z), \qquad (5.34)$$

which has the same form as the wave equation for a homogeneous medium, except that $v (= \beta)$ varies with z (slowly).

For the *SV* waves, the "rotation" is

$$\nabla \times \mathbf{u} = (0, \xi, 0), \quad \xi \equiv \frac{\partial u_x}{\partial z} - \frac{\partial u_z}{\partial x} \quad \text{with} \quad \xi = \frac{\chi}{\mu(z)},$$

where χ satisfies (5.34) with $v = \beta(z)$.

For the *P* waves, the dilatation is

$$\mathcal{D} \equiv \nabla \cdot \mathbf{u} = \chi \Big/ \big[\lambda(z) + 2\mu(z)\big],$$

where χ again satisfies (5.34) but with $v = \alpha(z)$.

Hence, for instance, Equation (5.34) could be used to approximately describe acoustic waves ($\mu = 0$) in a slowly varying vertically inhomogeneous medium ($v = \alpha(z)$). The acoustic wave pressure p would be $p = -\lambda \mathcal{D} = -\chi$, where χ is the solution of (5.34) (see Equation 2.35).

For a slowly varying vertically inhomogeneous medium, there is no lateral inhomogeneity. Consequently, we may assume a solution to (5.34), which is harmonic in x (and t) but whose amplitude varies with depth z in some as yet unknown way $U(z)$, i.e.,

$$\chi = U(z)e^{i\omega(px - t)}, \tag{5.35}$$

where p is the horizontal component of slowness, i.e., the **apparent slowness** in the x direction. $c = 1/p$ is the wave's apparent velocity in the x direction, and $\lambda_x = 2\pi c/\omega$ is the wave's apparent wavelength in the x direction. The word "apparent" refers to the value that the quantity *appears* to have if it is measured along a given direction (in this case, the x direction). Substituting (5.35) into (5.34) yields the Helmholtz equation:

$$\frac{d^2 U}{dz^2} + \omega^2 \eta(z)^2 U = 0, \quad \eta(z)^2 = \frac{1}{v(z)^2} - p^2. \tag{5.36}$$

If the medium is highly inhomogeneous, so that the velocity changes substantially over a seismic wavelength, then the very concept of a wavelength loses its meaning. We will assume then that the velocity changes only by a very small amount over the span of a wavelength. If η were constant (independent of z), the solution for, say, an upgoing wave would be $U = A \exp(-i\omega\eta z)$ with $\eta > 0$, where A is a constant. For η varying slowly with z, this suggests that we try a solution of the form

$$U = Ae^{i\omega T(z)}, \qquad A = \text{constant}, \tag{5.37}$$

where $T(z)$ is nearly linear in z (for $\eta = $ constant, $T = -\eta z$, with $\eta > 0$). Substituting this into Equation (5.36) gives

$$\left(\frac{dT}{dz}\right)^2 = \eta(z)^2 + \frac{i}{\omega}\frac{d^2T}{dz^2}. \tag{5.38a}$$

Since T is nearly linear in z, we have $d^2T/dz^2 \approx 0$. So, in (5.38a),

$$\frac{1}{\omega}\left|\frac{d^2T}{dz^2}\right| \ll \eta(z)^2. \tag{5.38b}$$

Consequently, we then have

$$\frac{dT}{dz} \approx \pm\eta(z), \quad\Longrightarrow\quad T(z) \approx -\int_0^z \eta(z)\,dz, \tag{5.39}$$

where we have assumed that $T(0) = 0$ and where the minus sign has been chosen so that in the homogeneous limit, $T(z)$ reduces to the correct result for an upgoing wave ($T = -\eta z$, with $\eta > 0$). Hence, to a first approximation ($d^2T/dz^2 \approx 0$), we have, for an upgoing wave,

$$U = A\exp\left[-i\omega\int_0^z \eta(z)\,dz\right]. \tag{5.40}$$

A second approximation can be obtained by iteration. Letting $T' \equiv dT/dz$, we have, from Equation (5.39), $T'' = -\eta'$. Substituting this into Equation (5.38a) gives $(T')^2 = \eta^2 - i\eta'/\omega$, which becomes

$$T' \approx -\eta + \frac{i\eta'}{2\omega\eta}, \tag{5.41}$$

where Equation (5.5) was used. Integrating with respect to z gives

$$T = -\int_0^z \eta(z)\,dz + \frac{i}{\omega}\ln\sqrt{\frac{\eta(z)}{\eta(0)}} \tag{5.42}$$

$$\Longrightarrow\quad U = A\sqrt{\frac{\eta(0)}{\eta(z)}}\exp\left[-i\omega\int_0^z \eta(z)\,dz\right], \tag{5.43a}$$

which is the solution of (5.36) to a second approximation. As stated previously, the *SH* displacement u_y for an upgoing wave is then given by (for $\eta > 0$)

$$u_y = A\sqrt{\frac{\eta(0)}{\mu(z)\eta(z)}}\exp\left[i\omega\left(px - \int_0^z \eta(z)\,dz - t\right)\right]. \tag{5.43b}$$

The preceding outlined method is known as the WKBJ method (after Wentzel, Kramers, Brillouin, and Jeffreys). It has been developed much further, and has been very useful for calculating seismic wave amplitudes in heterogeneous media, among other things.

The preceding solutions for U are valid if (5.38b) is true, i.e., if $|\eta'| \ll \omega\eta^2$ is approximately true (where Equation 5.39 has been used). For instance, for a vertically upgoing wave, we have $p = 0$, which implies $\eta(z) = 1/v(z)$, and hence the condition $|\eta'| \ll \omega\eta^2$ becomes

$$\left|\frac{dv}{dz}\right| \ll \omega \quad (\text{for} \quad p = 0), \tag{5.44}$$

i.e., the velocity gradient must be small compared to seismic frequency values. As stated previously, this means that $v(z)$ cannot change by much over the span of a typical wavelength W if the approximate solutions are to be valid. To see this, substitute $\omega = 2\pi v/W$ into (5.44) to obtain $|dv/v| \ll 2\pi(dz/W)$. This states that if $dz \sim W$, i.e., if the small distance dz through which the wave propagates is on the order of a wavelength, then the fractional change in the velocity $v(z)$ over this distance must be very small (i.e., $\ll 2\pi$).

The condition (5.44) is usually satisfied in the Earth. For example, in areas where $v(z) = v_0 + az$ applies (where a is a constant), the velocity gradient a has typical values around 1 sec^{-1}. For a low frequency, say $f = 10$ Hz, we have $\omega \approx 63$ Hz, and the condition clearly holds ($1 \ll 63$).

Conversely, we can say that the solutions are valid only for frequencies much higher than the velocity gradient. In an area with a very large velocity gradient, the solutions are valid only for high frequencies. Hence, the WKBJ solution is a **high frequency approximation**. It is also sometimes called the **geometrical optics approximation** because the concept of energy propagating along the ray paths of geometrical optics is meaningful only for frequencies much larger than the optical velocity gradient (the laws of geometrical optics apply only to high enough frequencies).

For nonvertically traveling waves, i.e., for $p \neq 0$, the condition that the velocity gradient must satisfy (obtained from $|\eta'| \ll \omega\eta^2$) for the approximate solutions to be valid is more restrictive than (5.44), i.e., (5.44) is the minimum requirement for the velocity gradient dv/dz.

The first approximation to U (Equation 5.40) can be obtained in a way that is perhaps more familiar. If η is constant, then the Helmholtz differential equation for U (Equation 5.36) can be written in the "factored" form

$$\left(\frac{d}{dz} + i\omega\eta\right)\left(\frac{d}{dz} - i\omega\eta\right)U = 0. \tag{5.45}$$

In this formula, the inner differential operator is applied to U first, then the outer one is applied to the result. If we apply each operator individually, we obtain equations whose solutions correspond to upgoing and downgoing waves: for $\eta > 0$, we get

$$\frac{dU}{dz} + i\omega\eta U = 0 \qquad \text{for upgoing waves,} \qquad (5.46)$$

$$\frac{dU}{dz} - i\omega\eta U = 0 \qquad \text{for downgoing waves.} \qquad (5.47)$$

For an upgoing wave, the solution is $U = Ae^{-i\omega\eta z}$. If η varies slowly with z, the upgoing waves are approximately described by Equation (5.46) with $\eta = \eta(z)$. Rewriting this differential equation, and solving, we have

$$\frac{dU}{U} = -i\omega\eta(z)\,dz, \qquad \Longrightarrow \qquad \ln U = -i\omega\int^z \eta(z)\,dz + \text{const}, \qquad (5.48)$$

$$\Longrightarrow \qquad U = A\exp\left[-i\omega\int_0^z \eta(z)\,dz\right], \qquad (5.49)$$

where A is a constant. Equation (5.49) agrees with Equation (5.40). The usefulness of this approach is that a second-order differential equation has been reduced to a first-order differential equation, which is easier to work with and solve. It has been used in seismic migration theory (see, e.g., Stolt and Benson, 1986, p. 109).

5.5 The Eikonal Equation

Suppose a medium is both vertically and laterally heterogeneous, or heterogeneous in all three directions, x, y, and z. In that case, $\lambda = \lambda(x, y, z)$ and $\mu = \mu(x, y, z)$, meaning $\alpha = \alpha(x, y, z)$ and $\beta = \beta(x, y, z)$. In deriving the equation of motion, λ and μ could no longer be treated as constant. The equation of motion (2.40) would then contain additional terms involving the space derivatives of λ and μ, i.e., the equation of motion would then become

$$\rho\frac{\partial^2 u_i}{\partial t^2} = (\lambda + \mu)\frac{\partial\mathcal{D}}{\partial x_i} + \mu\nabla^2 u_i + \mathcal{D}\frac{\partial\lambda}{\partial x_i} + \sum_{j=1}^3 \frac{\partial\mu}{\partial x_j}\left[\frac{\partial u_i}{\partial x_j} + \frac{\partial u_j}{\partial x_i}\right]. \qquad (5.50)$$

If the heterogeneity is weak, i.e., if the frequencies are high enough, and the wavelengths small enough (so that many wavelengths could be fit into a space interval over which the velocity $v = v(x, y, z)$ changes substantially), then approximate solutions to this equation can be obtained in some cases using a technique similar to the one in the previous section.

If the medium were homogeneous, then the harmonic wave solution would be

$$u_i = A_i\exp\left[i\omega(s_x x + s_y y + s_z z - t)\right],$$

where (s_x, s_y, s_z) is the slowness vector and $A_i = Ad_i$, with **d** being the polarization vector and A the amplitude, none of which vary with position for a plane wave traveling in a given direction. This suggests that, for the weakly heterogeneous medium, we try a solution of the form

$$u_i = A_i(x, y, z) \exp\left\{i\omega[T(x, y, z) - t]\right\}, \tag{5.51}$$

where $T(x, y, z)$ is called the **traveltime function** or the **phase function**. Note that (5.51) is not necessarily a plane wave, as the function $A_i(x, y, z)$ has not been specified. Substituting this trial solution into (5.50) leads to a quadratic equation in the frequency ω, i.e., $C_2\omega^2 + C_1\omega + C_0 = 0$. Assuming weak heterogeneity, i.e., high frequencies, means that the term C_0 can be dropped (since the other two terms will dominate at high ω), leading to $C_2\omega + C_1 = 0$. The coefficients C_j are independent of ω. Therefore, in order for this equation to hold for all ω, one must have $C_2 = 0$ and $C_1 = 0$, which after some math ultimately lead to partial differential equations (PDEs) for T and A_i, respectively. The PDE for T is

$$\left(\frac{\partial T}{\partial x}\right)^2 + \left(\frac{\partial T}{\partial y}\right)^2 + \left(\frac{\partial T}{\partial z}\right)^2 = \frac{1}{v(x, y, z)^2}, \tag{5.52}$$

where v can be either α or β. This is known as the **eikonal equation**. Note that for a homogeneous medium, $T = s_x x + s_y y + s_z z \Rightarrow \nabla T = (s_x, s_y, s_z)$ = the slowness vector. Similarly, in (5.51) and (5.52), ∇T is the slowness vector. This also means that the eikonal equation can be written as

$$(\nabla T) \cdot (\nabla T) = \left|\nabla T\right|^2 = s^2 = 1/v^2, \quad v = v(x, y, z), \tag{5.53}$$

which is consistent with the standard equation for slowness used in previous chapters, i.e., $s = 1/v$.

The PDEs for A_i are complicated, except in simple cases. For example, consider the equation

$$\nabla^2 \chi = \frac{1}{v(x, y, z)^2} \frac{\partial^2 \chi}{\partial t^2}, \tag{5.54}$$

which is the generalization of (5.34) to 3D heterogeneity. Equation (5.54) holds, for example, for acoustic waves ($\mu = 0$) waves in a constant density slowly varying 3D heterogeneous medium, with the acoustic pressure $p = -\chi$, where χ is the solution of (5.54) with $v = \alpha(x, y, z) = \sqrt{\lambda(x, y, z)/\rho}$. In Appendix 5A, it is shown that (5.54) is actually an exact equation for the acoustic pressure p and also for the P wave potential ϕ (with $\chi = -p$ or $\chi = \phi$) if the density ρ is constant.

If a trial solution,

$$\chi = A(x, y, z) \exp\left\{i\omega[T(x, y, z) - t]\right\}, \tag{5.55}$$

is substituted into (5.54), one obtains the equations $C_2 = 0$ and $C_1 = 0$ (see the discussion following Equation 5.51), with $C_2 = 0$ giving the eikonal equation in (5.53), and $C_1 = 0$ giving an equation for A, i.e.,

$$2\nabla A \cdot \nabla T + A\nabla^2 T = 0, \tag{5.56}$$

which is often called a **transport equation**. Once T is known (via solving the eikonal equation), the transport equation can be solved for $A(x, y, z)$, in principle. Note that (5.55) is similar to the trial solution (5.35) and (5.37) in the 1D *SH* case, except that here, A is not taken to be a constant, but is a function of **x**. This method for solving (5.54) can be applied to the 1D *SH* case as well – the second-order solution (5.43b) is obtained.

Appendix 5B provides more details on the preceding transport equation, as well as the derivation of a set of ray equations that can be solved numerically for tracing rays in a medium with an arbitrary velocity structure $v = v(x, y, z)$. Ray theory has been developed to a great depth in the seismic wave theory literature, and is a useful method in computing seismic wave travel times and amplitudes. See, e.g., Červený (2001) and Červený, Molotkov, and Pšenčík (1977).

5.6 Zero-Offset Ray Tracing

If a medium is inhomogeneous to a significant degree in both the vertical and lateral directions, then the computation of the wave motion, and travel times and amplitudes, is considerably more complicated. Often numerical methods, such as the finite difference method, are used (see the examples in Appendix 7A). Ray tracing methods can be used to obtain the travel time–offset curves for the medium. In particular, as we will see, zero-offset ray tracing can be used to simulate the travel time–offset curves observed on seismic stack sections (see, e.g., Graul and Hilterman, 1979; May and Covey, 1981; May and Hron, 1978; Sheriff and Geldart, 1995, p. 392; Taner, Cook, and Neidell, 1970).

Figure 5.9 shows some examples of zero-offset rays. For each zero-offset ray, the source and receiver are at the same location. The zero-offset rays corresponding to the reflection arrivals from a given interface are perpendicular to that interface (e.g., points *A*, *B*, *F*, *J*), which makes them easy to trace. If they pass through reflectors lying between the source and the given interface, then Snell's law determines the transmission angles (e.g., points *C*, *G*, *H*). A wave striking a sudden bend in a reflector (e.g., point *E*) is diffracted in all directions. The two-way travel time for each zero-offset ray is computed from knowledge of the velocities, and is plotted on the time section vertically beneath the receiver (e.g., the travel time for ray *YFY* is plotted at F', which is directly below the receiver at *Y*).

A seismic time section constructed in this way gives the arrival times of the various rays, but not the amplitudes. To construct synthetic seismograms, amplitudes must also be calculated. This involves computing reflection and transmission coefficients, geometrical spreading effects (spherical divergence), etc. Synthetic seismograms are often used in seismic modeling studies.

A seismic time section generated by tracing zero-offset rays through a model of a subsurface structure is a good approximation to the seismic CMP stack section that would be obtained by a conventional seismic survey of the structure, because when the NMO correction is applied to the traces of a CMP gather, they are essentially converted into zero-offset traces, and when they are stacked together, the output trace is positioned vertically beneath the midpoint (the location of the zero-offset receiver), just like the zero-offset trace.

Referring to Figure 5.9b, the receiver at X records two primary reflections, associated with ray paths XAX and $XABAX$. Two diffractions, associated with ray paths XEX and $XCDEX$, are also shown. Zero-offset ray path $XCDEX$ shows that when diffractions are involved, the upgoing and downgoing zero-offset ray paths do not necessarily coincide. Note the distortion of the reflectors on the time section, in particular the lower reflector, which is *flat* on the depth model. The distortion of the flat reflector is due to the presence of the overlying dipping reflector. Distortion occurs because reflection points on the interfaces on the depth section have **migrated** to new positions on the time section (e.g., see points F and F'). This indicates that seismic stack sections are *not* "pictures" of the subsurface. To get a true geological "picture" of the subsurface, one must **migrate** the points on the time section back to their true spatial positions. A diffraction curve must be collapsed back to a point at the apex of the curve, since that is the spatial position of the diffracting point in the true depth section. The diffraction appearing along with the lower reflection on the time section is sometimes referred to as a **phantom diffraction**. It is slightly skewed (nonsymmetric) because the ray paths associated with the left and right sides of the diffraction curve are different, i.e., they have different travel paths and travel times (compare rays $XCDEX$ and $YGIEY$).

Figure 5.10a illustrates the **bow-tie effect** produced by a **buried focus**. The buried focus is not generally a point, but a region. The rays all focus at one point only if the syncline is perfectly circular. On the time section, the segments AB and CD generally have low amplitudes because of the geometrical divergence of the associated rays, and the segment BC generally has a high amplitude because of the focusing or convergence of the associated rays. Hence, if BC is a strong event, and if AB and CD are masked by noise, say, then event BC might be incorrectly interpreted as an *anticline*.

For the low-curvature syncline of Figure 5.10b, the focus is above the surface, and so a syncline in depth is a syncline in time as well, although the shape of the

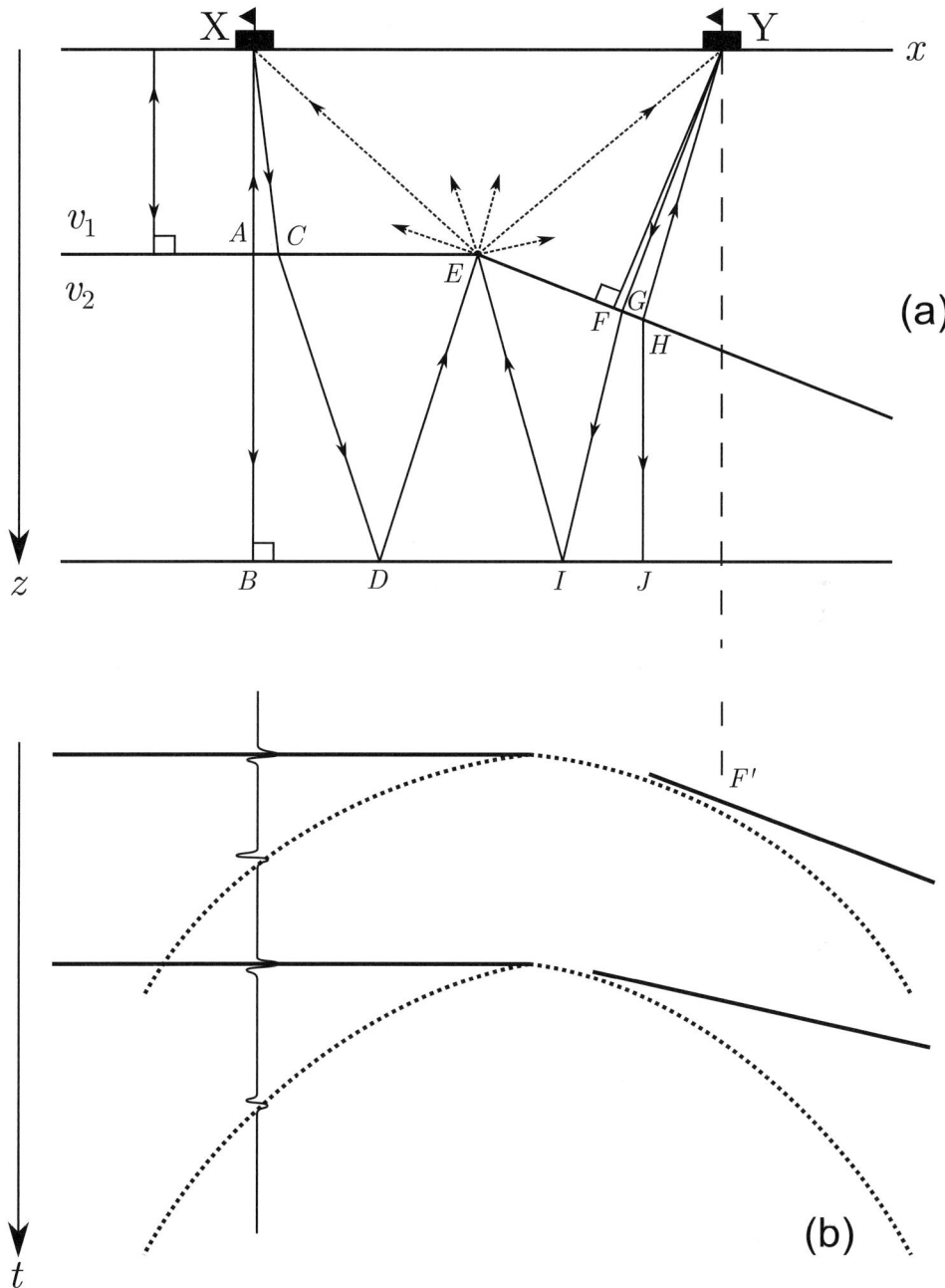

Figure 5.9 Schematic diagrams of zero-offset rays (primaries only) for a depth model (a), with $v_2 > v_1$, and the corresponding time section (b). The dotted lines represent diffractions.

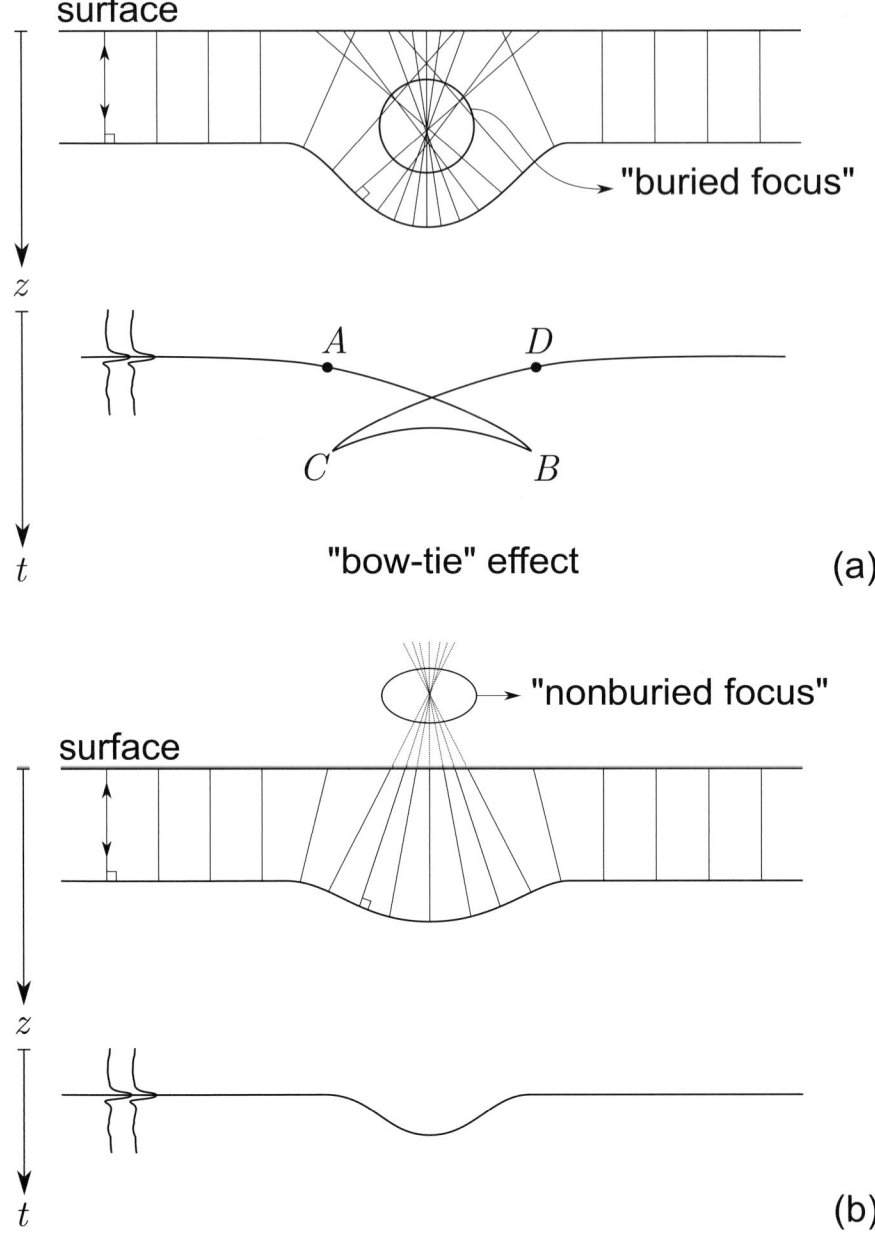

Figure 5.10 Schematic diagrams of zero-offset rays and the time section for a syncline that focuses the rays (a) below the surface and (b) above the surface.

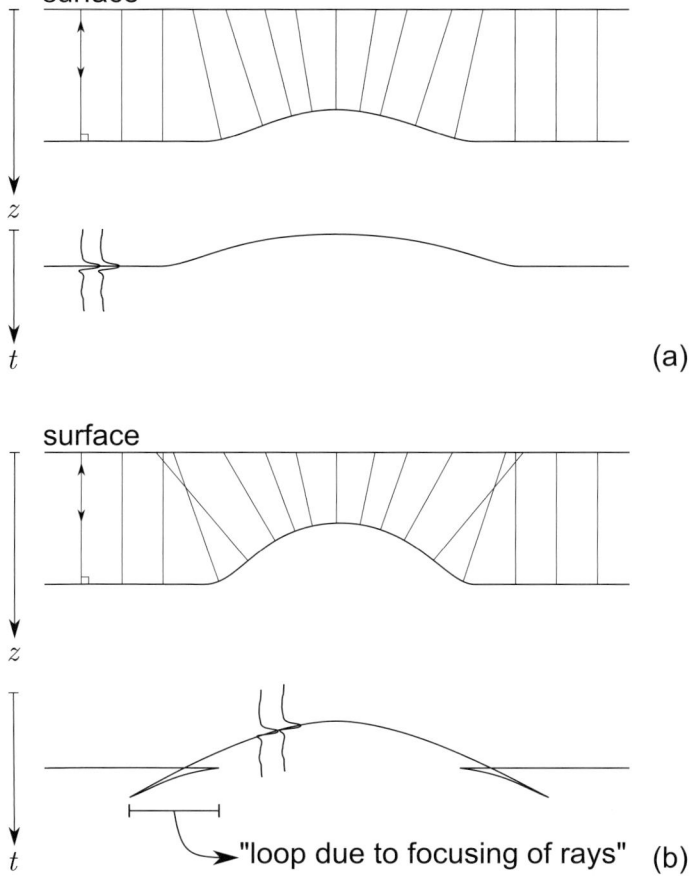

Figure 5.11 Schematic diagrams of zero-offset rays and the time section for an anticline of (a) low curvature and (b) high curvature.

time syncline is *not* the same as the shape of the depth syncline. The time syncline should be a moderately strong event since the rays do not diverge too much. A high-curvature syncline, such as the one in Figure 5.10a, can have a focus that is above the surface as well if the syncline is close enough to the surface.

Referring to Figure 5.11, the crest of a time anticline will have a relatively low amplitude (compared to the flanks) because of the geometrical divergence of the rays.

Referring to Figure 5.12, the flat reflector on the depth section is curved on the time section because rays passing through the anticline, such as the ray at *B*, spend more time in the high-velocity medium and less time in the low-velocity medium than do rays not passing through the anticline, such as the ray at *A*. This effect is known as **velocity pull-up**. The second reflection on the time section can of course

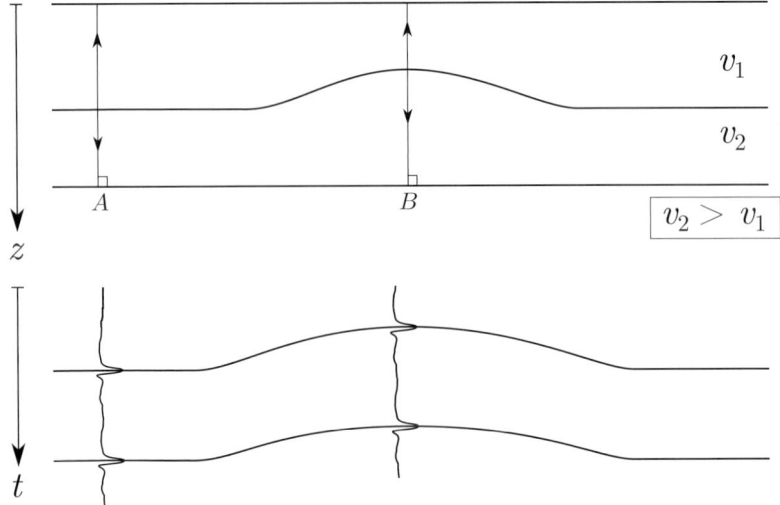

Figure 5.12 The velocity pull-up effect.

be artificially flattened by isotiming; however, this then changes the depths of the other reflections. The depths to reflectors should not be calculated from an isotimed seismic section. If the first reflector were a syncline instead of an anticline, the second reflection on the time section would have a synclinal appearance — this would be a **velocity pull-down** effect.

5.7 Seismic Diffractions

If a continuous reflector has zones where the curvature changes substantially over a short distance (on the order of a seismic wavelength), then diffracted waves with significant amounts of energy will be generated. The characteristics of these waves cannot be described by the usual geometric ray path theory. However, at distances from a diffraction zone that are large compared to a seismic wavelength, Huygens' principle can be used to construct the diffracted wavefronts. Drawing lines perpendicular to these wavefronts then permits a geometrical ray path interpretation for diffracted waves (e.g., Figures 4.7 and 5.9). The higher the frequency, the better the ray path interpretation will be (if the curvature changes slowly enough, or if the frequency is high enough, then we basically have reflected waves that have a clear-cut ray path interpretation). The diffraction ray paths generally do not obey Snell's law (see Figure 5.13a), although Snell's law can be applied to determine the reflection and transmission angles for a diffracted wave incident upon a flat boundary. Diffractions can also be produced by "sudden" lateral changes in impedance along a flat continuous reflector, as will be seen later in this chapter.

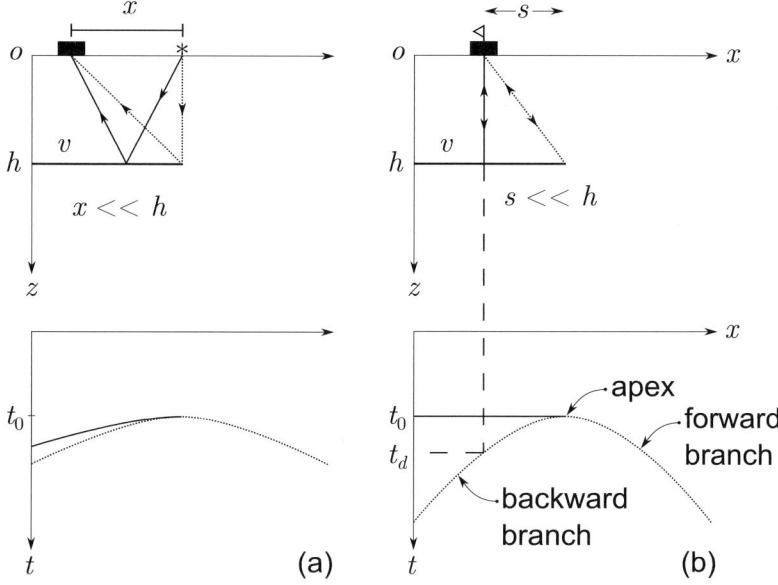

Figure 5.13 Schematic diagrams of a terminating reflector and the corresponding (a) shot record, and (b) CMP stack section, showing diffractions (the dotted lines).

Although seismic diffraction is discussed at a more or less qualitative level in the sections that follow, seismic diffraction theory can be quite mathematical. Details can be found, for example, in Klem-Musatov, Hoeber, Moser, and Pelissier, (2016a, 2016b). Physical modeling experiments have also be performed that demonstrate diffraction (see, e.g., Narod and Yedlin, 1986).

5.8 Diffraction Moveout on a Shot Record and a Stack Section

Consider the idealized case of an abruptly terminating reflector at a depth h (Figure 5.13a). The travel time t_r for the reflection is

$$t_r = \frac{2}{v}\sqrt{\left(\frac{x}{2}\right)^2 + h^2} = t_0\sqrt{1 + \left(\frac{x}{vt_0}\right)^2} \approx t_0 + \frac{x^2}{2v^2 t_0}, \qquad (5.57)$$

$$\text{for} \quad \frac{x}{vt_0} \ll 1 \quad (x \ll 2h),$$

where $t_0 = 2h/v$ and where Equation (5.5) was used. The travel time t_d for the diffraction is

$$t_d = \frac{h + \sqrt{x^2 + h^2}}{v} = \frac{t_0}{2}\left[1 + \sqrt{1 + \left(\frac{2x}{vt_0}\right)^2}\right] \approx t_0 + \frac{x^2}{v^2 t_0}, \qquad (5.58)$$

$$\text{for} \quad \frac{x}{vt_0} \ll 1 \quad (x \ll 2h).$$

Hence, we see that on the shot record, the moveout for the diffraction is larger than the moveout for the reflection – for small offsets, it is twice as large.

Consider Figure 5.13b. The CMP stack section for the abruptly terminating reflector can be obtained approximately from zero-offset ray tracing. The time t_d of the diffraction event when the source and receiver are at a horizontal distance s from the termination point is given by

$$t_d = \frac{2}{v}\sqrt{h^2 + s^2} = t_0\sqrt{1 + \left(\frac{2s}{vt_0}\right)^2} \approx t_0 + \frac{2s^2}{v^2 t_0} \quad \text{for} \quad \frac{2s}{vt_0} \ll 1 \quad (s \ll h).$$

$$(5.59)$$

The diffraction curve is a hyperbola. The deeper is the reflector, the smaller is the moveout (the smaller is the curvature of the diffraction). This can also be seen intuitively: as h gets bigger, the length (and therefore the travel time) of the zero-offset diffracted ray in the depth section approaches that of the zero-offset reflected ray, thereby reducing the diffraction curvature.

5.9 Diffraction Amplitudes and Phases

In the mathematical theory of diffraction, it is usually assumed that the subsurface behaves like an acoustic medium (S waves are neglected), so that a scalar wave equation can be used. Amplitudes and phases of diffraction signals are then worked out by solving the wave equation for the scalar field, i.e., the P-potential ϕ or the pressure (see, e.g., Hilterman, 1970; Trorey, 1970). Another simplifying assumption that is often made is that a continuous reflector can be treated as an ensemble of "elementary diffractors" (see Figure 5.14), and that the time section for the continuous reflector can be obtained by summing the time sections for each individual diffractor (this applies only to primary signals). This suggests that the signal at the receiver, which is usually thought of as coming from some point on the reflector, such as point B in Figure 5.14, actually comes from all points on the

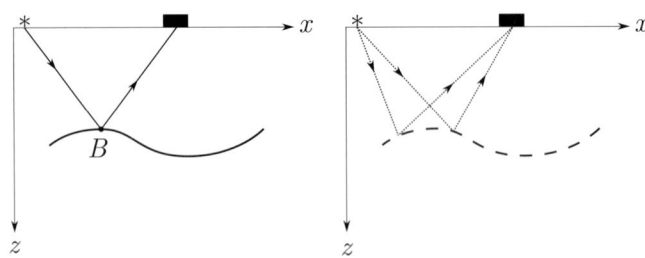

Figure 5.14 A continuous reflector modeled as an ensemble of elementary diffractors.

reflector (see Trorey, 1970). Several other simplifying assumptions are usually also made. The results obtained when the theory is applied to various simple geologic models are discussed later in this chapter.

5.10 A Terminating Reflector

Referring to Figure 5.15, the pulses on the reflection and the foreward branch of the diffraction have the same polarity (defined to be positive), but the pulses on the backward branch of the diffraction have the opposite polarity (defined to be negative). At the diffracting edge (termination point), the amplitude of the total response (diffraction plus reflection) is one-half that of the reflections alone. This can also be seen with simple algebra. Let r denote the reflection pulse and d the positive polarity diffraction pulse. Slightly to the left of the diffracting edge, the response is $r - d$, and slightly to the right it is d. To ensure continuity, we must have $r - d = d$ at the apex (termination point), which implies $d = \frac{1}{2}r$ at the apex. Hence, the total response at the apex is $\frac{1}{2}r$. It has the same polarity as the reflection pulse but half the amplitude. This effect makes it appear as if the amplitude of the reflection decreases as the diffracting edge is approached. We should also note that the amplitude of the diffraction pulse d decreases as one moves down the diffraction curve.

The $180°$ phase difference between the diffraction branches, predicted by the mathematical theory of diffraction, can be partially understood using a simple argument, illustrated by Figure 5.16: because of the opposite polarities on the diffraction branches, they cancel each other out by destructive interference to give the correct time section for the continuous reflector.

Because of the various complexities present in real data, the effects previously discussed may be difficult to observe in practice.

The rate of amplitude decay along the diffraction branches is faster than that implied by the usual geometrical spreading law ($1/D$, where D is the distance the

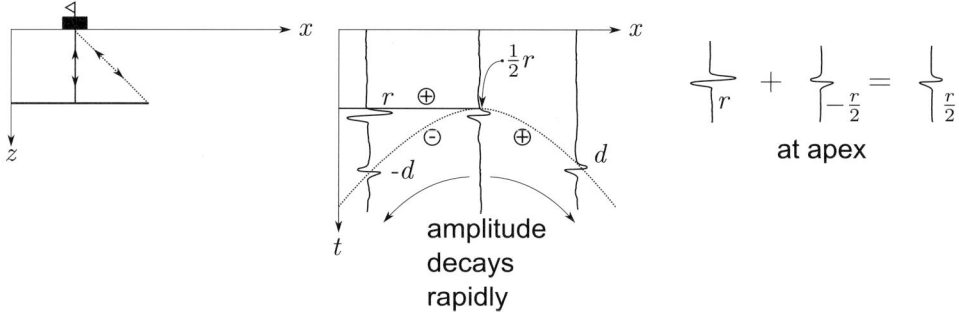

Figure 5.15 Schematic diagrams of the amplitude and polarity on a CMP stack section for a terminating reflector.

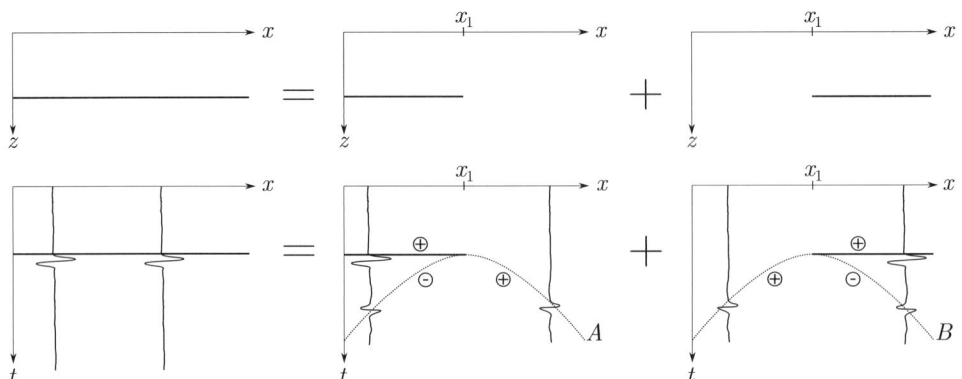

Figure 5.16 Schematic diagrams of a continuous reflector represented as a sum of terminating reflectors.

wave has traveled). This is because the diffracted wave does not emanate strictly from the termination point alone, but from the entire reflector. In Figures 5.15 and 5.16, and also Figures 5.17 and 5.18, it is assumed that a correction for geometrical spreading loss has been applied to the time section, so that amplitude decays are due to other (diffraction) effects.

5.11 A Flat Continuous Reflector with a Laterally Changing Reflection Coefficient

Referring to Figure 5.16, if the continuous reflector has a reflection coefficient R_1 to the left of x_1 and a smaller reflection coefficient R_2 to the right of x_1 ($R_2 < R_1$), then when the reflector is represented as the sum of two terminating reflectors, the reflected and diffracted waves from the terminating reflector on the right will have lower amplitudes than those from the terminating reflector on the left. Hence, the destructive interference of the diffraction branches is not quite complete, resulting in a diffraction curve for the continuous reflector. For example, the relatively high-amplitude branch A in Figure 5.16 added to the relatively low-amplitude branch B in Figure 5.16 gives the low-amplitude positive polarity branch C (see Figure 5.17). The reflection amplitudes will be high to the left of x_1 and low to the right of x_1. Note that in this example, the distinction between the terms "reflection" and "diffraction" is blurred, since it shows that diffractions can be produced even by a flat, perfectly continuous reflector. It is, however, possible to define "reflection" and "diffraction" more precisely for purposes of modeling (see Hilterman, 1970).

   If the reflection coefficient changes gradually rather than abruptly (or, if a reflector terminates gradually), then the diffraction pulses are generally too small to be visible.

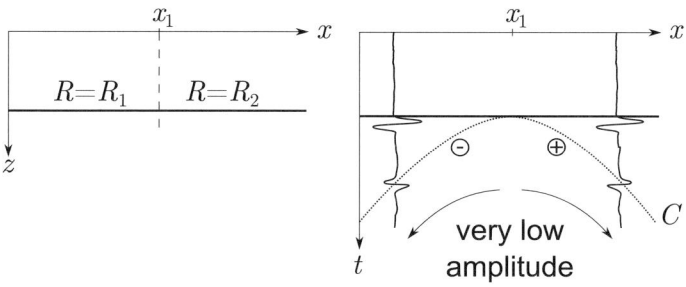

Figure 5.17 Schematic diagrams of a continuous reflector with a laterally changing reflection coefficient, with $R_2 < R_1$.

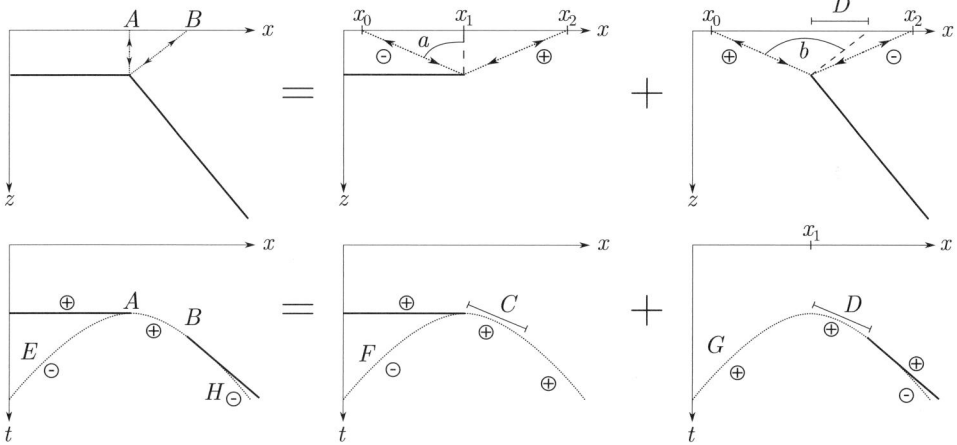

Figure 5.18 Schematic diagrams of diffractions from a sharp bend in a reflector.

5.12 A Bent Reflector

Referring to Figure 5.18, the diffraction segment AB has a positive polarity and relatively high amplitudes (but still lower than the reflection amplitudes) because it is due to the constructive interference of the two positive polarity diffraction segments C and D. The diffraction branch E has a relatively low amplitude and negative polarity for the following reasons: (i) destructive interference of the negative polarity segment F with the positive polarity segment G, and (ii) F has a higher amplitude than G because $a < b$ (i.e., as the angle, a or b, from the reflector normal increases, the amplitude of the diffracted ray decreases). By similar reasoning, we can say that the diffraction segment H has a relatively low amplitude and negative polarity.

This type of diffraction pattern may be used to locate faults (see, e.g., Dobrin, 1976, p. 267).

Referring to Figure 5.9, the event on the time section associated with the deeper reflector shows that diffraction patterns can occur even for flat continuous reflectors above which the medium is laterally homogeneous. In that case, the diffraction patterns are due to the presence of complex structure in the overburden. As mentioned previously, such patterns are sometimes called **phantom diffractions**.

Finally, more "exotic" amplitude-phase effects (e.g., phase changes other than 180°) associated with more complex subsurface structures can be studied as well, both theoretically and experimentally (with a physical modeling facility) – see, e.g., Hilterman (1970, 1975).

5.13 Appendix 5A: Acoustic Wave Equations for Heterogeneous Media

In this appendix, various versions of the acoustic wave equation are derived.

5.13.1 *Equation for the Potential for Constant Density*

Start with the equation of motion for a heterogeneous medium, Equation (5.50):

$$\rho \frac{\partial^2 u_i}{\partial t^2} = (\lambda + \mu)\frac{\partial \mathcal{D}}{\partial x_i} + \mu \nabla^2 u_i + \mathcal{D}\frac{\partial \lambda}{\partial x_i} + \sum_{j=1}^{3} \frac{\partial \mu}{\partial x_j}\left[\frac{\partial u_i}{\partial x_j} + \frac{\partial u_j}{\partial x_i}\right] + f_i. \quad (5.60)$$

Then set $\mu = 0$ so that we have a heterogeneous fluid. We get, in vector form,

$$\rho \frac{\partial^2 \mathbf{u}}{\partial t^2} - \mathbf{f} = \lambda \nabla \mathcal{D} + \mathcal{D}\nabla\lambda = \nabla(\lambda \mathcal{D}) = \nabla(\lambda \nabla \cdot \mathbf{u}), \qquad \mathcal{D} \equiv \nabla \cdot \mathbf{u}. \quad (5.61)$$

Now let $\mathbf{u} = \nabla\phi$ where ϕ is the P wave potential. Also let $\mathbf{f} = \nabla F$. We get

$$\rho \frac{\partial^2}{\partial t^2}\nabla\phi - \nabla F = \nabla(\lambda \nabla \cdot \nabla\phi) = \nabla(\lambda \nabla^2\phi). \quad (5.62)$$

∇ and $\partial^2/\partial t^2$ can be switched on the left side. So, if we assume the density ρ is a constant, we get

$$\nabla\left(\rho \frac{\partial^2 \phi}{\partial t^2} - F - \lambda \nabla^2\phi\right) = 0 \quad (5.63)$$

or

$$\nabla^2\phi = \frac{1}{v^2}\frac{\partial^2 \phi}{\partial t^2} - \frac{F}{\lambda}, \quad v^2 = \lambda/\rho, \quad \lambda = \lambda(x, y, z), \quad v = v(x, y, z), \quad \rho = \text{const}. \quad (5.64)$$

This wave equation for the P wave potential ϕ is exact for a fluid with constant density.

5.13.2 Equation for the Pressure for Constant Density

Consider now the dilatation $\mathcal{D}$. Take the divergence ($\nabla\cdot$) of Equation (5.61). Since the order of partial differentiation can be switched, and assuming the density ρ again to be constant, we get

$$\rho\nabla\cdot\frac{\partial^2\mathbf{u}}{\partial t^2} - \nabla\cdot\mathbf{f} = \rho\frac{\partial^2}{\partial t^2}\nabla\cdot\mathbf{u} - \nabla\cdot\mathbf{f} = \nabla\cdot\nabla(\lambda\nabla\cdot\mathbf{u}) = \nabla^2(\lambda\nabla\cdot\mathbf{u}) \quad (5.65)$$

or,

$$\rho\frac{\partial^2\mathcal{D}}{\partial t^2} - \nabla\cdot\mathbf{f} = \nabla^2(\lambda\mathcal{D}). \quad (5.66)$$

The pressure $p = -\lambda\mathcal{D}$ for a fluid, i.e., $\mathcal{D} = -p/\lambda$. Substituting this in gives

$$\frac{\rho}{\lambda}\frac{\partial^2 p}{\partial t^2} + \nabla\cdot\mathbf{f} = \nabla^2 p \quad (5.67)$$

or

$$\nabla^2 p = \frac{1}{v^2}\frac{\partial^2 p}{\partial t^2} + \nabla\cdot\mathbf{f}, \quad v^2 = \lambda/\rho, \quad \lambda = \lambda(x,y,z), \quad v = v(x,y,z), \quad \rho = \text{const}. \quad (5.68)$$

This wave equation is therefore exact for a fluid with constant density.

5.13.3 Equation for the Pressure for Nonconstant Density

For both $\rho = \rho(x,y,z)$ and $\lambda = \lambda(x,y,z)$, first rewrite (5.61) as

$$\frac{\partial^2\mathbf{u}}{\partial t^2} = \frac{1}{\rho}\nabla(\lambda\nabla\cdot\mathbf{u}) + \frac{\mathbf{f}}{\rho}. \quad (5.69)$$

Now take the divergence ($\nabla\cdot$). Switching the order of differentiation, this gives

$$\nabla\cdot\left[\frac{\partial^2\mathbf{u}}{\partial t^2}\right] = \frac{\partial^2}{\partial t^2}\nabla\cdot\mathbf{u} = \frac{\partial^2\mathcal{D}}{\partial t^2} = \nabla\cdot\left[\frac{1}{\rho}\nabla(\lambda\mathcal{D})\right] + \nabla\cdot\left(\frac{\mathbf{f}}{\rho}\right). \quad (5.70)$$

Substituting $p = -\lambda\mathcal{D}$ and $\mathcal{D} = -p/\lambda$ gives

$$\frac{1}{\lambda}\frac{\partial^2 p}{\partial t^2} = \nabla\cdot\left[\frac{1}{\rho}\nabla p\right] - \nabla\cdot\left(\frac{\mathbf{f}}{\rho}\right), \quad \lambda = \lambda(x,y,z), \quad \rho = \rho(x,y,z). \quad (5.71)$$

This equation for the pressure p is exact for spatially varying λ and ρ.

Suppose λ is constant and ρ varies with x, y, z. Then it is still possible to represent a spatially variable velocity via $v(x,y,z) = \sqrt{\lambda/\rho(x,y,z)}$. Equation (5.71) then becomes, when mulitiplied through with λ,

$$\frac{\partial^2 p}{\partial t^2} = \nabla\cdot\left[v^2\nabla p\right] - \nabla\cdot\left(v^2\mathbf{f}\right), \quad \lambda = \text{const}, \quad \rho = \rho(x,y,z). \quad (5.72)$$

In this way, only the values of v need to be given, and not density, for numerical modeling using (5.72).

5.13.4 A Point Source

It is useful to have a time-varying point source for Equation (5.71). Is it possible to have a body force $\mathbf{f}$ that satisfies

$$\nabla \cdot \left(\frac{\mathbf{f}}{\rho} \right) = \delta(\mathbf{x} - \mathbf{x}_0) s(t)? \tag{5.73}$$

One possible solution is

$$\frac{\mathbf{f}}{\rho} = \frac{1}{3} s(t) \left[H(x - x_0)\delta(y - y_0)\delta(z - z_0), \ H(y - y_0)\delta(x - x_0)\delta(z - z_0), \right.$$

$$\left. H(z - z_0)\delta(x - x_0)\delta(y - y_0) \right]. \tag{5.74}$$

Although the Heaviside step functions suggest an infinitely long line source, $\rho = \rho(\mathbf{x})$ can be used to taper them so that the source is basically concentrated at (x_0, y_0, z_0). For instance, if $\rho_m(\mathbf{x})$ is the actual density of the medium, then we could let $\rho = G + \rho_m$, where G is a tall ($G_{\max} \gg (\rho_m)_{\max}$) and very narrow Gaussian function centered at (x_0, y_0, z_0). The Gaussian would go to zero very quickly as one moves away from (x_0, y_0, z_0), leaving only $\rho = \rho_m$. In the limit of an infinitesimally narrow G, this would be a source that is nonzero at (x_0, y_0, z_0) only. Or G could be a narrow 3D boxcar.

5.14 Appendix 5B: Eikonal, Transport, and Ray Equations

In this appendix, we will use q, instead of s, for the slowness, and $\mathbf{q}$ for the slowness vector, so that we can use s for the arc length along a ray path.

5.14.1 Eikonal and Transport Equations

As discussed in the last part of Section 5.5, the wave equation with $v = v(x, yz)$ and its trial solution, i.e.,

$$\nabla^2 \chi = \frac{1}{v(x, y, z)^2} \frac{\partial^2 \chi}{\partial t^2} \quad \text{and} \quad \chi = A(x, y, z) \exp\left\{ i\omega \left[T(x, y, z) - t \right] \right\}, \tag{5.75}$$

lead to, for high frequencies, the eikonal and transport equations, i.e.,

$$(\nabla T) \cdot (\nabla T) = 1/v(x, y, z)^2 \quad \text{and} \quad 2\nabla A \cdot \nabla T + A\nabla^2 T = 0. \tag{5.76}$$

Assuming high frequencies is equivalent to assuming that the heterogeneity is "weak," i.e., that the characteristic distance over which the wave velocity changes appreciably is much larger than the wavelengths in the wave. The heterogeneity might actually be strong, but geometrical ray theory will still apply for high enough frequencies if this criterion is satisfied.

Also, we know that a wavefront is a surface of constant phase, so setting the phase in the preceding trial solution to be a constant gives $T(x, y, z) - t = $ const. Taking t and T to be zero when $\mathbf{x} = (x, y, z) = \mathbf{0}$ (which makes the constant zero), we obtain the equation of a wavefront, i.e.,

$$T(x, y, z) = t \qquad \text{i.e.,} \qquad T(\mathbf{x}) = t. \qquad (5.77)$$

$T(x, y, z) = t$ is the equation of a set of mathematical surfaces, each with a different t value, and each surface corresponds to the wavefront at a different time t. From vector analysis, we know that the vector ∇T is normal to the surface $T(x, y, z) = t$ for a fixed t value, i.e., ∇T is normal to the wavefront. As a simple example, consider the surface of a sphere of radius a, i.e., $T = x^2 + y^2 + z^2 = a^2$. Then $\nabla T = 2(x, y, z) = 2\mathbf{x} = 2\mathbf{r}$, which is the radial position vector, which is perpendicular to the surface of the sphere.

If we consider the corresponding ray, which is also normal to the wavefront, then the vector ∇T is tangential to the ray at every point on the ray path, which further confirms that $\mathbf{q} = \nabla T$.

Figure 5.19 shows the basic geometry of rays and wavefronts. s is the arc length along the ray. Note that $\mathbf{n} \equiv d\mathbf{x}/ds$ is a unit vector in the direction of the ray and the slowness, i.e., $\mathbf{q} = q\mathbf{n} = (1/v)\mathbf{n}$. Substituting $\mathbf{q} = \nabla T$ into the first term of the preceding transport equation gives

$$2\nabla A \cdot \nabla T = 2\mathbf{q} \cdot \nabla A = \frac{2}{v}\mathbf{n} \cdot \nabla A = \frac{2}{v}\nabla A \cdot \frac{d\mathbf{x}}{ds} = \frac{2}{v}\sum_j \frac{dA}{dx_j}\frac{dx_j}{ds} = \frac{2}{v}\frac{dA}{ds}, \qquad (5.78)$$

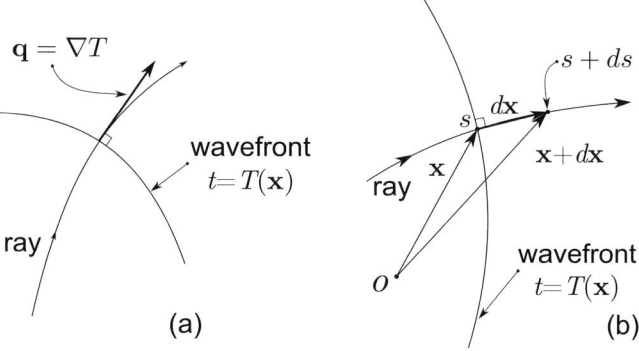

Figure 5.19 The ray corresponding to the wavefront $T(\mathbf{x}) = t$. The slowness vector $\mathbf{q}$ is shown in (a) and the position vector $\mathbf{x}$ and arc length s in (b). The length of $d\mathbf{x}$ is ds.

where the ray path is described parametrically as $x_j = x_j(s)$, $j = 1, 2, 3$. Noting that $\nabla^2 T = \nabla \cdot \nabla T = \nabla \cdot \mathbf{q}$, the transport equation then becomes

$$\frac{2}{v} \frac{dA}{ds} + A \nabla \cdot \mathbf{q} = 0, \tag{5.79}$$

whose solution, $A = A(s)$, describes how the amplitude A varies with arc length s along the ray.

More generally, when the trial solution for u_i in (5.51) is substituted into the equation of motion (5.50) for an isotropic heterogeneous medium, three eikonal equations are obtained: one with $v = \alpha$, the P wave speed; and two with $v = \beta$, the S wave speed (two are obtained because of the two polarizations for an S wave, SV and SH). In that calculation, one can also obtain the following results:

$$\mathbf{u} \times \nabla T = \mathbf{u} \times \mathbf{q} = \mathbf{0} \qquad \text{for a } P \text{ wave}, \tag{5.80}$$

$$\mathbf{u} \cdot \nabla T = \mathbf{u} \cdot \mathbf{q} = 0 \qquad \text{for an } S \text{ wave}, \tag{5.81}$$

showing that the particle displacement is parallel to the slowness vector $\mathbf{q}$ for a P wave and perpendicular to the slowness vector $\mathbf{q}$ for an S wave. The transport equation, however, is more complicated in this case.

One may also obtain eikonal and transport equations for heterogeneous anisotropic media.

For general heterogeneity, waves cannot be clearly identified as being either P or S – they are a mix of P and S. But assuming high enough frequencies, or "weak" heterogeneity, waves that are *predominantly* P or S can be identified, and they can be studied with the eikonal and transport equations.

5.14.2 Ray Equations

We now derive a set of equations that can be solved numerically in order to trace the path of a ray in a medium with an arbitrary velocity structure, $v = v(x, y, z)$.

The ray in Figure 5.19 can be represented by a set of parametric equations, using the arc length s as the parameter, i.e., $x_i = x_i(s)$, $i = 1, 2, 3$. From Figure 5.19 we have $|d\mathbf{x}| = ds$, i.e., $|d\mathbf{x}/ds| = 1$, which implies that $d\mathbf{x}/ds$ is a unit vector normal to the wavefront (and tangential to the raypath). However, the eikonal equation can also be written as $(v \nabla T) \cdot (v \nabla T) = 1$, i.e., $|v \nabla T| = 1$, and since ∇T is normal to the wavefront, $v \nabla T$ is also a unit vector normal to the wavefront. Therefore, we must have

$$\frac{d\mathbf{x}}{ds} = v \nabla T = v \mathbf{q} \qquad \text{i.e.,} \qquad \frac{dx_i}{ds} = v \frac{\partial T}{\partial x_i} = v q_i, \quad i = 1, 2, 3, \quad v = v(x, y, z).$$
$$\tag{5.82}$$

These three equations for $x_i(s)$ are called the **normal equations** for the ray. We would, however, like to have an equation for $\mathbf{x}$ that does not involve T, but only v, for the geometry of the raypath. So we use the normal equations to obtain

$$\frac{d\mathbf{q}}{ds} = \frac{d}{ds}\nabla T = \nabla\left[\frac{dT}{ds}\right] = \nabla\left[\sum_j \frac{\partial T}{\partial x_j}\frac{dx_j}{ds}\right] = \nabla\left[\frac{1}{v}\sum_j \frac{dx_j}{ds}\frac{dx_j}{ds}\right] = \nabla\left[\frac{1}{v}\right],$$

(5.83)

since $\sum_j(dx_j/ds)(dx_j/ds) = |d\mathbf{x}/ds| = 1$. Therefore, the desired equation is

$$\frac{d\mathbf{q}}{ds} = \frac{d}{ds}\left[\frac{1}{v}\frac{d\mathbf{x}}{ds}\right] = \nabla\left[\frac{1}{v}\right].$$

(5.84)

Equations (5.82) and (5.84) can be called the **ray equations**, and they can be used to determine ray paths in a heterogeneous medium.

For example, consider a homogeneous medium. In this case, v is a constant, and (5.84) gives

$$v\frac{d\mathbf{q}}{ds} = \frac{d^2\mathbf{x}}{ds^2} = v\nabla\left[\frac{1}{v}\right] = \mathbf{0} \quad\Longrightarrow\quad \mathbf{x} = \mathbf{a}s + \mathbf{b},$$

(5.85)

where $\mathbf{a}$ and $\mathbf{b}$ are constant vectors. In other words, the rays are straight lines, as expected.

For another example, consider a vertically heterogeneous medium, i.e., $v = v(z)$, and a ray in the xz plane, a case we have studied previously. The x component of (5.84) gives

$$\frac{dq_x}{ds} = \frac{\partial}{\partial x}\left[\frac{1}{v(z)}\right] = 0 \quad\Longrightarrow\quad q_x \text{ is independent of } s,$$

(5.86)

meaning q_x is constant along the ray path. But q_x is just p, the ray parameter. So we see that $p = [\sin\theta(z)]/v(z)$ is constant along the ray path, a result we discussed previously (see Equation 5.25b).

5.14.3 Another Form of the Ray Equations

The ray equations (5.82) and (5.84) can also be expressed in terms of t instead of s as follows. $T(x, y, z) = t$ is the equation of a wavefront at the time t. As t increases, the wavefront moves forward. Hence, we can write $ds = v\,dt$ (see Figure 5.19), and the ray equations become

$$\frac{d\mathbf{x}}{dt} = v^2\mathbf{q} \quad\text{and}\quad \frac{d\mathbf{q}}{dt} = v\nabla\left[\frac{1}{v}\right].$$

(5.87)

Note that $v\nabla(1/v)$ can also be written as $-(\nabla v)/v$ or as $-\nabla \ln v$. The preceding equations can be easily solved numerically, using schemes for solving ordinary differential equations, to trace rays and calculate their travel times for $v = v(x, y, z)$.

The preceding normal equations can also be used to derive the familiar equations (5.24c & d) for dx/dz and dt/dz in the $v = v(z)$ case: noting that $q_x = p$ and $\mathbf{x} = (x, y, z)$, we have

$$\frac{dx}{dt} = v^2 q_x, \quad \frac{dz}{dt} = v^2 q_z \quad \Rightarrow \quad \frac{dx}{dz} = \frac{q_x}{q_z} = \frac{p}{\pm\sqrt{v^{-2} - p^2}} = \frac{\pm pv(z)}{\sqrt{1 - p^2 v(z)^2}},$$
(5.88)

which is Equation (5.24d). Similarly, Equation (5.24c) can be obtained as follows:

$$\frac{dz}{dt} = v^2 q_z = v^2(\pm\sqrt{v^{-2} - p^2}) = \pm v\sqrt{1 - p^2 v^2} \quad \Rightarrow \quad \frac{dt}{dz} = \frac{\pm 1}{v(z)\sqrt{1 - p^2 v(z)^2}}.$$
(5.89)

Exercises

1. (a) Refer to the analysis of Equations (5.12)–(5.17) in the text. If simple hyperbolas, such as $t^2 = t(0)^2 + b_2 x^2$ where $b_2 = 1/\bar{v}^2$, do not fit the reflection curves well enough in certain cases (e.g., relatively large offsets), then it may be necessary to use a higher-order equation for t^2. Let's say that the reflection curves in a given case are approximated well enough by $t^2 = t(0)^2 + b_2 x^2 + b_4 x^4$. Extend the analysis to derive a formula for b_4. (Hint: work through and understand the analysis in the text *before* attempting this exercise. The general procedure is to start with a Taylor series expansion for $t(x)$ to fourth order, i.e.,

$$t(x) = t(0) + \cdots + \frac{1}{4!}\left(\frac{d^4 t}{dx^4}\right)_0 x^4 + \cdots,$$

calculate the coefficients, then square $t(x)$ to get t^2 and throw away terms of order higher than four. This is a generalization of the analysis in the text).

    (b) For the following three-reflector model, where v_n and h_n are interval velocities and thicknesses, compute b_2 and b_4 for each reflector:

n	v_n (km/s)	h_n (km)
1	1.0	0.2
2	2.0	0.4
3	3.0	0.6

    For small offsets x, the term $b_4 x^4$ in t^2 will be negligible, but for large enough x, it becomes important. Using the criterion that it becomes important when $|b_4 x^4 / b_2 x^2| > 0.1$, compute for each reflector the offset x at which the $b_4 x^4$ term becomes important.

2. Consider a medium in which the velocity v increases linearly with depth z, i.e., $v(z) = v_0 + az$, where $a > 0$ ($z = 0$ is the surface). Calculate the formula for $p(x)$, where p is the ray parameter and x is the source–receiver offset. Then calculate the formula for $t(x)$, where t is the travel time. $t(x)$ can, of course, be calculated by evaluating the integral for $t(p)$ for a continuous medium and using $p(x)$, but there is an easier way. Sketch or plot graphs of p vs. x and t vs. x. *Note: see the "Formula" section at the end of this chapter.*

3. Suppose that in a given area in which the velocity v changes continuously with depth z, the travel time curve $t = t(x)$ can be approximated by (see Figure 5.20)

$$t(x) \approx 1.1x - 0.1x^3 \quad \text{for} \quad 0 \le x \le 1.5 \,\text{km}.$$

(a) What is the velocity at the surface $z = 0$?
(b) Use the Wiechert–Herglotz formula involving $\cosh^{-1}$ to compute z_1 and $v(z_1)$ for $x_1 = 1$ km. Normally, the integral would have to be evaluated with a numerical method (e.g., Simpson's rule). However, in this case, there is a simple approximate way to evaluate it. To discover this way, plot a graph of the integrand $\cosh^{-1}\left[p(x)/p(x_1)\right]$ versus x for $0 \le x \le x_1$.
(c) Check the accuracy of your result by evaluating the integral in part (b) numerically to obtain the exact result (e.g., use Simpson's rule or the trapezoidal rule).

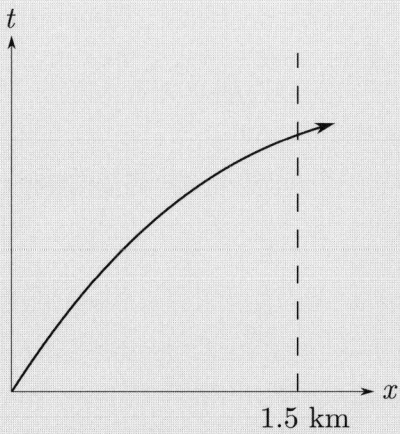

Figure 5.20 A schematic diagram of $t(x)$. See exercise 3.

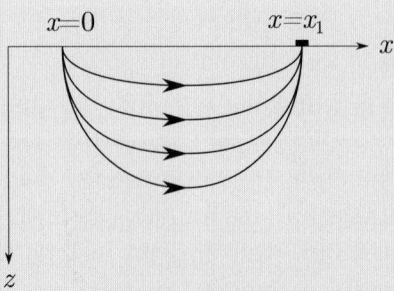

Figure 5.21 See exercise 4.

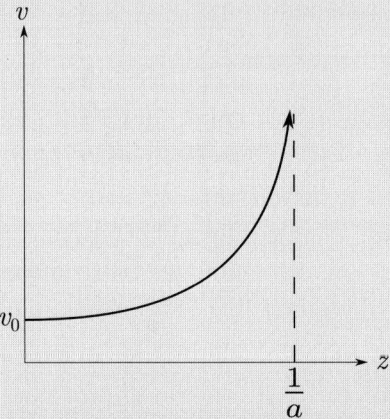

Figure 5.22 A schematic diagram of $v(z)$. See Exercise 5.

4. Consider the hypothetical vertically inhomogeneous medium shown in Figure 5.21 for which all the rays from a surface source emerge at the same point on the surface. Determine $v = v(z)$, the velocity–depth relationship, for this medium. Sketch or plot a graph of $v(z)$ vs. z. *Note: see the "Formula" section at the end of this chapter.*

5. Consider a hypothetical medium in which the velocity–depth relationship is (see Figure 5.22)

$$v(z) = \frac{v_0}{\sqrt{1 - az}}, \quad a = \text{positive constant.}$$

  (a) Obtain the equation $z = z(x)$ for a raypath in the medium (the source and receiver are on the surface). What kind of curve is it? *Note: see the "Formula" section at the end of this chapter.*

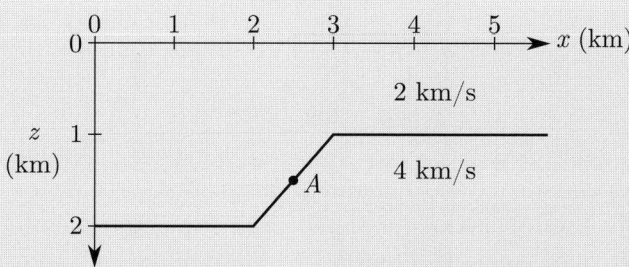

Figure 5.23 See exercise 6.

(b) Let $v_0 = 1$ km/s, $a = 0.2$ km^{-1}. If a ray leaves the source at an angle of 10° to the vertical, what is z_{max}, the maximum depth of penetration of the ray? What is x_r, the distance (from the source) at which the ray emerges on the surface?

6. (a) By tracing some appropriate zero-offset rays, construct the time model corresponding to the depth model given in Figure 5.23.
   (b) If there is a second reflector, flat and horizontal, at a depth of 4 km, calculate the travel time and the surface arrival point x of a zero-offset ray from this reflector which passes through the point A of the first reflector. Would this second reflector be flat in the time model?

7. Derive Equation (5.6b).

8. A two-layer model is proposed for a certain subsurface zone ($n = 2$ in Figure 5.1), with $h_1 = 0.3$ km, $h_2 = 0.5$ km, $v_1 = 1$ km/s, $v_2 = 2$ km/s. For the source–receiver offset $x = 1.0$ km, Equation (5.7) is solved for the p value of the primary ray via a root-finding method. The result is $p = 0.3129$ s/km. Using Equation (5.6a), compute the travel time for each segment of the ray, as well as the total two-way travel time t of the ray. Compute also the angle and horizontal offset, θ_j and x_j, for each ray segment. Verify that $x = 2 \sum_j x_j$.

9. Suppose that the value of the velocity $v(z)$ at some depth z is twice that of the surface velocity $v(0)$. For a vertically traveling wave, use Equation (5.43b) to calculate the ratio of the amplitude of the *SH* wave at the depth z to that at the surface. By how much does the amplitude change?

10. This problem involves deriving an approximate solution to the equation of motion for a heterogeneous medium in a relatively simple case: vertically traveling acoustic waves in a vertically heterogeneous medium.

    We have seen that the equation of motion for the pressure $\mathcal{P}$ in a heterogeneous medium is

$$\nabla^2 \mathcal{P} = \frac{1}{v^2} \frac{\partial^2 \mathcal{P}}{\partial t^2}, \quad v = v(x, y, z), \qquad (1)$$

where $\mathcal{P} = \mathcal{P}(x, y, z, t)$ and where the body force has been dropped, as we are interested only in plane wave solutions. This has the form of a standard classical wave equation, except that v is not a constant. For a vertically heterogeneous medium, and for plane waves traveling in the vertical direction only, the pressure $\mathcal{P}$ does not depend on x or y, meaning (1) reduces to

$$\frac{\partial^2 \mathcal{P}}{\partial z^2} = \frac{1}{v(z)^2} \frac{\partial^2 \mathcal{P}}{\partial t^2}, \quad \mathcal{P} = \mathcal{P}(z, t). \qquad (2)$$

We want to obtain an approximate wavelike solution to this equation, so we assume that the heterogeneity is weak, or equivalently, that we have high enough frequencies. To obtain a solution, we apply a relatively simple version of a general method known in geophysics as the WKBJ method:

(a) First, substitute the following trial solution into (2):

$$\mathcal{P} = A(z) \exp\{i\omega[T(z) - t]\}. \qquad (3)$$

For a homogeneous medium, $A = const$ and $T(z) = z/v$, but in this case, we must solve for the unknown functions $A(z)$ and $T(z)$. To save yourself some writing, you may want to write dT/dz as T', etc. After doing the math (taking the derivatives in (2), etc.), rearrange your result into the form

$$E_2 \omega^2 + E_1 \omega + E_0 = 0. \qquad (4)$$

What are the mathematical expressions that you get for E_2, E_1, and E_0? Are E_2, E_1, and E_0 independent of ω?

(b) Since we are looking for high frequency solutions, we keep only the first two terms on the left side of (4). This gives us two equations for the two unknowns A and T. Since E_2 and E_1 (and E_0) are independent of ω, in order for (4) to hold for all ω, we must have

$$E_2 = 0, \quad E_1 = 0. \qquad (5)$$

Verify that the first of these is the 1D eikonal equation, and the second is the 1D transport equation, by comparing them with (5.52) and (5.56) in the text.

(c) Solve the differential equations $E_2 = 0$ and $E_1 = 0$ for $T(z)$ and $A(z)$ and obtain the approximate solution for $\mathcal{P}(z, t)$. Express your answer for

$A(z)$ in terms of $A(0)$ and $v(0)$. Does your solution for $\mathcal{P}$ reduce to the correct solution for a homogeneous medium?

(d) What are $T(z)$ and $A(z)$ for the case $v = v_0 + az$? Sketch or plot graphs of them. Also, sketch or plot $\text{Re}(\mathcal{P})$ vs. z at the instant $t = 0$. Lastly, the solution is valid for frequencies satisfying $\omega \gg |dv/dz|$. If $a = 2$ km/s per km, for what frequencies would the solution hold?

5.15 Formulas That May Prove Useful in Doing Exercises 2, 4, and 5

C and a in the following are constants:

$$\int \frac{du}{\sqrt{1+u^2}} = \ln\left(u + \sqrt{1+u^2}\right) + C \equiv \sinh^{-1} u + C$$

$$\int \frac{du}{u\sqrt{1-u^2}} = \ln\left(\frac{u}{1+\sqrt{1-u^2}}\right) + C$$

$$\int \frac{u}{\sqrt{1-u^2}}\, du = -\sqrt{1-u^2} + C$$

$$\int \frac{du}{\sqrt{u^2-1}} = \ln\left(u + \sqrt{u^2-1}\right) + C \equiv \cosh^{-1} u + C$$

$$\int \frac{du}{u^2\sqrt{a^2-u^2}} = -\frac{\sqrt{a^2-u^2}}{a^2 u} + C$$

Also, $\ln\left(\dfrac{1}{u} + \sqrt{\dfrac{1}{u^2} - 1}\right) \equiv \text{sech}^{-1} u.$

6

Data Transformations

In this chapter, we briefly study the transformation of seismic data into the frequency domain, the frequency-wavenumber domain, and the τ-p domain, and the connection with seismic wave theory.

6.1 The Frequency Domain

The **Fourier time transform** is reviewed in the last section of Chapter 1. In seismology, it is used to transform seismic data from the time domain into the frequency domain. The Fourier forward and inverse transforms are defined by

$$\overline{g}(\omega) = \int_{-\infty}^{\infty} g(t) e^{i\omega t} \, dt \qquad \text{and} \qquad (6.1)$$

$$g(t) = \frac{1}{2\pi} \int_{-\infty}^{\infty} \overline{g}(\omega) e^{-i\omega t} \, d\omega \qquad (6.2)$$

respectively. $\overline{g}(\omega)$ is the **frequency spectrum** of $g(t)$. Since it is in general a complex function, it can be written as

$$\overline{g} = \text{Re}(\overline{g}) + i\text{Im}(\overline{g}) = |\overline{g}(\omega)| e^{i\psi(\omega)}, \qquad (6.3)$$

where $|\overline{g}(\omega)|$ is the **amplitude spectrum** of $g(t)$ and $\psi(\omega)$ is the **phase spectrum** of $g(t)$, and where

$$|\overline{g}(\omega)| = \sqrt{\left[\text{Re}(\overline{g})\right]^2 + \left[\text{Im}(\overline{g})\right]^2}, \qquad (6.4)$$

$$\psi(\omega) = \tan^{-1}\left[\text{Im}(\overline{g}) / \text{Re}(\overline{g})\right]. \qquad (6.5)$$

Consider the Gaussian function

$$g(t) = e^{-at^2}, \qquad (6.6)$$

where a is a real positive constant. Then

$$\overline{g}(\omega) = \int_{-\infty}^{\infty} e^{-at^2} e^{i\omega t} \, dt = \int_{-\infty}^{\infty} e^{-at^2} \cos(\omega t) \, dt = \sqrt{\frac{\pi}{a}} e^{-\omega^2/4a} \qquad (6.7)$$

where we have used the formula $e^{i\theta} = \cos\theta + i\sin\theta$ and the fact that $e^{-at^2}\sin(\omega t)$ is an odd function (meaning that its integral from $-\infty$ to $+\infty$ is zero). Hence, the frequency spectrum of a Gaussian function is also a Gaussian function.

Note also that the exponent or argument of the exponential in $g(t)$ is proportional to a, whereas in $\overline{g}(\omega)$, it is proportional to $1/a$. That means that as a increases, $g(t)$ gets narrower whereas $\overline{g}(\omega)$ gets wider, and vice versa. In other words, the narrower the signal, the wider is its frequency spectrum bandwidth, and vice versa.

What is the frequency spectrum of $\dot{g}(t)$ ($\equiv dg/dt$)? Using integration by parts, we obtain

$$\overline{\dot{g}}(\omega) = \int_{-\infty}^{\infty} \frac{dg}{dt} e^{i\omega t} \, dt = \left[e^{i\omega t} g(t) \right]_{-\infty}^{\infty} - \int_{-\infty}^{\infty} g(t) \left[i\omega e^{i\omega t} \right] dt = -i\omega \overline{g}(\omega), \quad (6.8)$$

where we have assumed that $[e^{i\omega t} g(t)]_{-\infty}^{\infty} = 0$ (which is obviously true for the Gaussian in Equation 6.6, as $g(t) \to 0$ as $t \to \pm\infty$). In words, the spectrum of dg/dt is just $-i\omega$ times the spectrum of g. We also obtained this result in Equation (1.150) using a different method. Applying this rule to the Gaussian pulse, we obtain

$$\dot{g}(t) = -2at e^{-at^2} \qquad \Longrightarrow \qquad \overline{\dot{g}}(\omega) = -i\omega \sqrt{\frac{\pi}{a}} e^{-\omega^2/4a} \qquad (6.9)$$

and

$$\ddot{g}(t) = 2a(2at^2 - 1)e^{-at^2} \qquad \Longrightarrow \qquad \overline{\ddot{g}}(\omega) = -\omega^2 \sqrt{\frac{\pi}{a}} e^{-\omega^2/4a}. \qquad (6.10)$$

The rule in (6.8) can be applied to obtain $\overline{\dot{g}}(\omega)$ and $\overline{\ddot{g}}(\omega)$ because $g(t)$ and $\dot{g}(t)$ both $\to 0$ as $t \to \pm\infty$.

$\ddot{g}(t)$ in Equation (6.10), or its negative, is known as a **Ricker wavelet** (Ricker, 1953). Figure 6.1 shows the amplitude and phase spectra of the Gaussian pulse and its first two derivatives.

Equation (6.8) shows that taking the time derivative of a wavelet g changes its phase by $90°$ (because $i = e^{i\pi/2} = e^{i90°}$). Since the frequency spectrum of dg/dt has the extra multiplicative factor "ω", the phase change depends on the sign of ω, and the spectrum of dg/dt is richer in high frequencies and poorer in low frequencies than the spectrum of g. Furthermore, the peak of the amplitude spectrum of dg/dt occurs at a higher frequency than does the peak of the amplitude spectrum of g.

Since reflection coefficients are complex numbers in general, having magnitude and phase, the shape of a wavelet can change upon reflection from an interface, due

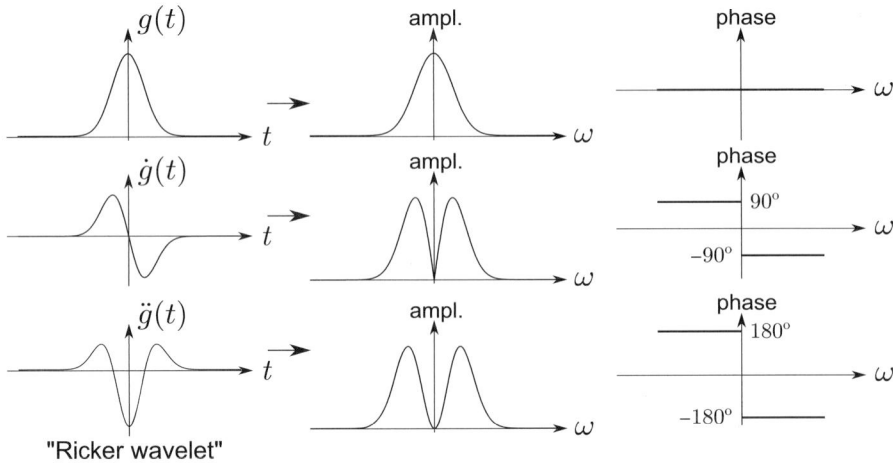

Figure 6.1 Graphs of $g(t) = e^{-at^2}$, dg/dt and d^2g/dt^2, and their amplitude and phase spectra. d^2g/dt^2 is called a "Ricker wavelet."

to a phase shift (although the shape of the amplitude spectrum would remain the same). The simplest example of this is a polarity reversal, which is a 180° phase shift. If reflection causes a 90° phase shift to occur for all frequencies, then a pulse that looks roughly like $\dot{g}(t)$, say, in Figure 6.1 before reflection, would look roughly like $\ddot{g}$ in Figure 6.1 after reflection. This effect can be seen on wide-angle reflections on seismic shot records.

6.2 The Frequency-Wavenumber Domain

The Fourier transform can be applied to a given trace $g(t)$ on a seismic section to obtain the spectrum $\bar{g}(\omega)$ of the trace – the trace is transformed from the time domain to the frequency domain. However, seismic data are functions of offset x as well as time t, and one can also do a Fourier transform in the x direction, so that the seismic section is transformed from the time-offset (t-x) domain to the frequency-wavenumber (ω-κ) domain. Some seismic processes can often be done more conveniently in the frequency-wavenumber domain, such as the removal of noise events such as ground roll and air waves.

Let $s(x, t)$ represent the seismic data, i.e., the sample values on a seismic section (e.g., a shot record, a stack section, a CMP gather, etc.). Then the Fourier time and space transforms are given by

$$\bar{s}(x, \omega) = \int_{-\infty}^{\infty} s(x, t) e^{i\omega t} \, dt \qquad \text{and} \qquad \widetilde{s}(\kappa, t) = \int_{-\infty}^{\infty} s(x, t) e^{-i\kappa x} \, dx. \qquad (6.11)$$

$\bar{s}(x, \omega)$ is the frequency spectrum of the data along a line of constant offset x (along the trace with offset x). $\widetilde{s}(\kappa, t)$ is the wavenumber spectrum of the data along a line of constant time t (across the traces). If we do a sequence of t-transforms of $s(x,t)$ for all values of x to obtain the 2D function $\bar{s}(x,\omega)$, and if we follow these by a sequence of x-transforms of $\bar{s}(x,\omega)$ for all values of ω, we obtain the 2D Fourier transform $\tilde{\bar{s}}(\kappa,\omega)$ of the seismic data, i.e.,

$$\tilde{\bar{s}}(\kappa,\omega) = \int_{-\infty}^{\infty}\int_{-\infty}^{\infty} s(x,t)e^{i(\omega t - \kappa x)}\,dt\,dx. \tag{6.12}$$

$\tilde{\bar{s}}(\kappa,\omega)$ is the **frequency-wavenumber spectrum**, often written as "the ω-κ spectrum" of the seismic data (or "the f-k spectrum" if the true frequency and wavenumber are used). The wavenumber κ (or the true wavenumber k) can be thought of as a "spatial" frequency (with units of cycles/meter, say). The inverse 2D Fourier transform is given by

$$s(x,t) = \frac{1}{(2\pi)^2}\int_{-\infty}^{\infty}\int_{-\infty}^{\infty} \tilde{\bar{s}}(\kappa,\omega)e^{i(\kappa x - \omega t)}\,d\omega\,d\kappa. \tag{6.13}$$

Note that this states that a seismic data section $s(x,t)$ can be mathematically represented as a sum of harmonic steady-state horizontally traveling plane waves $e^{i(\kappa x - \omega t)}$ with different values of ω and κ, and where for each wave, we have, from Equation (2.56), $\omega = v\kappa$ (or $f = vk$), where v is the speed at which the plane wave travels in the x direction. This makes intuitive sense because the geophones that record the waves are located along a horizontal line on the surface, and as the geophones move, one after another, in response to a disturbance, it appears as if the disturbance is traveling in the horizontal direction. Since the actual waves in the subsurface travel in all directions, we can think of v as the *apparent velocity*. The amplitude and phase of each plane wave (obtained from $\tilde{\bar{s}}(\kappa,\omega)$) also vary with ω and κ.

For a shot record, this apparent velocity v is the same as the apparent velocity c defined in Equation (5.11), i.e., v is given by the slope dx/dt of an event on the shot record and also by $v_1/\sin\theta_1$ (see Figure 5.2b). However, for a stack section, v and c are different, as explained in the following paragraph.

Consider a stack section containing a single flat dipping event, produced by a flat geological reflector with dip angle δ, with the body wave speed being v_1 above the reflector. By tracing zero-offset rays from the reflector, we know that the event can be thought of as being generated by an upcoming plane wave incident upon the surface at the angle $\theta_1 = \delta$ (measured from the vertical). If we determine v by measuring the slope of the event on the stack section, i.e., $v = dx/dt$, and if we determine c from (5.11), i.e., $c = v_1/\sin\theta_1$, we find, using some algebra, that $c = 2v$. For example, if the event has a dip of 20 ms/trace, and if the trace interval is 40 m, then $v = 40/0.02 = 2{,}000$ m/s but $c = 4{,}000$ m/s. If the event is horizontal

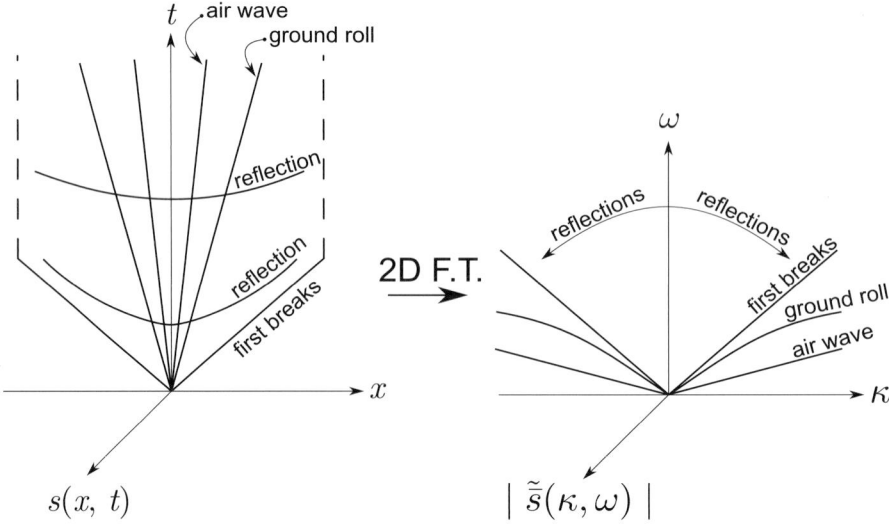

Figure 6.2 Schematic diagrams of a seismic shot record and its frequency-wavenumber amplitude spectrum. A linear event falls along the line $\omega = v\kappa$ on the spectrum, where v is the slope of the line, and is the apparent velocity of the event on the *x-t* section.

(no dip), then $dt = 0$ and $\theta_1 = 0$), meaning $c = v = \infty$. Also, note that c is always $>v_1$, but that v can be $<v_1$ for large enough θ_1.

Different types of events have different apparent velocities. One can use apparent velocities to discriminate between events on a plot of the frequency-wavenumber spectrum (see Figure 6.2).

The ground roll line on the ω-κ plot of Figure 6.2 is slightly curved, which indicates that ground roll is dispersive, i.e., $v = v(\omega)$ (the ω versus κ curve is not a straight line of constant slope v). A flat nondipping event on the seismic data whose waveform is bandlimited in frequency would appear as an event lying on the frequency axis between the frequency band limits on the ω-κ plot. A flat dipping bandlimited event with apparent velocity v would appear on the ω-κ plot along the line $\omega = v\kappa$ between the frequency band limits (and also between the wavenumber band limits). Since each hyperbolic reflection event in Figure 6.2 can be thought of as a superposition of linear events (a set of lines tangent to the reflection curve) with apparent velocities between some finite value and infinity, the reflection data appear in a conical region about the frequency axis on the ω-κ plot.

Unwanted events such as ground roll and air waves that contaminate the reflections on a shot record can be eliminated by zeroing them out on the ω-κ plot and transforming back to the *x-t* domain. Also, the migration of seismic data can be conveniently performed in the frequency-wavenumber domain.

6.3 The τ-p **Domain**

Consider a subsurface model consisting of a sequence of flat nondipping homogeneous layers (Figure 5.1a). The travel time t for the primary ray with offset x whose ray parameter is p ($= p(x)$) that reflects off the nth reflector is given by Equation (5.6b), i.e.,

$$t = \tau + px \tag{6.14}$$

where

$$\tau = \sum_{j=1}^{n} \tau_j = \sum_{j=1}^{n} \frac{2h_j}{v_j}\sqrt{1 - p^2 v_j^2} \equiv \tau(p). \tag{6.15}$$

Consider a new data plane, the p-τ plane, where p and τ are the horizonal and vertical coordinates, respectively. The preceding travel time curve for the nth reflector can be mapped onto the p-τ plane by plotting τ as a function of p. The corresponding curve in the p-τ plane would consist of the continuous set of points $[p, \tau(p)]$. What kind of curve is this?

Note first that the equation for τ_j is the equation for an ellipse in the p-τ_j plane. This can also be seen by rearranging the equation for τ_j into the form

$$v_j^2 p^2 + \left(\frac{v_j}{2h_j}\right)^2 \tau_j^2 = 1, \tag{6.16}$$

which is the equation of an ellipse in the p-τ_j plane with semi-axis lengths $2h_j/v_j \equiv \tilde{t}_j$ and $1/v_j$. Since τ is just the sum of the τ_js, we see that (a) for a single reflector, $\tau(p)$ for the reflection travel time curve, which is a hyperbola in the x-t plane, is an ellipse in the p-τ plane, and (b) for n reflectors, $\tau(p)$ for the reflection travel time curve for the nth reflector, which is approximately a hyperbola in the x-t plane, is the sum of n ellipses in the p-τ plane, i.e., the reflection curve (6.14) maps onto the set of points $[p, \tau(p)]$, where $\tau(p)$ is given by (6.15) – see Figure 6.3.

Note that in the sum of Equation (6.15), the expression under each square root must be positive ($1 - p^2 v_j^2 \geq 0$, $j = 1, \ldots, n$). Hence, for the nth reflector, the range of values that p is allowed to take is $0 \leq p \leq p_{\max}$, where $p_{\max}$ is the smallest of the numbers $1/v_1, \ldots, 1/v_n$. For example, $p_{\max} = 1/v_1$ for the first reflector, $p_{\max} = 1/\max(v_1, v_2)$ for the second reflector, etc.

Head wave refractions in a many-layered medium can also be mapped onto the p-τ plane. A head wave event on a shot record lies along a straight line tangent to the associated reflection event at the critical distance (Figures 4.5 and 6.3). From basic reflection seismology, we know that a head wave from the nth reflector has a travel time t given by

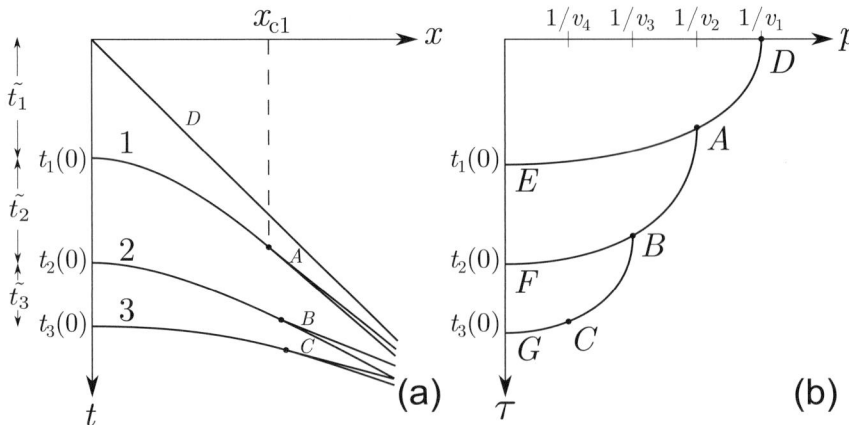

Figure 6.3 Schematic diagrams of the *x-t* plot (a) and *p-τ* plot (b) for a three-layer model for which $v_1 < v_2 < v_3 < v_4$. On the *x-t* plot, reflections 1, 2, and 3 are asymptotic to lines *D*, *A*, and *B*, respectively. Reflection 1 on the travel time plot maps onto ellipse *EAD* on the *p-τ* plot. Reflection 2 maps onto curve *FBA*, which is the sum of ellipses $\tau_1(p)$ and $\tau_2(p)$ in the range $0 < p < 1/v_2$ (see Equation 6.15). Similarly, reflection 3 maps onto curve *GCB*. The portion of reflection curve 1 that lies past the critical distance x_{c1} maps onto segment *AD*. Similarly, the supercritical portions of reflections 2 and 3 map onto segments *BA* and *CB*, respectively. The direct wave *D* maps onto the *point D* in the *p-τ* plot, and the head wave refractions *A*, *B*, and *C* associated with reflections 1, 2, and 3 map onto points *A*, *B*, and *C*, respectively. At point *A* on the *p-τ* plot, the slope of the ellipse *EAD* is $d\tau/dp = -x_{c1}$, with similar equations holding for points *B* and *C*. On the *x-t* plot, reflection 1 approaches the direct arrival *D* asymptotically, reflection 2 approaches the head wave *A* asymptotically, etc. After Diebold and Stoffa (1981), with permission from the Society of Exploration Geophysicists.

$$t = \frac{x}{v_{n+1}} + \sum_{j=1}^{n} \frac{2h_j}{v_j} \sqrt{1 - \frac{v_j^2}{v_{n+1}^2}}. \tag{6.17}$$

This has the form of Equation (6.14/15) with $p = 1/v_{n+1}$. Since this p is a constant, τ in (6.15) is also a constant, and hence Equation (6.14), $t = \tau + px$, is the equation of a straight line in the *x-t* plane with slope p and *t*-intercept τ, i.e., the head wave travel time line on a shot record. Since p and $\tau(p)$ are constants, the head wave travel time line maps onto the *point* (p, τ) in the *p-τ* plane, where

$$p = \frac{1}{v_{n+1}}, \qquad \tau = \sum_{j=1}^{n} \frac{2h_j}{v_j} \sqrt{1 - \frac{v_j^2}{v_{n+1}^2}}. \tag{6.18}$$

Similarly, the direct wave event line on the shot record, being also a straight line, maps onto a point on the p axis (see Figure 6.3).

We have seen that the slope of a reflection travel time curve on a shot record is the ray parameter, i.e., $p = dt/dx$ (Equation 5.8, Figure 5.2a). Similarly, for the slope $d\tau/dp$ of the corresponding curve in the p-τ plane, Equation (6.15) can be differentiated to give $d\tau/dp = -x$.

In Equation (6.14) for the travel time t along the nth reflection curve, p varies with x as we move along the curve. However, if p is fixed as x varies, then Equation (6.14) becomes the equation for the tangent line to the reflection curve. The slope of this tangent line is p at the offset x, with τ being the time at which the tangent line intersects the t axis, i.e., the t-intercept (Figure 5.2). As we move along the reflection curve, determining the slope p and intercept time τ at each point, we see that τ decreases as p increases (see Figure 6.3b), and if we also plot these determined points (p, τ) on the p-τ plane, we get the curves seen on the p-τ plane in Figure 6.3.

The preceding discussions suggest a general way of mapping any curve $t = t(x)$ in the x-t plane onto the p-τ plane. At each point on the curve $t = t(x)$, we measure the slope p and intercept time τ of the tangent line at that point. This gives us a series of points (p, τ) on the p-τ plane, which represent the desired mapping. As we have seen, hyperbolic-type reflection curves map onto curves which are sums of ellipses. A straight line in the x-t plane maps onto a single point in the p-τ plane, because each point on the straight line gives the same slope p and intercept time τ. If a straight line event on the x-t plane is considered to be noise, this noise can be compressed onto a single point in the p-τ plane. This is one of the advantages of transforming data into the p-τ plane.

Referring again to Figure 5.2a, it turns out that the tangent line of slope p is also the travel time curve of the reflection signal that would be recorded by the geophone spread if the wave emanating from the source was a single *plane* wave with ray parameter p. This can be seen by drawing an analogy with the upgoing head wave of Figure 4.4, a plane wave (in 2D) whose recorded signal lies along the straight line H on the travel time plot of Figure 4.5. Another way to see it is to note that a plane wave from the source will produce a reflected wave that is also a plane wave. For a single frequency, this reflected plane wave has the waveform $\exp[-i\omega(t - px - \xi z)]$, meaning that its travel time curve is $t = px + \xi z$, which is a straight line in the x-t plane. The reflection travel time curve in Figure 5.2a is the response to a source wave that is a *superposition* of plane waves with different p values, such as a spherical wave generated by a point source. The travel time curve itself is the envelope of all the different straight line travel time curves (the tangent lines) produced by the different plane waves in the superposition. For a point source generating a spherical wave, the reflected wave, in the case of a single reflector at depth h, has the form $\exp[-i\omega(t - r/v)]/r$, where $r = \sqrt{x^2 + 4h^2}$, i.e., the travel time curve is $t = r/v = \sqrt{x^2 + 4h^2}/v$, which is a hyperbola in

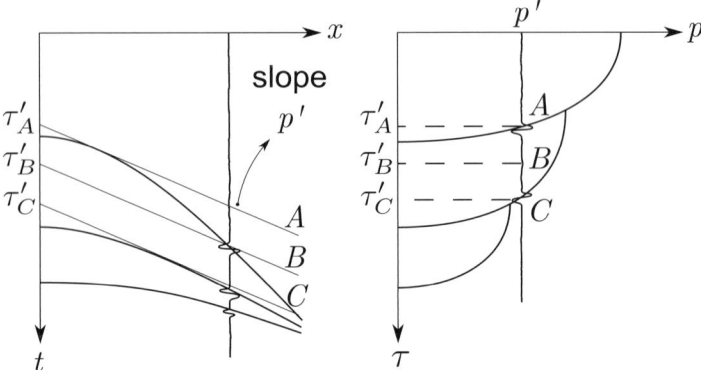

Figure 6.4 An illustration of the slant-stack process. The three slant lines A, B, and C on the travel time plot $s(x, t)$ all have the same slope p'. The sums of the samples along the slant lines appear at points A, B, and C on the p-τ plot $S(p, \tau)$.

the x-t plane. To summarize, a point source that generates a spherical wave results in a hyperbolic-type reflection travel time curve, whereas a source that generates a single plane wave results in a straight-line reflection travel time curve. In field practice, though, plane wave sources are rarely used.

Consider the following operation. On a shot record, sum the amplitudes of the data samples that lie along the straight line $t = \tau + px$, and plot the sum at the point (p, τ) on the p-τ plane. This sum is essentially the sum of the geophone pulses for the plane wave component (of the source wave) whose ray parameter is p. Also, sum along all other possible parallel lines (which have the same slope p but different intercept times τ) and plot the sums on the p-τ plane. These plotted values form a trace in the p-τ plane that represents, in an average sense, the response of the subsurface to the plane wave component (of the source wave) whose ray parameter is p. Repeat the whole procedure for all other possible values of p to obtain a set of traces in the p-τ plane (see Figure 6.4). This process is known as **slant-stacking**. The main contributions to the sums along the lines will obviously come from the zones where the lines are tangent to a reflection. Slant-stacking is a **plane wave decomposition** technique, i.e., the slant-stack section (or **p-gather**) represents a decomposition of the recorded seismic travel time data into plane waves.

Mathematically, the slant-stacking process is represented as follows. If $s(x, t)$ is the function representing the recorded seismic data on a shot record, and if $S(p, \tau)$ is the function representing the data on the slant-stack section, then the slant-stacking procedure described in the previous paragraph is expressed mathematically as

$$S(p, \tau) = \sum_{i=1}^{N} s(x_i, \ \tau + px_i), \qquad (6.19)$$

where the x_is are the offsets of the N traces on the shot record. In the continuous sense, the slant-stack is given by

$$S(p, \tau) = \int_{-\infty}^{\infty} s(x, \ \tau + px) \, dx, \tag{6.20}$$

which is also called the **Radon transform** of $s(x, t)$. To go from the p-τ domain to the x-t domain, one must perform an inverse slant-stack, or inverse Radon transform, on the p-τ data. The inverse Radon transform is given mathematically by

$$s(x, t) = \int_{-\infty}^{\infty} \frac{d}{dt} \mathcal{H}\Big[S(p, \ t - px) \Big] \, dp, \tag{6.21}$$

where $\mathcal{H}$ denotes the **Hilbert transform**, defined by

$$\mathcal{H}\Big[f(t) \Big] = \frac{1}{\pi} \int_{-\infty}^{\infty} \frac{f(t')}{t' - t} \, dt' = f(t) * \left(-\frac{1}{\pi t} \right), \tag{6.22a}$$

where "$*$" denotes the convolution operation. To handle the singularity at $t' = t$, the Cauchy principal value of the integral is evaluated (see any advanced calculus text).

Another convenient formula for determining $\mathcal{H}\Big[f(t) \Big]$, the Hilbert transform of $f(t)$, is

$$f(t) + i\mathcal{H}\Big[f(t) \Big] = \frac{1}{\pi} \int_{0}^{\infty} \overline{f}(\omega) e^{-i\omega t} \, d\omega, \tag{6.22b}$$

(Červený and Ravindra, 1971, pp. 94–95; Červený, 2001, equations A.3.2, A.3.3). Using this formula, one may actually determine both $f(t)$ and its Hilbert transform, if one knows $\overline{f}(\omega)$, the frequency spectrum (Fourier transform) of $f(t)$. The expression on the right-hand side of Equation (6.22b) is sometimes called the **analytic signal**.

It can be shown that the Hilbert transform of $f(t)$ advances the phase of each Fourier component of $f(t)$ by $90°$. The Hilbert transform of the Dirac delta function $\delta(t)$ is the function $-1/\pi t$.

The Radon transform and the 2D Fourier transform are closely related. We have seen that a linear event with apparent velocity v in the x-t domain transforms into a linear event along the line $\omega = v\kappa = \kappa/p$ in the ω-κ domain. Consider then the 2D Fourier transform $\widetilde{\overline{s}}(\kappa, \omega)$ of the seismic data $s(x, t)$. If we evaluate $\widetilde{\overline{s}}$ along the line $\omega = \kappa/p$, we obtain

$$\begin{aligned}
\widetilde{\overline{s}}(\omega p, \ \omega) &= \int_{-\infty}^{\infty} \int_{-\infty}^{\infty} s(x, t) e^{i(\omega t - \omega p x)} \, dt \, dx \\
&= \int_{-\infty}^{\infty} \left[\int_{-\infty}^{\infty} s(x, \ \tau + px) \, dx \right] e^{i\omega\tau} \, d\tau = \int_{-\infty}^{\infty} S(p, \tau) e^{i\omega\tau} \, d\tau,
\end{aligned} \tag{6.23}$$

where $\tau = t - px$ has been substituted in the second line. Equation (6.23) simply states that the sequence of data values along the line $\omega = \kappa/p$ in the ω-κ domain can be obtained by doing a 1D Fourier transform of the trace in the p-τ domain, whose ray parameter value is p.

Various seismic processes, such as deconvolution, dip-filtering, inversion for velocities, trace interpolation, multiple suppression, analysis of normal modes, etc., can often be done more conveniently in the τ-p domain. For more details and examples, see Diebold and Stoffa (1981); Robinson (1983); Schultz and Claerbout (1978); Stoffa, Buhl, Diebold, and Wenzel (1981); Yilmaz (1987); among others.

Exercises

1. The frequency spectrum $\bar{s}(\omega)$ of a signal $s(t)$ is given by $\bar{s}(\omega) = i\omega \exp(-a|\omega|)$, where a is a positive constant (see Figure 6.5).

   (a) Verify that Figure 6.5 is correct (use curve-sketching techniques from basic calculus).
   (b) Calculate $s(t)$.
   (c) Calculate the Hilbert transform $h(t)$ of the wavelet $s(t)$ by advancing the phase of each Fourier component $e^{-i\omega t}$ by 90° (in the inverse Fourier transform, multiply the positive-frequency Fourier components by the phase factor $e^{-i\pi/2}$ and the negative ones by $e^{i\pi/2}$ and then integrate). Note that in this case the 90° phase shift (advance) applied to $s(t)$ cancels its phase, i.e., $h(t)$ is a zero-phase wavelet.
   (d) Sketch or plot $s(t)$ and $h(t)$ on a graph.
   (e) Sketch or plot graphs of the time-derivative ds/dt of the wavelet and its amplitude and phase spectra.
   (f) Sketch or plot graphs of the time-integral $\int s(t)\,dt$ of the wavelet and its amplitude and phase spectra (the wavelet for the head wave, i.e., refraction arrival, is basically the time-integral of the reflection-arrival wavelet).

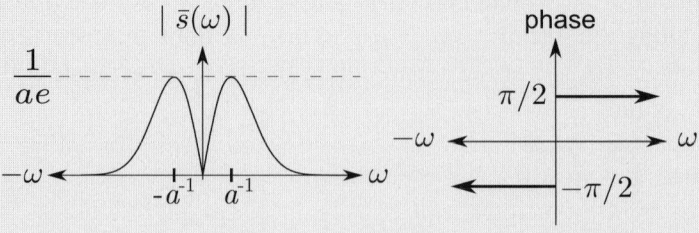

Figure 6.5 See exercise 1a.

(g) Compute $s(t)$ and its Hilbert transform $h(t)$ by using Equation (6.22b). Confirm that you obtain the same results as in parts (b) and (c).

2. Derive the formula for the Hilbert transform $\mathcal{H}[f(t)]$ of the boxcar $f(t)$ given by

$$f(t) = \begin{cases} 1, & -T < t < T \\ 0, & \text{all other } t \end{cases}.$$

Plot a graph of $\mathcal{H}[f(t)]$ vs. t.

Note: For those values of t for which the integrand of the Hilbert transform (equation 6.22a) contains a singularity ($t' = t$) within the range of integration, you will have to evaluate the integral by taking the Cauchy principal value (*CPV*): if $g(x)$ is singular at $X(g(X) = \pm\infty)$, then for $a < X < b$,

$$CPV\left(\int_a^b g(x)\,dx\right) = \lim_{\epsilon \to 0}\left(\int_a^{X-\epsilon} g(x)\,dx + \int_{X+\epsilon}^b g(x)\,dx\right).$$

3. A subsurface medium consists of two homogeneous horizontal layers over a half-space, with the P wave speed increasing with depth. A shot record for the medium shows the two primary P wave reflections, the direct P wave, the two P head waves, the air wave, and ground roll. Carefully sketch these data in the t-x plane, the p-x plane, the p-τ plane, and the ω-κ plane.

7

Synthetic Seismograms

For a given model of the subsurface, a theoretical seismogram, or **synthetic seismogram**, showing the various reflection arrivals (body waves) can be computed for a given source–receiver offset by tracing rays through the model from source to receiver. Synthetic seismic sections can be compared with the actual seismic data to see if the assumed subsurface model is realistic or accurate. If not, the model can be refined and a new synthetic section can be computed and compared with the data. This **modeling** procedure can be repeated iteratively until an accurate model is obtained. Synthetic sections can also be helpful in identifying unknown events in the recorded data.

For each ray that is traced, the amplitude and phase at the receiver are calculated. For an ideal (perfectly elastic) medium, this generally involves the calculation of reflection and transmission coefficients, and the effects of geometrical spreading (spherical divergence of wavefronts). If the medium is significantly nonideal, then the effects of other processes and physical conditions, such as anelastic absorption and anisotropy, may have to be calculated. Also, some type of waveform function or wavelet $w(t)$ is usually chosen for the synthetic traces (such as a Ricker wavelet – see Figure 6.1), in order to facilitate comparison with the recorded field data.

7.1 The Subsurface Model and Synthetic Trace

Consider a model consisting of a sequence of N flat horizontal homogeneous layers, with ρ_j, α_j, and h_j being the density, P wave speed and thickness, respectively, for the jth layer (Figure 5.1a). Assume we are dealing with vertically incident P waves (the zero-offset case) and primary reflections only (e.g., no multiples). Assume the shot goes off at $t = 0$, and that the shot is located at or very near the surface.

Consider first a single frequency ω (harmonic waves) and assume perfect elasticity. For perfectly elastic (nondissipative) media, the wave velocities are independent of frequency.

Let A_n denote the displacement amplitude at the receiver of the primary reflection from the nth interface. As mentioned previously, A_n contains the effects of reflection, transmission, geometrical spreading, etc. Let A_0 denote the displacement amplitude of the source wave at a small distance r_0 from the source. A_n and A_0 have units of length. We can arbitrarily set $A_0 \equiv 1$, but we will retain A_0 for clarity in the formulas that follow.

Let $\tau_n \equiv t_n(0)$ denote the total two-way travel time of the normal-incidence primary reflection from the base of layer n (in Chapter 5, $t_n(x)$ denoted the two-way travel time for offset x). Then,

$$\tau_1 = \frac{2h_1}{\alpha_1}, \quad \tau_2 = \frac{2h_1}{\alpha_1} + \frac{2h_2}{\alpha_2}, \quad \dots, \quad \tau_n = 2\sum_{j=1}^{n} \frac{h_j}{\alpha_j}, \quad \dots . \quad (7.1a)$$

The displacement $\mathbf{u}^{(G)}$ of the geophone due to the nth primary reflection is given by

$$\mathbf{u}^{(G)} = W(\omega)A_n \exp\left[-i\omega(t - \tau_n)\right]\mathbf{d}^{(G)}, \quad (7.1b)$$

where $W(\omega)$ is the (complex) frequency spectrum of the source pulse $w(t)$, and $\mathbf{d}^{(G)}$ is the polarization vector, modified to include the free surface effect. $\mathbf{d}^{(G)}$ is given by (3.32), i.e.,

$$\mathbf{d}^{(G)} = [\mathbf{d}^{(IP)} + R_{PP}\mathbf{d}^{(RP)} + R_{PS}\mathbf{d}^{(RS)}]. \quad (7.1c)$$

For normal incidence, $R_{PP} = -1$, $R_{PS} = 0$, and $\mathbf{d}^{(IP)} = -\mathbf{e}_z = -\mathbf{d}^{(RP)}$ (for the z axis pointing downward into the Earth). Consequently, $\mathbf{d}^{(G)} = -2\mathbf{e}_z$, and the z component of displacement is

$$u_z^{(G)} = -2W(\omega)A_n \exp\left[-i\omega(t - \tau_n)\right]. \quad (7.1d)$$

To obtain the full waveform, an integration over frequency must be performed (i.e., an inverse Fourier transform, e.g., equation 1.133b). The nth reflection pulse is then given by

$$u_z^{(G)} = -2\int_{-\infty}^{\infty} W(\omega)A_n \exp\left[-i\omega(t - \tau_n)\right] d\omega = -4\pi A_n w(t - \tau_n), \quad (7.1e)$$

where A_n has been taken outside of the integral since A_n is independent of frequency for nondissipative media. The waveform in (7.1e) is essentially the source wavelet $w(t)$ shifted to the arrival time of the pulse, τ_n, and multiplied by A_n.

Ignoring the "-4π" (it is only a scale factor), the trace $s(t)$ then consists of the superposition of all such reflection arrivals, i.e.,

$$s(t) = \sum_{n=1}^{N} A_n w(t - \tau_n) \tag{7.1f}$$

$$= \left[\sum_{n=1}^{N} A_n \delta(t - \tau_n) \right] * w(t) = A(t) * w(t), \tag{7.1g}$$

where the asterisk stands for the convolution operation, and where we have used the convolution rule $\delta(t - \tau) * w(t) = w(t - \tau)$, in which $\delta(t)$ is the Dirac delta function, and where

$$A(t) \equiv \sum_{n=1}^{N} A_n \delta(t - \tau_n). \tag{7.1h}$$

$A(t)$ is a series of Dirac delta function "spikes." The nth spike has the amplitude A_n and is located at the time τ_n. Note that $s(t)$ is the *convolution* of the reflection amplitude spike sequence $A(t)$ with the wavelet $w(t)$. Equation (7.1g) is referred to as the **convolutional model of a seismic trace** (for normal incidence).

7.2 Reflection Losses

Let us construct a synthetic trace in which only reflection losses are considered in the computation of the ray amplitude (other effects, such as transmission losses and geometrical spreading, will be considered later). In that case, the amplitude of the reflection from the first reflector is $A_1 = A_0 R_1$, where R_1 is the reflection coefficient (for particle displacement), i.e., $R_1 = (Z_2 - Z_1)/(Z_2 + Z_1)$ where $Z_j = \rho_j \alpha_j, j = 1, 2$. Similarly, $A_2 = A_0 R_2$, where $R_2 = (Z_3 - Z_2)/(Z_3 + Z_2)$. In general,

$$A_n = A_0 R_n, \qquad R_n \equiv \frac{Z_{n+1} - Z_n}{Z_{n+1} + Z_n}, \tag{7.2a}$$

where R_n is the reflection coefficient for interface n (for an incident wave in layer n), and $A(t)$ in (7.1h) then becomes

$$A(t) = A_0 \sum_{n=1}^{N} R_n \delta(t - \tau_n) \equiv A_0 r(t). \tag{7.2b}$$

$r(t)$ is often called the **reflectivity function**. Note then that the seismic trace $s(t)$ is essentially the *convolution* of the reflectivity function with the wavelet $w(t)$ – see Figure 7.1a.

7.3 Transmission Losses

Let us examine the effects of transmission losses on the ray amplitudes. Consider the ray in Figure 7.1b. Its amplitude will then involve not only the reflection

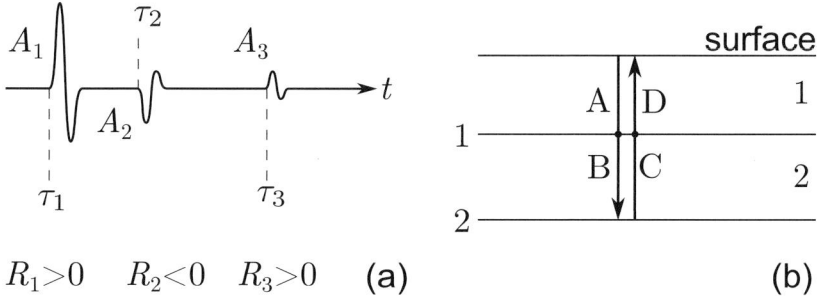

Figure 7.1 (a) A schematic example of a zero-offset synthetic seismic trace in which only reflection losses are included, with $A_n = A_0 R_n$, and (b) a zero-offset ray path showing two points on interface 1 where transmission losses occur (the down- and upgoing raypaths, which overlap, are drawn horizontally separated so that they can be seen).

coefficient R_2, but also the transmission coefficients at interface 1. Note, from Equation (3.20a), that the normal incidence P wave transmission coefficient is $1 - R$, where R is the reflection coefficient. Hence, for the downgoing wave A in Figure 7.1b, the transmission coefficient is $1 - R_1$. Now, the *reflection* coefficient for the upgoing wave C is $(Z_1 - Z_2)/(Z_1 + Z_2)$, which is just $-R_1$. Hence, the transmission coefficient for the upgoing wave C is $1 - (-R_1) = 1 + R_1$. Therefore, referring to Figure 7.1b, ray segments A, B, C, and D have the following amplitudes:

$$\text{A}: \quad A_0$$
$$\text{B}: \quad A_0(1 - R_1)$$
$$\text{C}: \quad A_0 R_2(1 - R_1)$$
$$\text{D}: \quad A_0 R_2(1 - R_1)(1 + R_1) = A_0 R_2(1 - R_1{}^2)$$

with the amplitude of D being the amplitude A_2 of the wave arriving at the receiver, i.e.,

$$A_2 = A_0 R_2(1 - R_1{}^2).$$

In general, when both reflection and transmission losses are included, the amplitude at the receiver of the primary reflection from the base of layer n is

$$A_n = A_0 R_n\left(1 - R_{n-1}{}^2\right)\left(1 - R_{n-2}{}^2\right) \cdots \left(1 - R_1{}^2\right), \tag{7.3}$$

with $A_1 = A_0 R_1$. The trace $s(t)$ is given by (7.1f), with A_n given by (7.3). For example, if $N = 2$, $\rho_1 = \rho_2 = \rho_3 = 2.5 \, \text{g/cm}^3$, $\alpha_1 = 1{,}000 \, \text{m/s}$, $\alpha_2 = 2{,}500 \, \text{m/s}$, $\alpha_3 = 4{,}000 \, \text{m/s}$, $h_1 = 200 \, \text{m}$ and $h_2 = 400 \, \text{m}$, and if we arbitrarily set $A_0 = 1$, then $\tau_1 = 0.400 \, \text{sec}$, $\tau_2 = 0.720 \, \text{sec}$, $R_1 = 0.4286$, $R_2 = 0.2308$, $A_1 = A_0 R_1 = 0.4286$ and $A_2 = A_0 R_2(1 - R_1{}^2) = 0.1884$. Note that the amplitude of the second reflection is reduced from 0.23 to 0.19 when transmission losses are included. For many layers, transmission losses can significantly affect the amplitudes of seismic signals.

7.4 Geometrical Spreading Losses

Consider a spherical wave emanating from a point source explosion at $r = 0$ in an infinite homogeneous medium. The constant amount of energy in the wave is distributed over an ever-expanding spherical wavefront, meaning that the energy per unit area on the wavefront, and hence the amplitude A of the wave, decreases with time, i.e., with the distance r traveled by the wave. More precisely, the amplitude of the wave at r is $A(r) = A_0 r_0 / r$, where A_0 and r_0 are constants with units of length. $A_0 = A(r_0)$ is the amplitude of the wave at some small reference distance r_0 from the source. The key thing to note is that $A(r) \propto 1/r$ (e.g., see Equations 2.54 and 2.91). The loss of amplitude is due to the **geometrical spreading** of the wavefront.

The rule "$A(r) \propto 1/r$" holds only for a homogeneous medium, where the ray paths are straight lines and the wavefronts are perfectly spherical. In an *inhomogeneous* medium, where ray paths are bent due to refraction, $A(r) \propto 1/L$, where L is called the **geometrical spreading factor**, and is a more complicated function of the medium parameters, i.e., L no longer given by the total distance traveled by the wave. For our primary two-way zero-offset P ray reflecting off the base of layer n, it can be shown that

$$L = L_n = \frac{2}{\alpha_1}\left(\alpha_1 h_1 + \alpha_2 h_2 + \cdots + \alpha_n h_n\right). \qquad (7.4)$$

As mentioned previously, even for a zero-offset ray, L_n is *not* simply given by the total distance traveled by the ray, as is the case for a homogeneous medium, i.e., $L \neq 2(h_1 + \cdots + h_n)$. This is because the wavefront is refracted at the interfaces and is no longer perfectly spherical. However, for $n = 1$ (the primary from the base of the first layer), the wave is contained in a single homogeneous medium (no refraction occurs), and so $L = 2h_1$ *is* the total distance traveled by the wave.

Hence, when reflection, transmission, and geometrical spreading losses are taken into account, the amplitude A_n at the receiver of the zero-offset primary reflection from the base of layer n is

$$A_n = \frac{A_0 r_0}{L_n} R_n \left(1 - R_{n-1}^{\,2}\right) \cdots \left(1 - R_1^{\,2}\right), \qquad (7.5)$$

and the trace $s(t)$ is given by (7.1f).

For example, the numerical values for the medium parameters given after Equation (7.3) yield $L_1 = 2h_1 = 400$ m, and $L_2 = (2/\alpha_1)(\alpha_1 h_1 + \alpha_2 h_2) = 2,400$ m.

Note that L_2 is *twice* as large as the total distance traveled by the second primary, i.e., 1,200 m. Using these, and setting $A_0 \equiv r_0 \equiv 1$, the relative amplitudes of the first two primary reflections are

$$A_1 = \frac{A_0 r_0}{L_1} R_1 = 1.071 \times 10^{-3}, \tag{7.6a}$$

$$A_2 = \frac{A_0 r_0}{L_2} R_2 \left(1 - R_1{}^2\right) = 7.849 \times 10^{-5}. \tag{7.6b}$$

The value of the ratio A_2/A_1 shows how the amplitude decays down the trace more and more rapidly as more loss effects are included. For instance, for reflection losses only, $A_2/A_1 = R_2/R_1 = 0.54$. Similarly, for reflection and transmission losses, $A_2/A_1 = (R_2/R_1)(1 - R_1{}^2) = 0.44$, and for reflection, transmission, and spreading losses, we have from (7.6), $A_2/A_1 = 0.073$, which is considerably smaller than 0.44 because of the inclusion of geometrical spreading losses. In general, the amplitude of a deep reflection is very small compared to the amplitude of a shallow one.

7.5 Absorption Losses

So far, we have assumed that we are dealing with a perfectly elastic medium. However, the Earth is generally **anelastic** or **viscoelastic** in nature. Due to **internal friction**, seismic waves lose energy as they propagate, which results in a loss of amplitude. The higher the frequency, the greater is the energy loss over a given distance traveled by a wave. Seismic wave propagation in anelastic or viscoelastic media has been treated in great mathematical detail in the literature (see, e.g., the texts by Aki and Richards, 1980, 2002; Ben-Menahem and Singh, 2000; Borcherdt, 2009; Carcione, 2001, 2007; Červený, 2001; Hudson, 1980). A brief introduction is also provided in Chapter 10 of this book. However, in this section and the next one, we treat it only from a relatively simple and approximate 1D viewpoint.

Consider a traveling spherical sinusoidal wave of frequency ω generated by a point source at $r = 0$ in a homogeneous medium. Let $A(r)$ be the amplitude of the wave at the distance r. The amplitude attenuation of the wave, due to anelastic absorption, in going from some distance r_0 to the distance r is expressed by

$$\frac{A(r)}{A(r_0)} = \exp\left[-\frac{\omega(r - r_0)}{2vQ}\right], \tag{7.7a}$$

where v is the speed of the wave and Q is the **quality factor**. The quantity $1/Q$ is sometimes called the **loss factor**. For a sinusoidal wave passing through a volume of material, the loss factor can be expressed as $1/Q = -\Delta E/2\pi E$, where E is the peak strain energy stored in the volume per cycle, and $-\Delta E (> 0)$ is the energy lost per cycle in the volume due to anelasticity. Sometimes other definitions of Q are used (e.g., *average* strain energy, rather than peak strain energy, is sometimes used), but in practice, the differences in definition are usually negligible. As $E \propto A^2$, where A is amplitude, a definition in terms of amplitude is $1/Q = -\Delta A/\pi A$ (from which (7.7a) can be derived, as discussed later). For a perfectly elastic medium, $Q = \infty$ ($Q^{-1} = 0$), which gives $A(r) = A(r_0)$, i.e., there is no anelastic attenuation.

Typical values of Q are $Q = 5\text{-}20$ for near surface layers, $Q = 20\text{--}100$ for the upper crust, $Q = 50\text{--}150$ for the lower crust, $Q = 100\text{--}500$ for the upper mantle, and $Q \approx 10,000$ for water. For most materials, $Q \gg 1$.

The derivation of the exponential amplitude decay factor in (7.7a) from the definition of Q goes as follows. Consider a photograph or snapshot of an absorbing medium taken while a wave is traveling in the r direction. Consider a small distance δr over which the amplitude changes by a small amount δA. Let λ be the wavelength and v the wave speed. Assume $Q \gg 1$ (weak absorption), which also means that A and λ do not change much over a cycle, which in turn means that the concepts of "amplitude" and "wavelength" are still physically meaningful. We then have

$$\Delta A = \frac{\text{change}}{\text{cycle}} = \frac{\text{change}}{\text{distance}} \cdot \frac{\text{distance}}{\text{cycle}} = \frac{\delta A}{\delta r}\lambda \to \frac{dA}{dr}\lambda = \frac{dA}{dr}\frac{2\pi v}{\omega}. \tag{7.7b}$$

Substituting this into the definition $1/Q = -\Delta A/\pi A$ then gives

$$\frac{dA}{dr} = -\left[\frac{\omega}{2vQ}\right]A \quad \Rightarrow \quad A(r) = A(r_0)\exp\left[-\frac{\omega(r - r_0)}{2vQ}\right]. \tag{7.7c}$$

A more careful treatment of seismic waves in absorbing media involves using, in the equation of motion, a stress–strain relation modified to account for anelasticity. The resulting equation of motion is then solved. For more details, see the references at the beginning of this section.

Anelasticity not only causes the amplitude to decay with distance, but also causes the waveform to change its shape because the higher frequencies in a pulse experience more attenuation than the lower frequencies over the same distance. As the wave propagates, any oscillations in the waveform tend to get "damped out," and the waveform tends to become broader because of the loss of the higher frequencies (see Figure 7.2).

In general, in anelastic media, v and Q are functions of frequency, i.e., *dispersion* occurs. This is different from the kind of dispersion that occurs for surface waves in a perfectly elastic medium. Studies have shown that the frequency dependence of Q is very weak over seismic frequency bands, i.e., Q is effectively constant, and that the phase velocity $v = v(\omega)$ is given approximately by

$$v(\omega) = v(\omega_1)\left[1 + \frac{1}{\pi Q}\ln\left(\frac{\omega}{\omega_1}\right)\right], \tag{7.8}$$

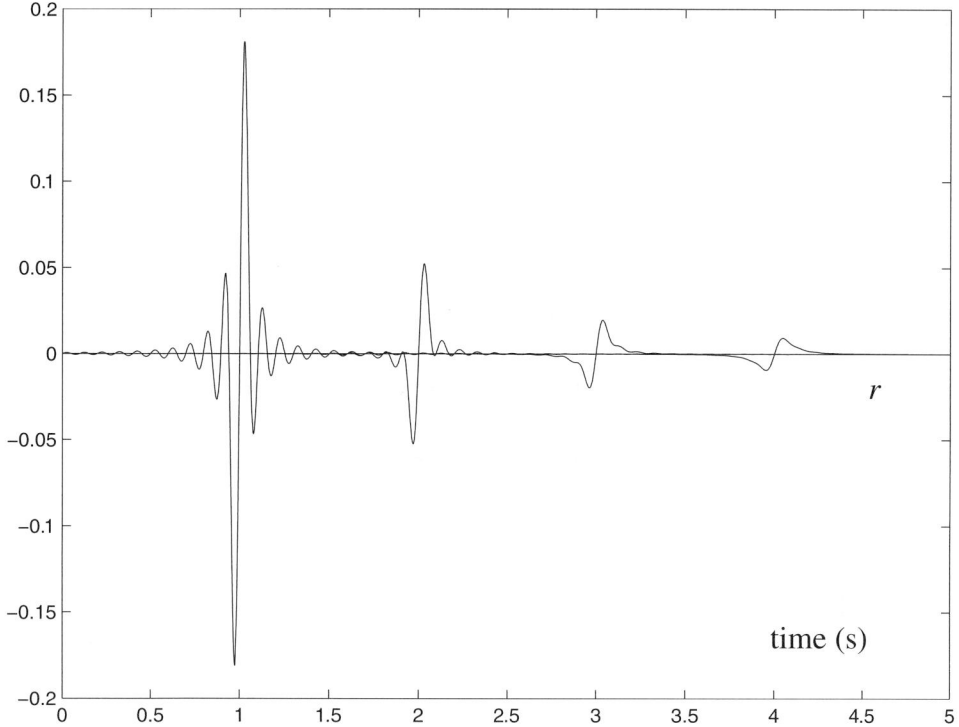

Figure 7.2 An example of a waveform recorded at the successive distances $r = 1$, 2, 3, 4 km, for a wave traveling in the r direction in a homogeneous dissipative or anelastic medium. The waveform is the one used by Krebes and Hron (1980, eqs. 35–40), with $\eta = 3$, $f_0 = 10$ Hz, $Q = 20$, and a wave speed of 1 km/s.

where ω_1 is some arbitrary reference frequency. Note that the velocity increases with frequency, although the increase is very slow because of the logarithm function. For example, if $\omega_1 = 1$ Hz and $Q = 20$, then $v(100) = 1.073\, v(1)$. Dispersion will further distort the waveform since higher frequencies travel at higher speeds. This also tends to broaden the waveform.

We can include absorption losses in our synthetic trace by calculating $(r - r_0)/2vQ$ in (7.7a) for each ray segment and summing the results. For the first ray segment, we assume $r_0 \ll h_1$ so that $r - r_0 = h_1 - r_0 \approx h_1$, and for each remaining ray segment, $r - r_0$ is simply the thickness of the layer containing the ray segment. Hence, the amplitude of the nth primary reflection will be attenuated by an amount $\exp[-a_n\omega]$, where

$$a_n = \frac{h_1}{\alpha_1 Q_1} + \cdots + \frac{h_n}{\alpha_n Q_n}, \tag{7.9a}$$

where the velocities $\alpha_1, \ldots, \alpha_n$ are frequency-dependent, as mentioned previously (and the Q values as well, in general). To derive this, consider the ray in Figure 7.1b.

Let r_1 and r_2 be the depths of interfaces 1 and 2, respectively. Then, applying (7.7a) to the first two (downgoing) ray segments for example, we have

$$A(r_2) = A(r_1) \exp\left[-\frac{\omega h_2}{2\alpha_2 Q_2}\right] = A(r_0) \exp\left[-\frac{\omega h_1}{2\alpha_1 Q_1}\right] \exp\left[-\frac{\omega h_2}{2\alpha_2 Q_2}\right]$$

$$= A(r_0) \exp\left[-\frac{\omega}{2}\left(\frac{h_1}{\alpha_1 Q_1} + \frac{h_2}{\alpha_2 Q_2}\right)\right]. \tag{7.9b}$$

If (7.7a) is applied to the entire ray, the "2" disappears because the upgoing rays contribute an exponential factor identical to the last one in the preceding equation. Hence, when reflection, transmission, spreading, and absorption losses are taken into account, the amplitude at the receiver of the Fourier component of frequency ω of the zero-offset primary reflection from the base of layer n is given by

$$A_n = e^{-a_n\omega} \frac{A_0 r_0}{L_n} R_n \left(1 - R_{n-1}^2\right) \cdots \left(1 - R_1^2\right). \tag{7.10}$$

Note that in order to compute A_n, we now need to know the angular frequency ω and the values of the wave speeds α_n and A_0 at that frequency. In general, we need to know these for a wide range of frequencies, because a wave pulse consists of a superposition of frequencies. For example, using the numerical values of the medium parameters that follow Equation (7.3), together with $Q_1 = 70$ and $Q_2 = 100$, and assuming that these numerical values are those at a frequency of $f = \omega/2\pi = 20$ Hz (which could be the 20 Hz Fourier component of a pulse), we can calculate, from Equation (7.10), the amplitude ratio A_2/A_1 (introduced following equation 7.6) for reflection, transmission, spreading, and absorption losses. The result is $A_2/A_1 = 0.060$. This is only a little less than 0.073, the value obtained for A_2/A_1 when absorption losses are not included (see the text that follows Equation 7.6), indicating that absorption losses are comparatively small. Absorption will be significant when long wave propagation paths are involved, but in general, geometrical spreading causes an amplitude reduction that is much greater than that caused by absorption.

Lastly, Equations (7.1f/g) no longer give the seismic trace when absorption or dissipation is included, because the rightmost expression in Equation (7.1e) no longer is correct. Since A_n is now a function of frequency (see Equation 7.10), A_n must be left inside the integral in (7.1e). In addition, the τ_n in the integral are also functions of frequency (because the α_n are – see Equation 7.8). This integral, in general, can no longer be evaluated analytically – it must be evaluated numerically (usually by the fast Fourier transform algorithm) to obtain the nth reflection waveform. It is this complication that causes the waveform to change as the wave propagates – each of the N reflection waveforms are different (although the differences may be small).

The trace is then given by the superposition of the individual waveforms obtained by numerical integration.

If desired, selected multiply reflected waves (multiples) can also be included in the normal-incidence calculations discussed up to now. A discussion of a signal analysis–based approach to normal-incidence synthetic seismogram calculations that accounts for all multiples is given by Robinson and Treitel (1980, chapter 13).

7.6 The General Case

In general, ray tracing can also be used to construct synthetic seismic traces for the case of nonnormal incidence, or nonzero offset ($x \neq 0$). Multiply reflected waves, often called "multiples" and **converted waves**, can be included (some of the segments of the ray path of a converted wave are P waves and some are SV waves). At each point where a ray encounters a boundary, a reflection or transmission coefficient (one of those in Equation 3.28e) must be computed. A more general formula for the geometrical spreading factor L, such as (7.11b) later in this section for horizontal layers, must be used. Absorption losses can be estimated as explained earlier, i.e., by calculating the quantity $(r - r_0)/2vQ$ for each ray segment and summing these quantities – if the sum is written as a, then the absorption amplitude factor for the Fourier component of frequency ω is $e^{-a\omega}$ (see the formula for a near the end of this section). If absorption losses are included, the amplitude is a function of frequency, hence, a Fourier sum or integration over frequency (an inverse Fourier transform) must be done to obtain the waveform at the receiver, as for the normal-incidence case previously discussed. Absorption losses can also be calculated more carefully by using the solutions for the equation of motion for a dissipative medium.

The theory for computing the individual waveforms composing the seismic trace can be developed in a way similar to Equation (7.1a)–(7.1e) (with A_n and $\mathbf{d}^{(G)}$ staying inside the integral in Equation 7.1e if absorption is included). The seismic trace is then obtained as usual by superimposing the individually computed waveforms.

Since phase changes can occur upon reflection and transmission, and since absorption and dispersion can alter the frequency spectrum of the propagating pulse, the waveform of a given reflection is, in general, substantially different from the waveform of the source pulse.

Head waves and surface waves can also be included in the synthetic seismograms. Furthermore, dipping interfaces, curved interfaces, and inhomogeneous layers can also be treated, although technical complications abound for complex structures.

Consider the ray path shown in Figure 7.3. Let v_j be the seismic velocity (P or S) associated with the jth ray segment (not the jth layer). Similarly, let h_j be the

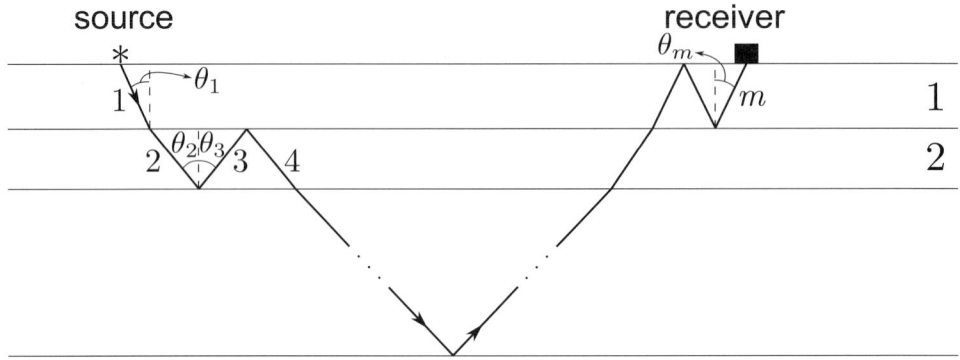

Figure 7.3 A typical ray path in a medium consisting of a sequence of flat horizontal homogeneous layers. Each ray segment can be either a *P* or *SV* wave (or all segments are *SH* waves). The ray segments are numbered from 1 to *m*.

thickness of the layer in which the *j*th ray segment is found (in Figure 7.3, this means that $h_2 = h_3 = h_4 = h_{m-3}$). Let θ_j be the angle that the *j*th ray segment makes with the vertical. Let *m* be the total number of ray segments in the ray. For such a case of horizontal layers, *p* is no longer zero for a ray with a nonzero offset. Equations (5.6) and (5.7) can be applied to this case, with slight modifications – the leading "2" is deleted and the sum is replaced by a sum over the ray segments, not the layers, i.e., a sum from $j = 1$ to $j = m$. For horizontal layers, *p* for a ray with offset *x* can be computed by numerically solving the modified Equation (5.7), and the travel time of the ray can then be computed by using this value of *p* in the modified (5.6). The takeoff angle θ_1 can be determined from *p* and the other angles θ_j as well. For more complicated structures, *p*, or the takeoff angle of the ray, must be determined by trial-and-error "ray shooting."

If the ray starts off with an amplitude of A_0 at a small distance r_0 from the source, then at the receiver, the ray has the amplitude

$$A_m = \frac{A_0 r_0}{L} U_1 U_2 \cdots U_m, \tag{7.11a}$$

due to reflection, transmission, and spreading losses, where one usually sets $A_0 \equiv 1$ and $r_0 \equiv 1$, and where U_j is the amplitude of the *j*th ray segment due to reflection or transmission, i.e., $U_1 = 1$, U_2 is the transmission coefficient between ray segments 1 and 2, U_3 is the reflection coefficient between ray segments 2 and 3, etc., and where *L* is the geometrical spreading factor given by

$$L = \frac{\cos \theta_1}{v_1} \sqrt{\left(\sum_{j=1}^{m} \frac{v_j h_j}{\cos \theta_j} \right) \left(\sum_{j=1}^{m} \frac{v_j h_j}{\cos^3 \theta_j} \right)} \tag{7.11b}$$

(see Červený and Ravindra, 1971, eq. 2.143). For one ray segment ($m = 1$), $L = h_1/\cos\theta_1$, which is just the total distance traveled – this is expected since the whole ray is in a single homogeneous medium. For the case of zero-offset and m ray segments, $\theta_j = 0$ for all j, implying

$$L = (1/v_1)(v_1h_1 + \cdots + v_mh_m). \tag{7.11c}$$

Equation (7.4) is a special case of this equation, i.e., (7.4) is for $v_j = \alpha_j$. Note also that (7.4) has a leading "2" because (7.4) is for a *primary* zero-offset symmetric reflection with the subscripts referring to the layers, whereas (7.11c) is for a primary or multiple generally nonsymmetric zero-offset reflection, with the subscripts referring to the ray segments.

Absorption can be approximately included by multiplying the amplitude A_m by the absorption amplitude factor $e^{-a\omega}$, with $a = \sum_{j=1}^{m} \ell_j/(2v_jQ_j)$ where ℓ_j is the length of the jth ray segment. A more accurate treatment would include reflection coefficients modified to include absorption, and other details – see the references at the beginning of the previous section.

The method discussed in this section, for computing synthetic seismograms for the nonzero-offset case, is not accurate for all offsets. Ray amplitudes computed in this way are quite accurate for subcritical angles, and moderately accurate for supercritical angles, but not accurate at all for rays containing near-critical angles of incidence.

Other methods besides ray tracing can be used to construct synthetic seismograms for elastic and anelastic media. For instance, the **finite difference method** or the **finite element method** can be used to solve the wave equation or the equation of motion and/or to calculate travel times (see, e.g., Carcione, 2001, 2007; Dai, Kanasewich, and Vafidis, 1993; Eaton, 1993; Emmerich and Korn, 1987; Kay and Krebes, 1999; Krebes and Quiroga-Goode, 1994; Moczo, Bystrický, Kristek, Carcione, and Bouchon, 1997; Moczo, Kristek, and Galis, 2014.) See Appendix 7A for an example of synthetics produced by the finite difference method.

Finite difference forward modeling has also been done for solid media (elastic or anelastic) containing interfaces that are in non-welded contact with each other (e.g., Carcione, 2001, 2007; Cui, Lines, Krebes, and Peng, 2018; Martinez Fernandez, 2014; Slawinski and Krebes, 2002a, 2002b). Nonwelded contact means that the displacement **u** is no longer continuous across an interface, although the traction **T** still is. Media containing joints, fractures, and faults may in some cases fall into this category.

Other methods for solving partial differential equations can be used. Furthermore, numerical methods can be combined with analytical methods (such as integral transform methods) to generate synthetic seismograms for complex media. Such **full-wave methods** are generally accurate at all offsets, but are more expensive and time consuming.

7.7 The Reflectivity Function

In order to calculate normal-incidence synthetic seismograms for a medium modeled by a sequence of N flat horizontal homogeneous layers, it is necessary to know the acoustic impedance Z as a function of depth. This information can be obtained from sonic logs combined with density logs. If density logs are not available, the density is typically assumed to be constant (this is also done sometimes simply for convenience). To make the calculations simpler, the acoustic impedance log is sometimes smoothed or averaged over different zones of depth, and R_n, the reflection coefficient for the nth interface, is computed from $R_n = (Z_{n+1} - Z_n)/(Z_{n+1} + Z_n)$ for all n (see Figure 7.4). The travel times τ_n can be computed from the known thicknesses and interval velocities (see Equation 7.1a), and the synthetic trace $s(t)$ can be computed from Equations (7.1f/g).

Consider the simple case in which only reflection losses (Equation 7.2a) are considered (normal-incidence synthetic traces are often compared with stack sections to identify reflectors, and amplitudes are often scaled on stack sections, meaning detailed amplitude calculations are unnecessary). In that case, from (7.2a/b), the synthetic trace is given by

$$s(t) = r(t) * w(t), \qquad \text{where} \qquad r(t) = \sum_{n=1}^{N} R_n \delta(t - \tau_n), \qquad (7.12\text{a,b})$$

is the reflectivity function for a layered medium.

On the other hand, smoothing can be omitted, and the log can be finely sampled, so that the medium is modeled by a sequence of many discrete thin layers. In this way, most of the information in the log can be used (Figure 7.5b).

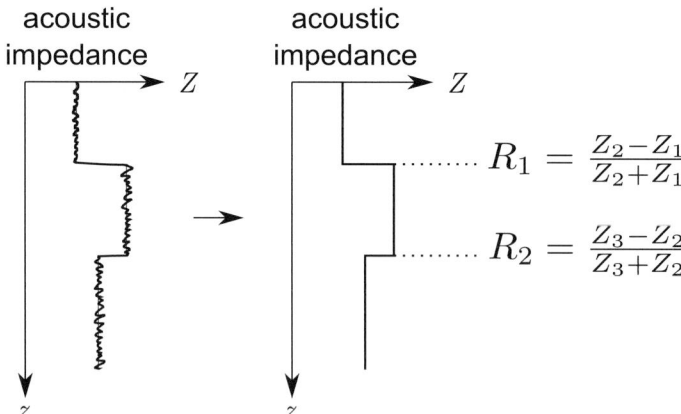

Figure 7.4 A schematic diagram of the smoothing of an acoustic impedance log and the calculation of the normal-incidence reflection coefficients.

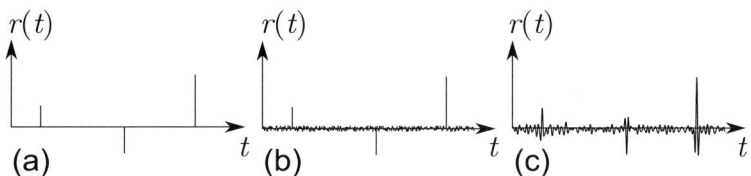

Figure 7.5 Schematic diagrams of the reflectivity function for (a) the discrete case in which the acoustic impedance log is smoothed, (b) the discrete case in which the log is not smoothed but finely sampled, and (c) the continuous case ($r(t)$ is the time derivative of $\ln Z$). For the discrete cases, the nth spike is located at $t = \tau_n$.

Now consider the medium to be continuous, rather than discretely layered. If the log is very finely sampled, the impedance will change very little from sample to sample. In other words, $Z_{n+1} = Z_n + \Delta Z_n$ with $\Delta Z_n \ll Z_n$. Hence, from (7.2a),

$$R_n = \frac{\Delta Z_n}{2Z_n + \Delta Z_n} \approx \frac{\Delta Z_n}{2Z_n}. \tag{7.13a}$$

Now, convert the log from depth to time by computing all the values of τ_n in (7.1a) and replacing the depth values on the log with their corresponding time values τ_n. Let τ denote the general time variable on the converted log. R_n is then the value of the reflection coefficient at $\tau = \tau_n$. Or, depth could also be converted to time by integrating the recorded sonic log curve, as discussed in Chapter 5, and extracting $Z = Z(\tau)$. Let $\Delta \tau_n \equiv \tau_{n+1} - \tau_n$ be the time sample interval at sample n on the converted log. Then, using Equation (7.13a), Equation (7.12b) can be written as

$$r(t) = \sum_{n=1}^{N} \frac{1}{2Z_n} \frac{\Delta Z_n}{\Delta \tau_n} \delta(t - \tau_n) \Delta \tau_n.$$

Passing to the infinitesimal limit, i.e., letting $\Delta \tau_n \to d\tau$, gives

$$r(t) = \int_{-\infty}^{\infty} \frac{1}{2Z(\tau)} \frac{dZ(\tau)}{d\tau} \delta(t - \tau) \, d\tau = \frac{1}{2Z(t)} \frac{dZ(t)}{dt} = \frac{1}{2} \frac{d}{dt} \Big[\ln Z(t) \Big], \quad (7.13b)$$

where (1.135c) has been used. Equation (7.13b) is the formula for the reflectivity function for a continuous medium (see Figure 7.5c). The synthetic seismic trace is the convolution of the wavelet $w(t)$ with the reflectivity function.

7.8 Interference Effects

Suppose we want to construct the synthetic trace for a model consisting of two flat horizontal layers with impedances Z_1 and Z_2 overlying a half-space with impedance Z_3 such that $Z_1 = Z_3$ and $Z_2 > Z_1$ (Figure 7.6a). This also implies that $R_2 = -R_1$, where R_1 and R_2 are the reflection coefficients at interfaces 1 and 2 for an incident

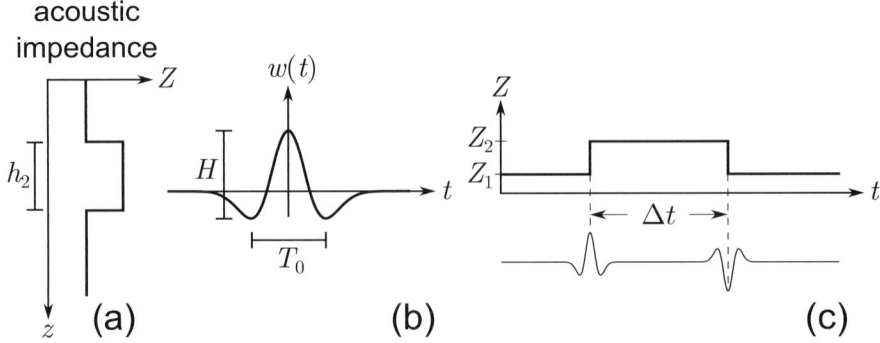

Figure 7.6 (a) The acoustic impedance model, (b) the wavelet, and (c) the synthetic trace.

wave in the upper layer, resulting in a polarity reversal in the two reflections, as can be seen in Figure 7.6c. Let Δt be the difference between the two-way vertical travel times of the two primary reflections. Choose, for example, a symmetrical zero-phase wavelet like the Ricker wavelet with a basic period of T_0, i.e., with a time separation of T_0 between the two minor peaks in the case of a Ricker wavelet, such as in Figure 7.6b. T_0 is approximately the dominant period in the spectrum of the wavelet. Since we are interested in the effects of interference of the two waveforms due to the second layer being a thin bed, it is sufficient to consider only reflection losses.

If the thickness of the second layer is large, i.e., if $\Delta t \gg T_0$, then the trace consists of two distinct noninterfering wavelets (the two primaries). The second primary has a polarity opposite to that of the first because $Z_3 < Z_2$ (Figure 7.6c).

However, if Δt is small enough, the wavelets will interfere with each other. If $\Delta t = \frac{1}{2}T_0$, there is a maximum **tuning effect** due to constructive interference – the resultant pulse (the sum of the two interfering wavelets) has a relatively high amplitude (Figure 7.7a). If $\Delta t \ll T_0$, the resultant pulse has, approximately, the shape of the time derivative dw/dt of the wavelet $w(t)$ (Figure 7.7b), which can be explained as follows. Let us arbitrarily shift the time origin to the center of the second wavelet (the second reflection). The waveform for the second primary is then $-w(t)$ ("$-$" because of the polarity reversal), and the waveform for the first primary is $w(t+\Delta t)$. Hence, the resultant pulse, i.e., "SUM" in Figure 7.7b, is given by, for small Δt,

$$\text{SUM} = w(t + \Delta t) - w(t) \approx \frac{dw}{dt}\Delta t, \qquad (7.14)$$

where the definition of a derivative has been used. Figure 7.9 shows that the tuning effect is largest when $\Delta t = \frac{1}{2}T_0$. The tuning effect occurs for thin beds of relatively

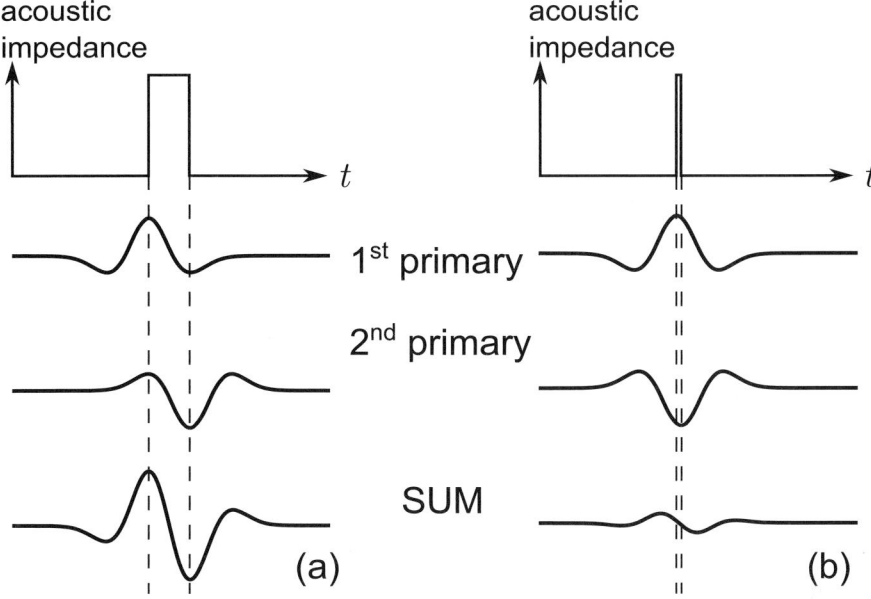

Figure 7.7 A schematic diagram showing the interference of two primary reflections for (a) $\Delta t = \frac{1}{2}T_0$, and for (b) $\Delta t \ll T_0$. In (b), if the uppermost pulse is $w(t + \Delta t)$, then the middle one is $-w(t)$, meaning that the lowermost pulse (the sum) is approximately $(dw/dt)\Delta t$ for small Δt.

high velocity. In terms of the basic wavelength $\lambda_0 = v_2 T_0$, the frequency $f_0 = 1/T_0$, and the thickness h_2 of the second layer in the model, the tuning formula $\Delta t = \frac{1}{2}T_0$ becomes

$$h_2 = \frac{v_2 \Delta t}{2} = \frac{v_2 T_0}{4} = \frac{1}{4}\lambda_0 = \frac{v_2}{4f_0}, \tag{7.15}$$

indicating how thick (i.e., thin) the layer must be for maximum tuning to occur. For example, if $v_2 = 2400$ m/s and $f_0 = 30$ Hz, then $h_2 = 20$ m.

The preceding result is also consistent with the familiar rule of thumb that in order to be able to vertically resolve two horizontal flat reflectors, they need to be separated by a distance of at least $\lambda_0/4$.

Another example of the tuning effect shown in Figure 7.7 can be seen later in Figure 7.8.

Suppose now that the acoustic impedance Z increases linearly with depth z in the second layer, forming a "ramp" in the $Z(z)$ function. Just as the response to a spike in the acoustic impedance is proportional to the time derivative of the wavelet, the response to a ramp in the acoustic impedance is proportional to the time integral of the wavelet (see Figure 7.10). If the ramp has a relatively large width, i.e., if

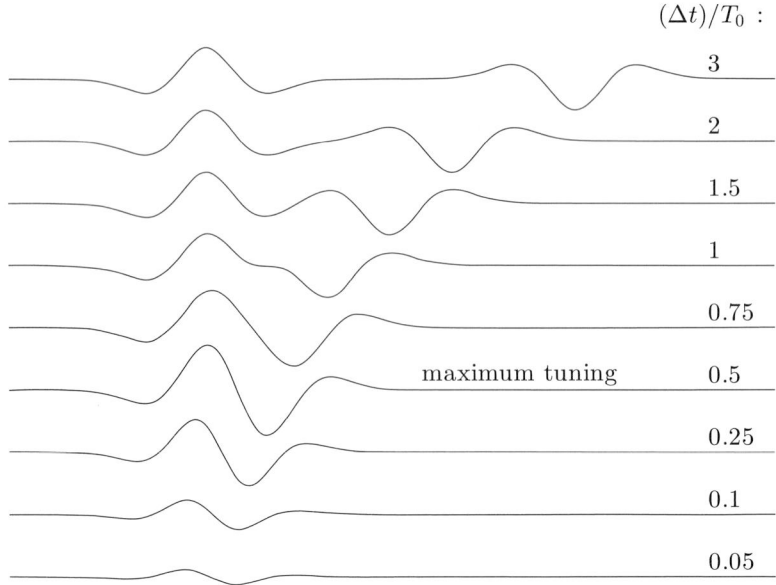

Figure 7.8 This shows how tuning occurs as Δt decreases. A Ricker wavelet (Equation 6.10) was used to compute the signals.

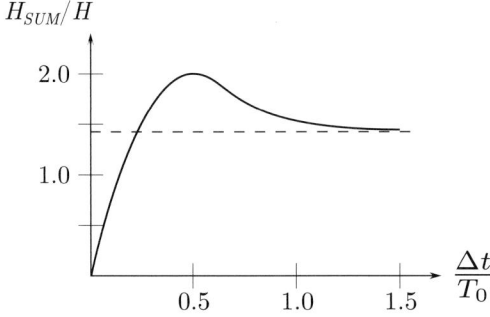

Figure 7.9 A schematic diagram of the height H_{SUM} of the resultant pulse SUM in Figure 7.7 relative to the height H of the wavelet for different values of $\Delta t/T_0$. Maximum tuning occurs at $\Delta t/T_0 = 0.5$, where the height of the resultant pulse is twice that of the wavelet $w(t)$. As $\Delta t/T_0$ increases, the curve approaches the relative height of the *sum* of the two separated noninterfering pulses, i.e., the sum of the value of the large positive peak of the leading pulse and the absolute value of the large negative peak of the trailing pulse, and the relative height of this sum is greater than the relative height 1.0 of the wavelet.

$\Delta t \gg T_0$ where T_0 is the basic period of the ramp wavelet, then the trace consists of two distinct noninterfering wavelets (the two primaries) with opposite polarities (see Figure 7.11a). However, if Δt is small enough, the two wavelets will interfere

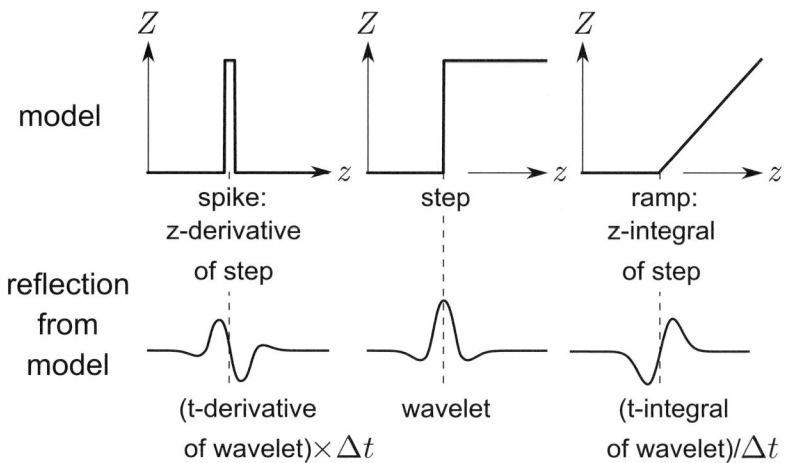

Figure 7.10 The response to a spike, step, and ramp in the acoustic impedance.

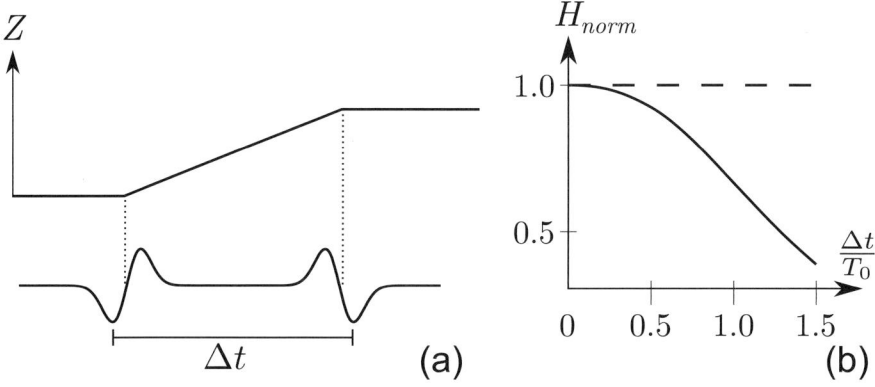

Figure 7.11 Schematic diagrams of (a) the response to a wide ramp in the acoustic impedance ($\Delta t \gg T_0$), and (b) the normalized height H of the resultant pulse for the impedance ramp for different values of $\Delta t/T_0$.

with each other. For $\Delta t = \frac{1}{2}T_0$, tuning occurs, and for $\Delta t \ll T_0$, the total response has again the shape of the time derivative of the ramp wavelet, which is the original wavelet (Figure 7.6b). This is also consistent with the fact that for $\Delta t \ll T_0$, the ramp model essentially reduces to the step model. The resultant pulse also reaches its maximum height in this case (Figure 7.11b).

Synthetic seismograms that are constructed from acoustic impedance logs in the previously described manner are often compared with field records in order to identify reflections on the data.

For more details and further reading, see Sengbush (1961, 1983) and Widess (1973).

7.9 Appendix 7A: Synthetic Shot Records and Wavefronts

This appendix is excerpted and modified from Krebes (2004), with permission from the Canadian Society of Exploration Geophysicists.

Figures 7.12–7.20 show examples of synthetic shot records and wavefronts computed with the finite difference method. They were produced by Chabot (2000) using a 2D finite difference code provided by G.T. Schuster, J. Xu, and Y. Luo of the University of Utah. Additional modifications to the code were made by Guevara (2001). The code was based on an algorithm described by Levander (1988). The code was also used in a study on single-well seismic imaging by Chabot, Henley, Brown, and Bancroft, (2001).

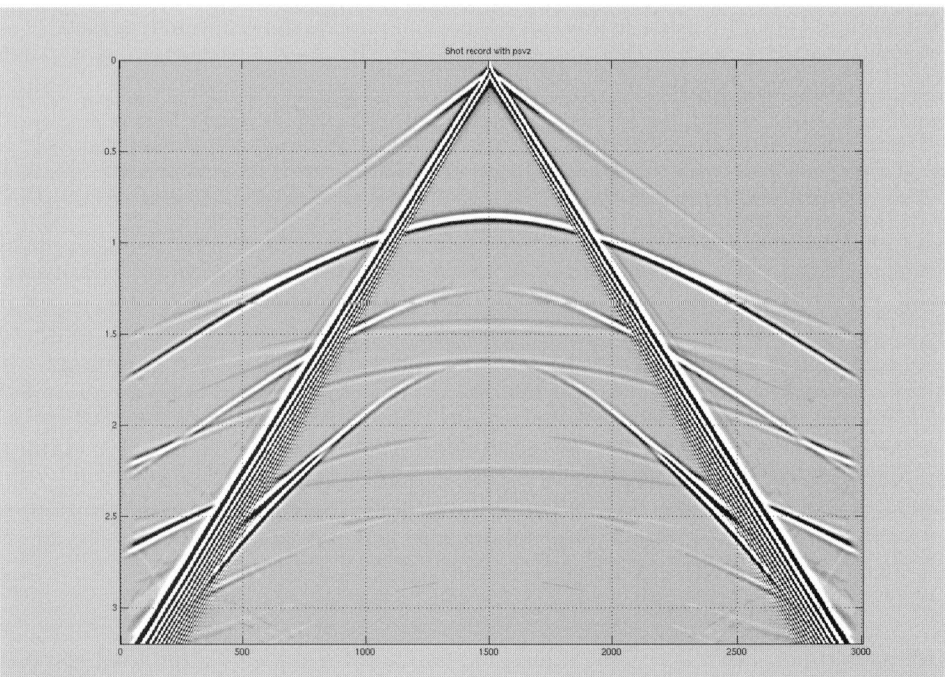

Figure 7.12 A 2D synthetic shot record for the vertical component of geophone motion produced by the finite difference method for a surface source. The horizontal axis is the source–receiver offset in meters and the vertical axis is the two-way travel time in seconds. The model is 3 km × 1 km. The geological model is a flat elastic layer with a free upper surface overlying an elastic half-space. The source is a vertically directed point force located on the free surface. The density, P, and S wave velocities in the elastic layer are 2.1 g/cm^3, and 1,000 m/s and 500 m/s, respectively, and in the half-space they are 2.3 g/cm^3, 1850 m/s, and 925 m/s. From Krebes (2004), with permission from the Canadian Society of Exploration Geophysicists.

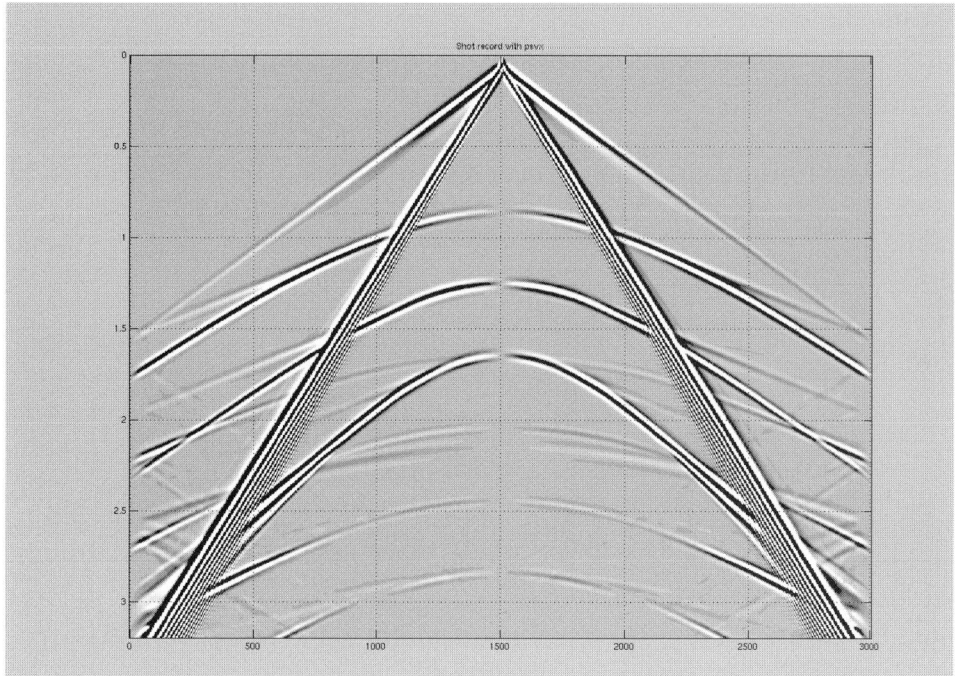

Figure 7.13 Same as Figure 7.12, except it is for the horizontal component of motion. From Krebes (2004), with permission from the Canadian Society of Exploration Geophysicists.

Figures 7.12 and 7.13 show 2D synthetic shot records for a source consisting of a force directed vertically downward at a point on the surface (e.g., a hammer blow). The nonsymmetric source generates both a *P* wave and an *SV* wave. The geological model is a horizontal homogeneous elastic layer overlying a homogeneous elastic half-space. The top of the elastic layer is a free surface. The synthetics show the expected events, i.e., the direct arrival, the reflected and converted waves, the head waves (refraction arrivals), and the prominent dispersive guided waves. They also show the amplitude and phase (polarity) relationships among the wave types. In addition, Figure 7.13, for the horizontal component of particle motion, shows a polarity reversal in all events as one goes from the left side of the record to the right side. This happens because the waves arriving at geophones left and right of center produce geophone motions whose *x* components of displacement have opposite signs. Some numerical artifacts can also be seen in Figure 7.13 on the lower-left and lower-right edges of the grid.

Figures 7.14–7.20 show a series of "snapshots" of the wavefronts corresponding to the shot records in Figures 7.12 and 7.13, at the following fixed times: 300 ms, 500 ms, 600 ms, 800 ms, 1,000 ms, 1,200 ms, and 1,400 ms. The upper half of each

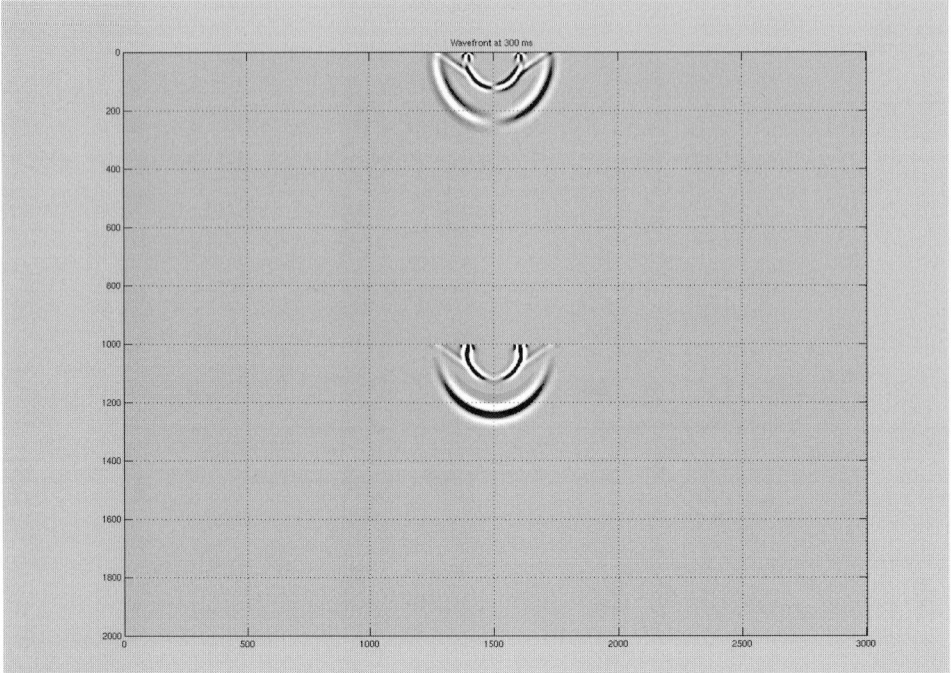

Figure 7.14 Fixed-time plot, or "snapshot," of the wavefronts corresponding to the shot records and geological model of Figures 7.12 and 7.13, produced by the finite difference method, at an elapsed time of 300 ms (the source is activated at time $t = 0$). The horizontal component of particle motion is shown in the upper half and the vertical component in the lower half. The locations of the free surface (where the source is) and the interface between the layer and half-space can be inferred from the locations of the incident and reflected wavefronts in this and the following figures. From Krebes (2004), with permission from the Canadian Society of Exploration Geophysicists.

figure shows the horizontal component of motion, and the lower half shows the vertical component. Several interesting wave propagation features can be seen in the figures – they show (a) that the "hammer blow" source produces both a downgoing P and SV wave; (b) that there is a downgoing head wave connecting the downgoing P and SV wavefronts emerging from the source point; (c) the upgoing P and SV head waves (refraction arrivals) produced by the transmitted P wavefront beyond the critical angle (when it breaks away from the other wavefronts at the interface); (d) the expected left-to-right polarity reversal in the horizontal component; (e) the variation of amplitude along some of the reflected and transmitted wavefronts due to the nonsymmetric "hammer blow" source; and (f) the complex nature of the wavefield produced by a nonsymmetric source, even for a simple one-layer model.

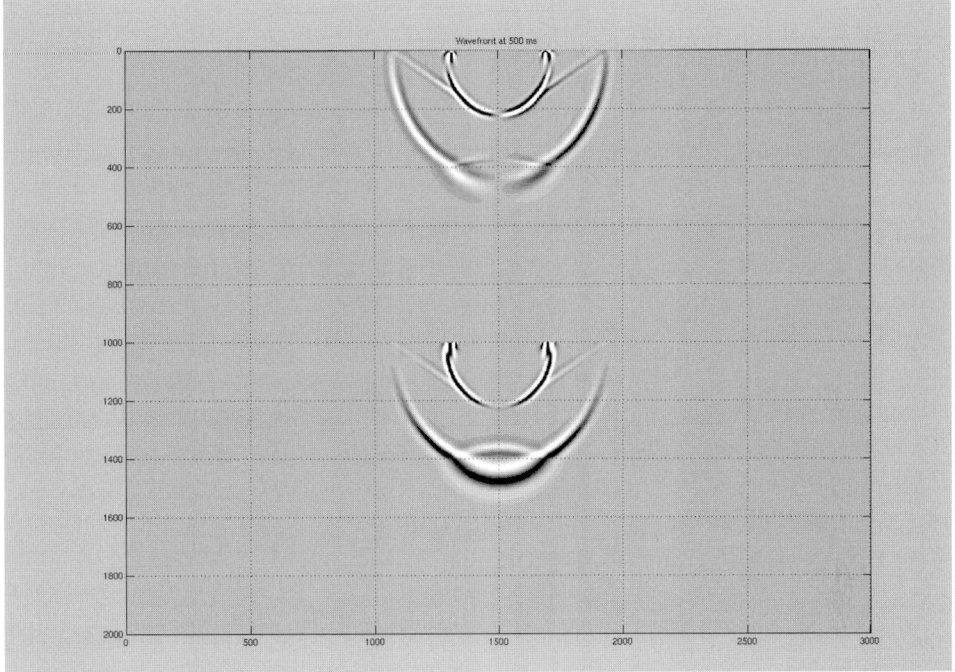

Figure 7.15 Same as Figure 7.14, except the elapsed time is 500 ms. From Krebes (2004), with permission from the Canadian Society of Exploration Geophysicists.

Regarding point (e) in the preceding paragraph, the variation of amplitude observed along the wavefronts is consistent with the radiation pattern for a vertically directed point force (see Figure 2.12). For example, this radiation pattern predicts that a downward hammer blow would result in mainly a large vertical (longitudinal) displacement of the medium at points vertically below the source point, with little or no horizontal (transverse) displacement at these points (which one might also predict using physical intuition). This prediction is confirmed, for example, in the transmitted *P* wavefront in Figure 7.16 (the lowermost wavefront in each half, upper and lower, of the figure) – for the horizontal component, the center of the transmitted *P* wavefront has zero amplitude, but for the vertical component, it has a high amplitude.

A good way to study Figures 7.14–7.20, with the intent of understanding wave propagation, is to select a certain wave and follow what happens to it from one snapshot to the next (as time progresses). For example, consider the first reflected *P* wave – the vertical component. Beginning with Figure 7.14 (lower half), which shows the downgoing high-amplitude incident *P* wave that has not yet reached the interface, we move to Figure 7.15 and note the beginnings of the first reflected

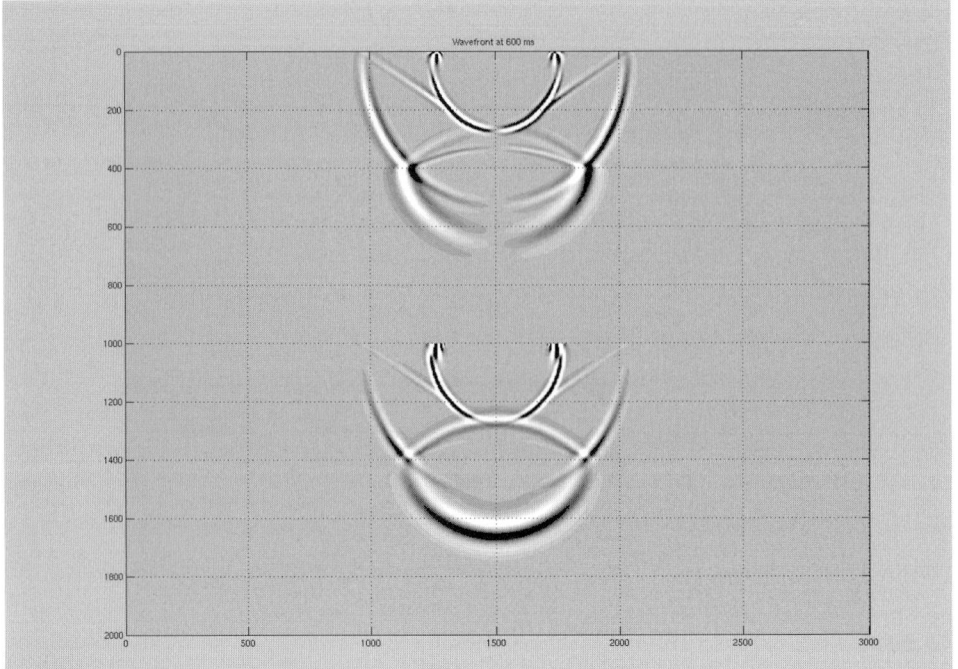

Figure 7.16 Same as Figure 7.14, except the elapsed time is 600 ms. From Krebes (2004), with permission from the Canadian Society of Exploration Geophysicists.

P wave. Since the incident *P* wave is reflecting off a medium (the half-space) which has a larger acoustic impedance (density × *P* wave velocity), the vertical component of displacement experiences a polarity reversal (the central peak of the reflected *P* wavelet is white, not black as for the incident *P* wave). This is just like the polarity reversal in displacement experienced, upon reflection, by an incident wave on thin string connected to a thick rope. We then follow the reflected *P* wave upward in Figures 7.16 and 7.17, where it then reflects off the free surface. This free surface reflection can be seen in Figure 7.18 – note that there is now no polarity reversal in the vertical component of displacement because the wave is reflecting off a medium with a lower acoustic impedance, i.e., the air layer (although there is a polarity reversal in the pressure at the free surface – a compression reflects as a rarefaction, i.e., a dilatation, off a medium with a lower acoustic impedance). We then follow the *P* wave reflected from the free surface back down to the interface in Figure 7.19, after which it reflects off the interface with another polarity reversal (due again to the larger impedance in the half-space), which can be seen in Figure 7.20. Note also the decreasing amplitude of the wave.

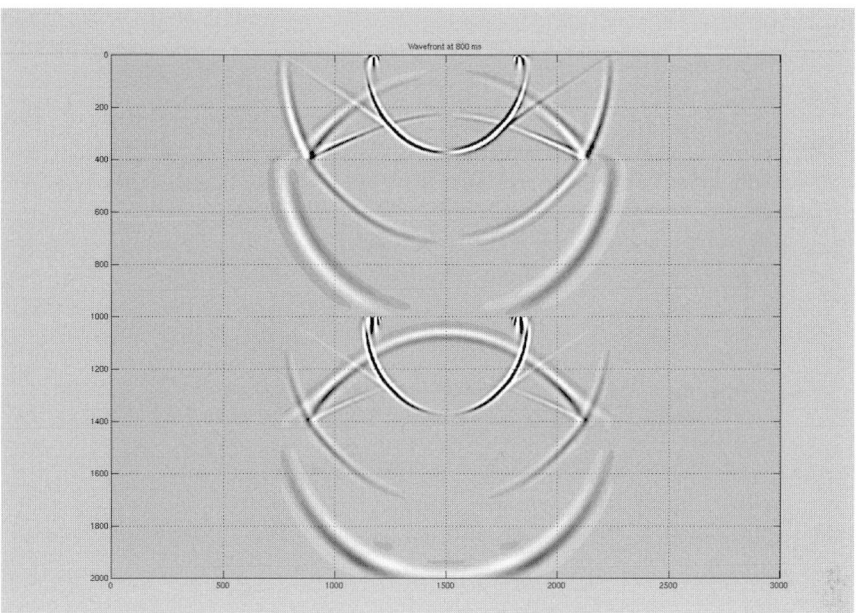

Figure 7.17 Same as Figure 7.14, except the elapsed time is 800 ms. From Krebes (2004), with permission from the Canadian Society of Exploration Geophysicists.

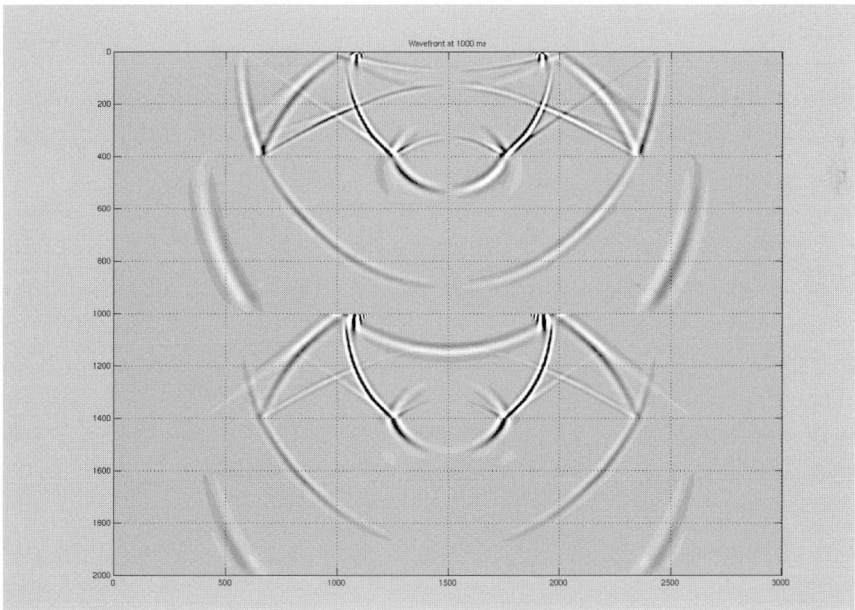

Figure 7.18 Same as Figure 7.14, except the elapsed time is 1,000 ms. From Krebes (2004), with permission from the Canadian Society of Exploration Geophysicists.

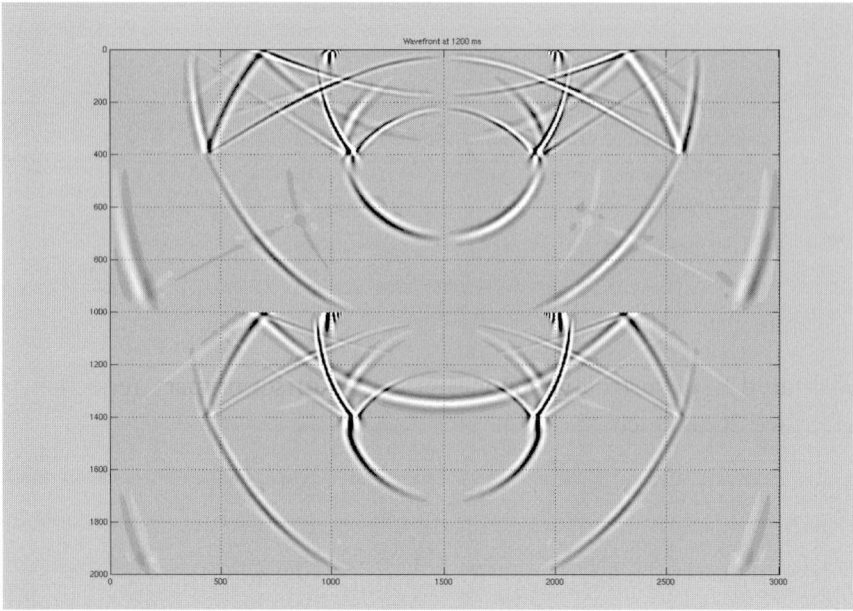

Figure 7.19 Same as Figure 7.14, except the elapsed time is 1,200 ms. From Krebes (2004), with permission from the Canadian Society of Exploration Geophysicists.

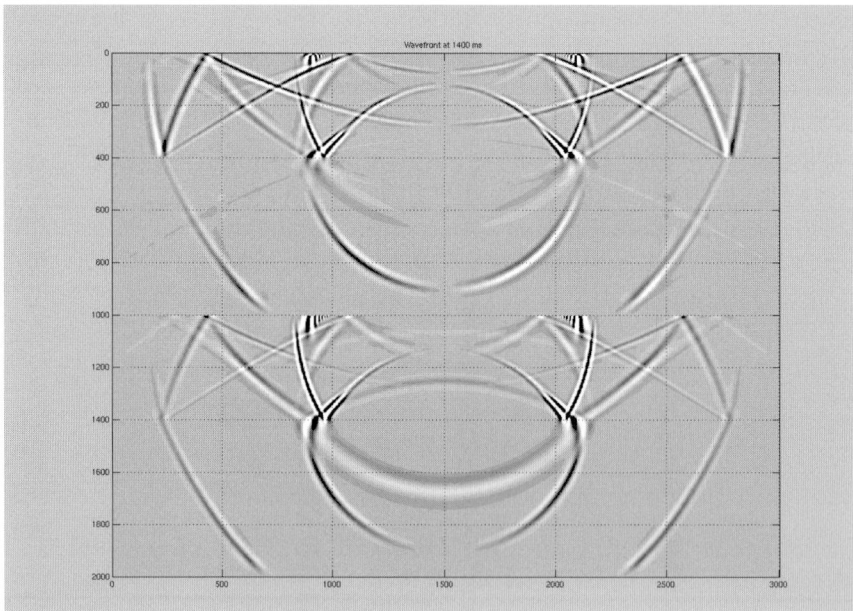

Figure 7.20 Same as Figure 7.14, except the elapsed time is 1,400 ms. From Krebes (2004), with permission from the Canadian Society of Exploration Geophysicists.

Exercises

1. (a) Referring to Figure 7.21, calculate the relative amplitude and phase of the reflected *SH* wave at the receiver. Account for losses due to reflection and geometrical spreading of wavefronts. The densities and velocities are $\rho_1 = \rho_2 = 2.0$ g/cm^3, $\beta_1 = 1,000$ m/s, and $\beta_2 = 1,500$ m/s.

   (b) By how much is the amplitude of the 30 Hz component of the wave further attenuated if $Q_S = 50$ in the first layer?

2. A depth model consists of 20 flat horizontal interfaces, each of which has a relative reflection amplitude coefficient of 0.3. What is the amplitude, measured by a surface geophone, of a zero-offset primary reflection from interface 20 if we consider

   (a) reflection losses only?

   (b) both reflection and transmission losses?

3. Construct two zero-offset synthetic seismic traces for the following depth model:

n	ρ (g/cm^3)	v (km/s)	h (m)
1	2.4	0.6	150
2	2.6	2.4	300
3	2.6	5.4	600
4	2.6	2.7	450
5	2.6	4.8	∞

   where n is the layer number, ρ is the density, v is the wave speed, and h is the layer thickness. The first trace should involve only reflection losses. The second trace should involve both reflection and transmission losses. Show all of your calculations of times and amplitudes. On your traces, label the pulses with their times and amplitudes.

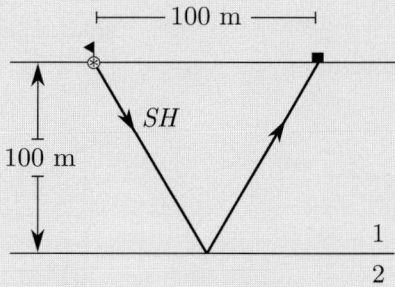

Figure 7.21 See exercise 1.

4. Construct a zero-offset synthetic seismic trace for the depth model given in the following table (three layers over a half-space) for the following three cases:

(a) Consider reflection losses only.
(b) Consider both reflection and transmission losses.
(c) Consider reflection, transmission, and geometrical spreading losses.

n	ρ (g/cm^3)	v (km/s)	h (m)
1	2.5	2.4	300
2	2.6	5.5	600
3	2.6	2.7	400
4	2.6	4.9	∞

In the preceding table, n is the layer number, ρ is the density, v is the wave speed, and h is the layer thickness.

5. A seismic trace shows a maximum tuning effect. Assuming it is due to a thin bed, as shown in Figure 7.22, calculate the thickness of the bed.

6. (a) For the *SH* ray shown in Figure 7.23, compute the offset x, the travel time t and the relative amplitude and phase of the signal recorded by the receiver. Include losses due to reflection, transmission, and geometrical spreading. The layer parameters are as follows:

n	ρ (g/cm^3)	v (km/s)	Q	h (m)
1	2.1	0.7	20	50
2	2.2	1.0	50	100
3	2.3	3.0	—	∞

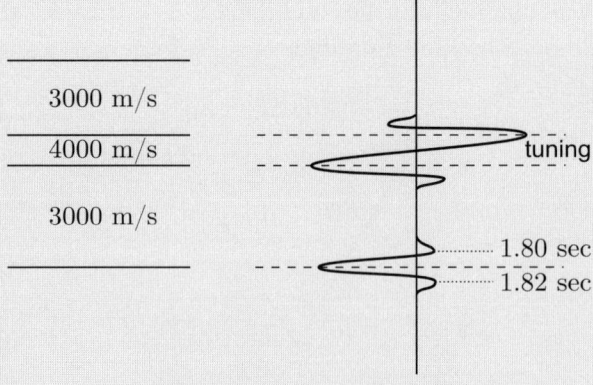

Figure 7.22 See exercise 5.

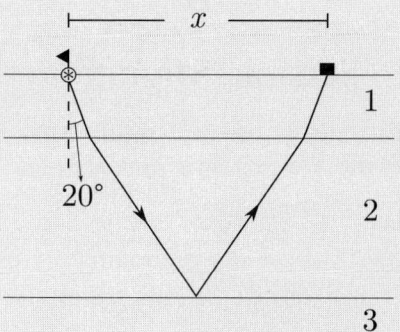

Figure 7.23 See exercise 6.

where n is the layer number, ρ is the density, v is the wave speed, Q is the quality factor, and h is the layer thickness.

(b) If absorption losses are also included, what is the amplitude and phase of the 30 Hz component of the signal at the receiver?

8

Seismic Migration

We have seen that on CMP stack sections (initial images of the subsurface) for complex subsurface structures, the reflectors are, in general, spatially mispositioned, and need to be **migrated** to their true spatial positions to obtain an accurate seismic image of the subsurface. In this chapter, we take an introductory look at how seismic wave theory can be used to develop some basic seismic migration methods, in particular, the basic methods of wave equation migration.

8.1 A Point Reflector

Consider the simple case of a homogeneous medium of velocity v containing a single **point reflector** or **point diffractor** at a depth z_0 and horizontal location $x = 0$ (Figure 8.1a). The CMP stack section corresponding to this model can be obtained from zero-offset ray tracing. It consists of a hyperbola whose apex is at $x = 0$ and has the two-way travel time $t_0 = 2z_0/v$ (Figure 8.1b). The CMP stack section is the given data set that we need to migrate to get a true image of the subsurface. To do the migration, we could first multiply the time axis values on the CMP stack section by $v/2$ to get the CMP depth section in Figure 8.1c. Then, the process of migration essentially consists of "collapsing" the hyperbola, i.e., moving each point on the hyperbola back to its true spatial position at the apex. The output section is called the **migrated depth section**. We could also obtain a **migrated time section** by converting depth back to time (or by collapsing the time hyperbola in Figure 8.1b).

Note that in order to position the point reflector correctly so that we can determine its true depth and horizontal location, i.e., in order to do an accurate migration, we need to know the medium velocity v – it is needed to convert the CMP stack section to depth (or the migrated time section to depth). If we use the wrong value of v, the point reflector would be vertically mispositioned on the migrated depth section – the value of z_0 in Figure 8.1c would be different from the true value of z_0 (the one in Figure 8.1a).

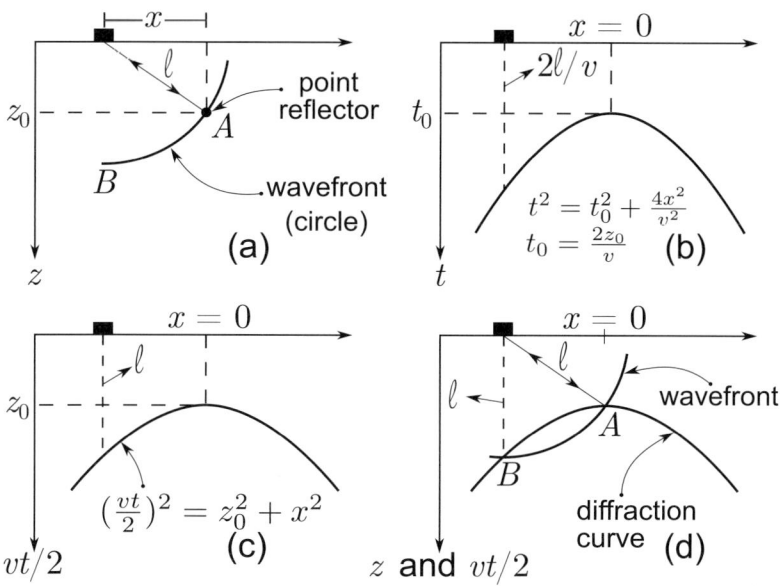

Figure 8.1 (a) A point reflector in a medium of constant velocity v. (b) The two-way travel time plot of the CMP stack section. (c) The stack section with time converted to depth. (d) A superposition of Figures 8.1a and 8.1c.

The velocity value v that we choose to perform the migration is called the **migration velocity**. In general, the accuracy of a migration depends on the accuracy of the migration velocity.

Figure 8.1d is a superposition of Figures 8.1a and 8.1c that shows the geometrical connection between the diffraction curve and the wavefront.

8.2 A Dipping Reflector

We can think of a continuous reflector as a series of closely spaced point reflectors. Figure 8.2 shows a CMP stack section with a flat dipping reflector before and after migration. Its migrated spatial position is given by the **envelope** of the wave-fronts (the line which is tangential to all the wavefronts simultaneously), and its unmigrated or apparent position is given by the envelope of the diffraction curves (e.g., the apparent reflector is tangent to the diffraction curve at point B).

Suppose we are given a CMP stack data section for a dipping reflector. Let v_1 be the velocity above the reflector. To migrate the data, we first choose a migration velocity v. If v_1 is known, we choose $v = v_1$. If v_1 is not known, we guess or estimate v (perhaps using other data). We then convert the CMP stack section to depth by multiplying the time axis by $v/2$, ensuring that the resulting offset and depth axes have the same scale. The converted CMP stack section is now the **unmigrated depth section** (Figure 8.2), which is to be migrated to obtain the migrated depth

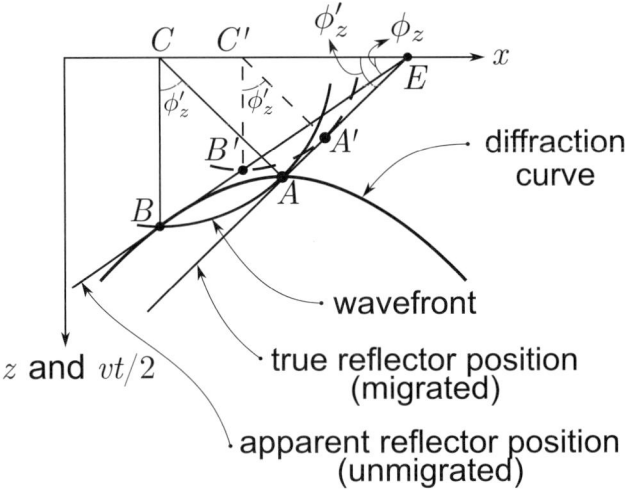

Figure 8.2 The unmigrated (apparent) and migrated positions of a flat dipping reflector on a CMP stack section. Migration would move the segment BB' to AA'. If $v = v_1$, AA' would be the true spatial position of BB'.

section. Or, the **unmigrated time section**, i.e., the CMP stack section, could be migrated to obtain the migrated time section, with time to depth conversion done afterward. Since the migration is performed on the stack section, i.e., after the stacking operation, using a constant velocity v, it is called a **post-stack constant velocity migration**.

To be more realistic, suppose that the apparent reflector in Figure 8.2 does not go all the way to the surface, but extends only from B to B'. The goal of migration is to determine the correct spatial position AA' of the segment BB'.

A simple geometrical way to migrate the data goes as follows. On the unmigrated depth section, for the point B on the apparent reflector, use the vertical distance to the surface ($\overline{CB}$ in Figure 8.2) as a radius to swing a circular arc (a wavefront) away from B (see Figure 8.2). Do the same for the point B'. Then, from the point E where the apparent reflector, when extended, intersects the surface, draw a line that is tangent to the two circular arcs. The migrated reflector would then lie on this line between the two tangency points A and A'.

Another way to migrate the data is to use the fact that $\tan \phi_z = \sin \phi_z'$, which is known as the **migrator's equation**, and is proven in (8.1a). ϕ_z and ϕ_z' are the dip angles of the reflectors on the unmigrated and migrated depth sections respectively. These angles can be measured on the two sections, but there is also a simple relation connecting them. Referring to Figure 8.2, note that the lines CB and CA have the same length. This means that

$$\tan \phi_z = \overline{CB} \,/\, \overline{CE} = \overline{CA} \,/\, \overline{CE} = \sin \phi_z'. \tag{8.1a}$$

This equation implies that $\phi'_z \geq \phi_z$ (e.g., $\tan 40° = \sin 57°$) i.e., the reflector has a steeper dip after migration. It also implies that as the angle ϕ'_z varies between $0°$ and $90°$, the angle ϕ_z varies between $0°$ and $45°$, i.e., the dip angle on the unmigrated depth section cannot exceed $45°$. The migrated dip angle ϕ'_z can be computed from the migrator's equation once the unmigrated angle ϕ_z has been measured on the unmigrated depth section. Referring to the apparent reflector in Figure 8.2, the goal of migration is to move the point B, whose coordinates (x_B, z_B) are known, to the point A. To do this, simple trigonometry can be used to determine the coordinates (x_A, z_A) of the point A in terms of (x_B, z_B). In effect, this is like fitting the diffraction curve $z^2 = z_A{}^2 + (x - x_A)^2$ to the apparent reflector, choosing x_A and z_A so that the diffraction curve is tangent to the apparent reflector at B, and then moving B to the apex A of the diffraction curve (see Hagedoorn, 1954). The process is repeated for every point on the apparent reflector BB' to obtain the migrated reflector AA'.

Yet another way is as follows. We can derive formulas for the coordinates of the endpoints of the (x_A, z_A) and $(x_{A'}, z_{A'})$ of the migrated reflector segment as follows: from Figure 8.2, noting that the distance $\overline{CA}$ is z_B, and that the distance $\overline{C'A'}$ is $z_{B'}$, we have

$$
\begin{aligned}
x_A &= x_B + z_B \sin \phi'_z, & z_A &= z_B \cos \phi'_z, \\
x_{A'} &= x_{B'} + z_{B'} \sin \phi'_z, & z_{A'} &= z_{B'} \cos \phi'_z.
\end{aligned}
\tag{8.1b}
$$

Using $\cos^2 \phi'_z = 1 - \sin^2 \phi'_z$, and using (8.1a) to substitute $\sin \phi'_z = \tan \phi_z \equiv m$, these become

$$
\begin{aligned}
x_A &= x_B + z_B m, & z_A &= z_B \sqrt{1 - m^2}, \\
x_{A'} &= x_{B'} + z_{B'} m, & z_{A'} &= z_{B'} \sqrt{1 - m^2}
\end{aligned}
\tag{8.1c}
$$

where $m = \tan \phi_z$ is the slope of the apparent reflector, and is taken to be positive ($m > 0$) in Figure 8.2. Hence, on the x-z unmigrated data section, the slope m and the points (x_B, z_B) and $(x_{B'}, z_{B'})$ are measured, and substituted into Equation (8.1c) to obtain the coordinates (x_A, z_A) and $(x_{A'}, z_{A'})$ of the migrated reflector segment. For a similar method, see Claerbout (1985, p. 8)

In each method described, the accuracy of the migration depends on the accuracy of the value chosen for the migration velocity v. If $v = v_1$, the velocity above the reflector, the migration would be perfectly accurate, i.e., the reflector would be correctly positioned on the migrated depth section. Otherwise, it is mispositioned.

If we convert the depth section in Figure 8.2 to a time section by multiplying the vertical scale by $2/v$, and let c denote apparent velocity on the stack time section, then we see that $c' < c$, i.e., migrating the dipping event lowers its apparent velocity. Remember also, from Chapter 6, that the apparent velocity c on a stack section (it was called "v" in Chapter 6) is different from the apparent velocity c on a shot record (on a stack, c can be less than v_1).

8.3 Diffraction-Summation Migration

This method of migration is suggested by Figure 8.1, and is often known as **diffraction-stack migration**, or just **diffraction migration**. The data of Figure (8.1b) can be migrated by summing the trace values along the diffraction curve and placing the result at the apex. This procedure can be applied more generally to a more complex CMP stack section. In such a case, the migration velocity is no longer constant, in general. Referring to Figure 8.3, use the appropriate velocity-time function, obtained from a velocity analysis technique, as the migration velocity function to calculate the diffraction curve at a given sample point on a given trace on the stack section. Then add up all the trace values along the diffraction curve and place the sum at the apex. Do this at each sample point on the stack section. The final result is a **migrated time section**. If desired, obtain the migrated depth section by converting time to depth, as discussed in Chapter 5 (e.g., Equation 5.20, or the methods discussed in relation to Figure 5.5).

The method is based on the assumption that a subsurface medium can be treated as a distribution of point reflectors. Referring to Figure 8.3, if the apex at $t = t_a$, for instance, corresponds to an actual reflection point in depth, then the energy along the diffraction curve will add up to produce a sample with a relatively high amplitude ($\sum_i a_i$) on the migrated section. Otherwise, the sample values a_i will be more or less random in value, and will tend to cancel each other out in the sum.

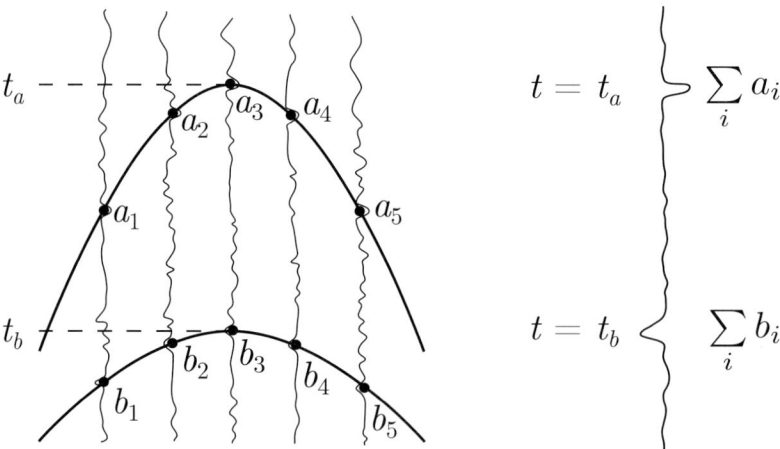

Figure 8.3 An illustration of the diffraction-summation migration of a stack section. a_i and b_i are the trace values at the points where the diffraction curves cross the traces. The trace at the far right is the output trace and it is placed in the same position as the middle trace of the input section, i.e., $\sum_i a_i$ is placed at the apex of the upper hyperbola, etc.

In simple applications of this method, the same velocity-time function is used for each *x*, i.e., each trace. In other words, the velocity is taken to be a function of depth (or time *t*) only, and not a function of *x*. In effect, this implies that vertical inhomogeneity is being assumed – the layers are flat and horizontal. One might then wonder how the method can possibly work if it is applied to typical data from media that are both vertically and laterally inhomogeneous (synclines, curved reflectors, dipping beds, etc.). The fact is that the method *does* work fairly well for such media (better than one might expect, anyway), and the reason seems to be that the errors produced by the assumption of flat and horizontal layers tend to cancel each other out in a statistical sense, due to the more or less random distribution of reflector dip values in the media (Robinson, 1983). Nevertheless, the generally wrong assumption of vertical inhomogeneity means that the migration will not be absolutely accurate, and some errors will be present.

Since simple applications of this method, and other conventional methods of migration as well, assume a 1D velocity function, but can be applied reasonably successfully to 2D data (from laterally and vertically inhomogeneous media), they are sometimes called $1\frac{1}{2}$D migration methods. Recent research efforts have come up with migration methods that are fully 2D – they have been called **depth migration** methods – and they are rapidly replacing the conventional $1\frac{1}{2}$D methods in complex media.

Finally, note that if the medium *is* in fact vertically inhomogeneous, then migration is actually not necessary because the reflection events are flat and horizontal. The correct depth section can be obtained from a simple time-to-depth conversion.

8.4 Wavefront Migration

This method is suggested by Figure 8.1d. Consider the data (diffraction curve) generated by a point reflector in Figure 8.4a. Choose a point *P* on the diffraction curve and construct the wavefront with its deepest point at *P*. Spread the energy at point *P* evenly along the wavefront. Repeat for all points (P_1, P_2, ...) on the diffraction curve. Since the wavefronts intersect at the apex *A*, and nowhere else, they will add constructively to produce a relatively high amplitude at point *A* (the location of the point reflector), but relatively low amplitudes everywhere else, resulting in a migrated depth section.

This can be applied to a CMP stack section, assuming that the subsurface medium can be treated as a distribution of point reflectors. Using an appropriate velocity function, convert time to depth, as discussed in Chapter 5 (e.g., Equation 5.20, or the methods discussed in relation to Figure 5.5), on the recorded data section, and construct the appropriate wavefront at a given sample point *P*. Spread

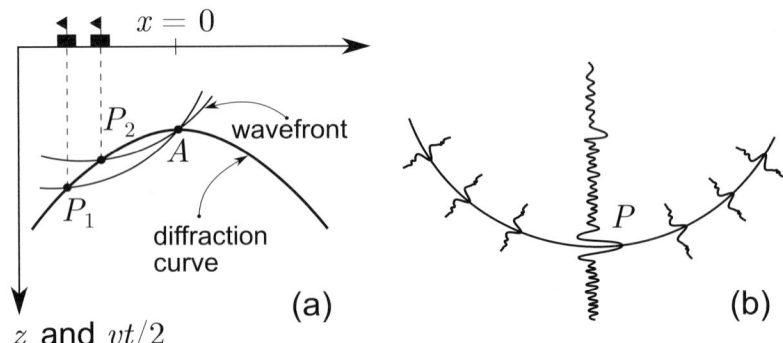

Figure 8.4 An illustration of wavefront migration. (a) Convergence of the wavefronts at the apex of the diffraction curve for a point reflector. (b) The amplitude of the sample at P is evenly spread out along the wavefront.

the energy at P evenly along the wavefront (Figure 8.4b). Repeat for all points P on the data section, then sum the wavefronts. The result is a migrated depth section.

For a continuous reflector, Figure 8.2 shows that the wavefronts are tangent to the migrated reflector and that it will lie along the envelope of the wavefronts. The wavefronts will intersect and add constructively along the envelope line. In zones where there are no reflectors, the wavefronts will tend to cancel each other out because of random interference effects.

If the subsurface is both vertically and laterally heterogeneous, but a 1D velocity function is used, then this method is a $1\frac{1}{2}$D migration method, and the migrated depth section will be correspondingly inaccurate.

8.5 Wave Equation Migration

The term "wave equation migration" generally refers to any migration method that is based on solving the wave equation. The *scalar* wave equation is normally used (no shear waves are considered), but research work is in progress on extending and generalizing this approach.

As we have seen previously, a CMP stack section is approximately the same as the section consisting of a collection of zero-offset recordings. The downgoing travel paths of the zero-offset rays are identical to the upgoing ones, i.e., they coincide (meaning they also have the same travel times). This means that we can think of the waves as originating at the reflectors themselves, rather than at the surface. This is, in effect, what we assume when we trace zero-offset rays. This is the basis of the **exploding reflector model** of seismic data, used in wave equation migration (see Loewenthal, Lu, Roberson, and Sherwood, 1976). In this model, we consider sources to be distributed along all reflectors, with source strength proportional to the reflection coefficient, and assume all sources are initiated at $t = 0$, with waves traveling upward at the half-velocity V, where $V = \frac{1}{2}v$. It is assumed that the wavefield generated by the model satisfies the scalar wave equation

$$\frac{\partial^2 u}{\partial x^2} + \frac{\partial^2 u}{\partial z^2} = \frac{1}{V^2}\frac{\partial^2 u}{\partial t^2}. \tag{8.2}$$

It is also assumed that the recorded seismic data (e.g., the CMP stack section) are the values of the wavefield at the surface, i.e., $u(x, 0, t)$. Knowing $u(x, 0, t)$, we can, in theory, calculate $u(x, z, t)$, i.e., the seismic section that would be obtained from the exploding reflector model if the recording plane was at depth z rather than at the surface. This is known as **downward continuation** or **extrapolation** of the data. Putting $t = 0$ in $u(x, z, t)$, i.e., extrapolating the wavefield backward in time to $t = 0$ when the sources were initiated, gives us the migrated depth section $u(x, z, 0)$. At $t = 0$, we know from Huygens' principle that the wavefronts associated with the wavefield will have the same shape as the reflectors, and that the wavefronts will become distorted as t increases.

Referring to Figure 8.5, we see that migration can be performed by calculating the seismic time section $u(x, z_i, t)$ for the recording plane at depth z_i from the stack section data, stripping off the uppermost part $u(x, z_i, 0)$, and repeating these steps for all depths z_i. The collection of strips is the migrated depth section.

There are limitations to the exploding reflector model (Claerbout, 1985). Certain zero-offset rays obtained from a model in which there is lateral velocity variation

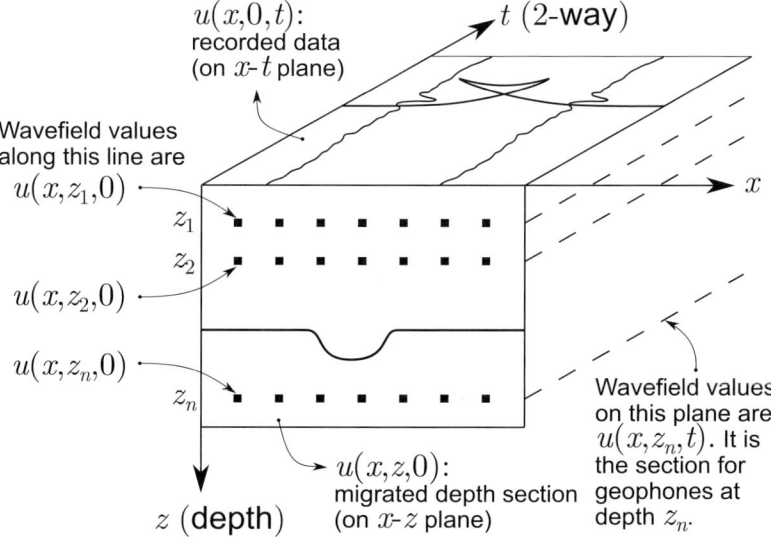

Figure 8.5 A schematic illustration of wave equation migration.

are not predicted by the exploding reflector model. For example, consider a **velocity lens**, e.g., an inclusion in the shape of a convex lens whose plane is horizontal and whose velocity is higher than the surrounding rock. The upgoing and downgoing parts of a zero-offset ray passing through such a lens would not coincide (except for a ray passing right through the middle of the lens, where the surface is horizontal). Another example is a single horizontal layer which also contains a vertical reflector – again, the up- and downgoing parts of the zero-offset rays do not coincide. The exploding reflector model also fails with multiple reflections. For example, for a single horizontal layer of velocity α and thickness h, zero-offset multiples would occur at times $4h/\alpha$, $6h/\alpha$, etc., whereas for an exploding reflector, they would occur at times $3h/\alpha$, $5h/\alpha$, etc. In the case of a multilayered medium, referring to Figure 7.3, the reflection between segments 3 and 4, and the reflection between the last two segments $m-1$ and m would have opposite polarities (for normal incidence) – the exploding reflector model would not predict this.

All migration methods require velocity v as input, and the accuracy of the migration depends on the accuracy of the velocity information. Also, only primary reflections are usually treated. Multiples are ignored – if they are present, they will be migrated incorrectly.

Three popular methods of conventional wave equation migration are (a) frequency-wavenumber (f-k_x or ω-κ_x) migration, (b) finite difference migration, and (c) Kirchoff migration.

8.6 Frequency-Wavenumber Migration

Consider a medium with constant velocity v. Wave motion in such a medium is described by the wave equation (8.2). If we apply a 2D Fourier transform with respect to x and t to Equation (8.2), we obtain

$$(i\kappa_x)^2 \, \bar{\tilde{u}}(\kappa_x, z, \omega) + \frac{\partial^2}{\partial z^2} \bar{\tilde{u}}(\kappa_x, z, \omega) = \frac{(-i\omega)^2}{V^2} \bar{\tilde{u}}(\kappa_x, z, \omega), \qquad (8.3)$$

where κ_x ($= \omega p$) is the radial horizontal wavenumber and where

$$\bar{\tilde{u}}(\kappa_x, z, \omega) = \int_{-\infty}^{\infty} \int_{-\infty}^{\infty} u(x, z, t) \exp\left[i(\omega t - \kappa_x x)\right] dx \, dt, \qquad (8.4)$$

is the 2D Fourier transform of the wavefield $u(x, z, t)$.

Equation (8.3) can also be more quickly and easily obtained as follows. From (1.150) or (6.8) we know that if the Fourier time transform of a function $g(t)$ is $\bar{g}(\omega)$, then the Fourier time transform of dg/dt is just $-i\omega\bar{g}(\omega)$ (assuming that $g(t) \to 0$ for $t \to \pm\infty$, which is typically true for seismic signals). Similarly, if the Fourier x transform of a function $h(x)$ is $\tilde{h}(\kappa_x)$, then the Fourier x transform of dh/dx is just

$+i\kappa_x\widetilde{h}(\kappa_x)$ (assuming that $h(x) \to 0$ for $x \to \pm\infty$). These rules also apply to partial derivatives. Since the x- and t-derivatives in (8.2) are second-order derivatives, we simply apply these rules twice to (8.2) to obtain (8.3). The 2D Fourier transforms of

$$\partial^2 u/\partial x^2 \qquad \text{and} \qquad \partial^2 u/\partial t^2,$$

in the wave equation (8.2) are

$$(i\kappa_x)^2\widetilde{\widetilde{u}} \qquad \text{and} \qquad (-i\omega)^2\widetilde{\widetilde{u}},$$

respectively. The transform of $\partial^2 u/\partial z^2$ is just $\partial^2\widetilde{\widetilde{u}}/\partial z^2$ because the integration is not over z, but only over x and t.

The horizontal and vertical radial wavenumbers κ_x and κ_z satisfy the equation

$$\kappa^2 = \kappa_x^2 + \kappa_z^2 = \frac{\omega^2}{V^2}. \tag{8.5}$$

Equation (8.3) can be rewritten as

$$\frac{\partial^2\widetilde{\widetilde{u}}}{\partial z^2} = -\kappa_z^2\,\widetilde{\widetilde{u}} \qquad \text{where} \qquad \kappa_z^2 = \frac{\omega^2}{V^2} - \kappa_x^2. \tag{8.6}$$

This is just a 1D Helmholtz equation whose solution, for upgoing or downgoing waves, is

$$\widetilde{\widetilde{u}}(\kappa_x, z, \omega) = \widetilde{\widetilde{u}}(\kappa_x, 0, \omega)\exp(i\kappa_z z) \quad \text{where} \quad \kappa_z = \pm\sqrt{\frac{\omega^2}{V^2} - \kappa_x^2}. \tag{8.7}$$

Note that $\widetilde{\widetilde{u}}(\kappa_x, 0, \omega)$ in Equation (8.7) is just the frequency-wavenumber (ω-κ_x) spectrum of the seismic data $u(x, 0, t)$ (see Equation 8.4). Equation (8.7) gives, in the κ_x-ω domain, the wavefield at depth z in terms of the known (recorded) wavefield at the surface.

To obtain $u(x, z, t)$, we transform $\widetilde{\widetilde{u}}$ back to the x-t domain via an inverse 2D Fourier transform, i.e.,

$$u(x, z, t) = \frac{1}{4\pi^2}\int_{-\infty}^{\infty}\int_{-\infty}^{\infty}\widetilde{\widetilde{u}}(\kappa_x, z, \omega)\exp\Big[i(\kappa_x x - \omega t)\Big]\,d\kappa_x\,d\omega. \tag{8.8a}$$

Substituting (8.7) into this gives

$$u(x, z, t) = \frac{1}{4\pi^2}\int_{-\infty}^{\infty}\int_{-\infty}^{\infty}\widetilde{\widetilde{u}}(\kappa_x, 0, \omega)\exp\Big[i(\kappa_x x + \kappa_z z - \omega t)\Big]\,d\kappa_x\,d\omega. \tag{8.8b}$$

How do we choose the sign of κ_z in this integral (see Equation 8.7)? We desire that this integral represent a superposition of *upgoing* plane waves, since we are using the exploding reflector model. If the complex exponential is to be an upgoing plane wave, then κ_z and ω must have opposite signs, i.e., if ω is positive then κ_z is negative, and vice versa (note that the integration is over negative values of ω as well as positive ones).

Putting $t = 0$ in this equation gives the *migrated depth section*:

$$u(x, z, 0) = \frac{1}{4\pi^2} \int_{-\infty}^{\infty} \int_{-\infty}^{\infty} \widetilde{\widetilde{u}}(\kappa_x, 0, \omega) \exp\left[i\kappa_x x + i\kappa_z z\right] d\kappa_x \, d\omega, \qquad (8.9a)$$

where

$$\kappa_z = -\text{sign}(\omega)\sqrt{\frac{\omega^2}{V^2} - \kappa_x^2} = -\frac{\omega}{V}\sqrt{1 - \left(\frac{V\kappa_x}{\omega}\right)^2}. \qquad (8.9b)$$

Given the velocity v, Equation (8.9) states that the process of migration consists of computing the frequency-wavenumber spectrum $\widetilde{\widetilde{u}}(\kappa_x, 0, \omega)$ of the data $u(x, 0, t)$, multiplying it by the complex exponential factor in the integrand of Equation (8.9a), and doing a double integral over κ_x and ω. Because multiplying the spectrum $\widetilde{\widetilde{u}}(\kappa_x, 0, \omega)$ by the complex exponential changes the phase of the spectrum, the method is called **phase-shift migration**. Note that if the integral over ω is done first, then the remaining integral over κ_x is merely an inverse Fourier transform over κ_x, which can be done efficiently by the fast Fourier transform (FFT) algorithm. Note that since the integration in Equation (8.9a) is over all possible values of κ_x and ω ($-\infty$ to ∞), the quantity under the square root in Equation (8.9b) is zero or negative for those values of κ_x and ω satisfying $|\kappa_x| \geq |\omega|/V$, i.e., the integration includes *evanescent* or *inhomogeneous* waves. However, in practical computations, the evanescent waves are left out of the integration for reasons of computational stability and because they do not affect the final result very much anyway. More specifically, the double integral (8.9a) is evaluated only over the region in the ω-κ_x plane for which $|\kappa_x| < |\omega|/V$, implying κ_z is real and nonzero. For more details, see Claerbout (1985, p. 30) and Stolt and Benson (1986, chapter 2).

The integration in Equation (8.9a) can be done more economically by replacing the integral over ω with an integral over κ_z. This effectively results in the migrated depth section being obtained by a 2D inverse Fourier transform, rather than an ordinary single integral followed by a 1D Fourier transform as in phase-shift migration (which is more expensive). From Equation (8.9b), remembering from earlier that κ_z is real and nonzero, we have

$$\omega = -\text{sign}(\kappa_z)V\sqrt{\kappa_x^2 + \kappa_z^2} \qquad (8.10a)$$

$$\implies \quad \frac{d\omega}{d\kappa_z} = -\text{sign}(\kappa_z)\frac{V\kappa_z}{\sqrt{\kappa_x^2 + \kappa_z^2}} = -\frac{V|\kappa_z|}{\sqrt{\kappa_x^2 + \kappa_z^2}}. \qquad (8.10b)$$

Substitution of this into Equation (8.9a) gives the migrated depth section, or **image**:

$$u(x, z, 0) = \frac{1}{4\pi^2} \int_{-\infty}^{\infty} \int_{-\infty}^{\infty} B(\kappa_x, \kappa_z) \exp\left[i\kappa_x x + i\kappa_z z\right] d\kappa_x \, d\kappa_z, \qquad (8.11a)$$

where

$$B(\kappa_x, \kappa_z) = -\frac{V|\kappa_z|}{\sqrt{\kappa_x{}^2 + \kappa_z{}^2}}\tilde{\tilde{u}}\left(\kappa_x, 0, -\text{sign}(\kappa_z)V\sqrt{\kappa_x{}^2 + \kappa_z{}^2}\right). \tag{8.11b}$$

Obtaining the image via Equation (8.11a) is the process which is normally referred to as frequency-wavenumber migration. It consists of the computation of $B(\kappa_x, \kappa_z)$ from the frequency-wavenumber spectrum of the data, followed by a 2D inverse Fourier transform of $B(\kappa_x, \kappa_z)$.

Alternatively, ω and $d\omega/d\kappa_z$ in Equation (8.10a) can be expressed as

$$\omega = -V\kappa_z\sqrt{1 + \frac{\kappa_x{}^2}{\kappa_z{}^2}} \quad \text{and} \quad \frac{d\omega}{d\kappa_z} = -\frac{V}{\sqrt{1 + \left(\kappa_x{}^2/\kappa_z{}^2\right)}}, \tag{8.12}$$

where the correct opposite-sign relationship between κ_z and ω is maintained without the use of "sign" or "| |." Hence, $B(\kappa_x, \kappa_z)$ can also be written as

$$B(\kappa_x, \kappa_z) = -\frac{V}{\sqrt{1 + \left(\kappa_x{}^2/\kappa_z{}^2\right)}}\tilde{\tilde{u}}\left(\kappa_x, 0, -V\kappa_z\sqrt{1 + \frac{\kappa_x{}^2}{\kappa_z{}^2}}\right). \tag{8.13}$$

The migrated *time* section can be obtained by dividing all values z on the depth axis of the migrated depth section $u(x, z, 0)$ by the half-velocity $V = v/2$ (time is two-way, depth is one-way). In the κ_x-κ_z domain, the corresponding operation is to multiply the values on the κ_z axis by $-V$, which changes it into a frequency axis with frequency values ω' given by

$$\omega' = -V\kappa_z = \omega\sqrt{1 - \left(\frac{V\kappa_x}{\omega}\right)^2} = \text{sign}(\omega)\sqrt{\omega^2 - V^2\kappa_x{}^2},$$

where Equation (8.9b) has been used. For a fixed κ_x, this represents a *reduction* in the magnitude of the frequency (from $|\omega|$ to $|\omega'|$) due to migration (see Figure 8.6). In other words, the wavelets on the migrated time section will be broader than the same wavelets on the unmigrated time section. Note that for small V, we have $\omega' \approx \omega$, i.e., there is very little change in frequency, but for larger V, the loss in resolution may be appreciable.

Referring to Figure 8.6, the angles ϕ_z and ϕ_z' are related by the equation $\sin \phi_z' = \tan \phi_z$, which can be used to obtain a relation between the reflection dips in the time sections, δ_t and $\delta_t{}'$, in the following way. First, write $\tan \phi_z$ in terms of $\tan \phi_z'$, i.e.,

$$\tan \phi_z = \sin \phi_z' = \frac{\tan \phi_z'}{\sec \phi_z'} = \frac{\tan \phi_z'}{\sqrt{1 + \tan^2 \phi_z'}}. \tag{8.14}$$

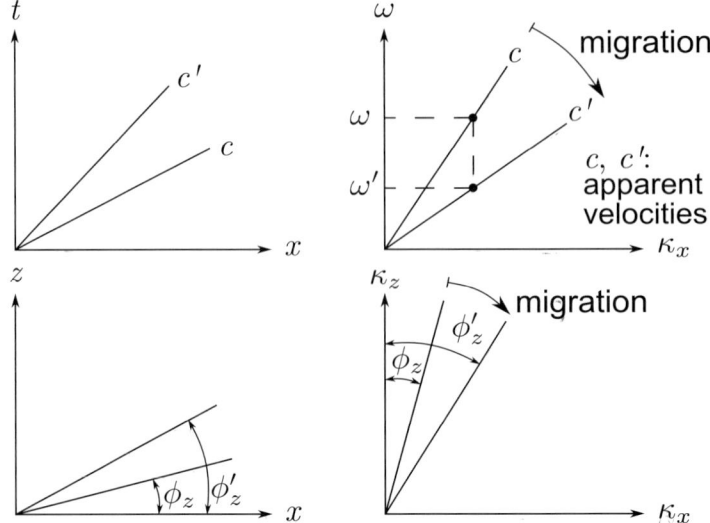

Figure 8.6 Schematic diagrams of the migration of a flat dipping event in various domains. c and c' are the apparent velocities of the associated events. The angles and apparent velocities shown satisfy the equations $\sin \phi'_z = \tan \phi_z$ and $c' = c\sqrt{1 - V^2/c^2}$, where $V = v/2$ with v being the velocity for the medium. The primes denote the migrated quantities.

Now, in the migrated depth section, if a reflector dips by an amount $\Delta z'$ in the vertical direction over a horizontal distance $\Delta x'$, then the dip angle ϕ'_z is given by

$$\tan \phi'_z = \frac{\Delta z'}{\Delta x'} = \frac{V \Delta t'}{\Delta x'} = V \delta_t', \tag{8.15a}$$

where $\delta_t' = \Delta t'/\Delta x'$ is the dip of the reflector in the migrated time section. Similarly, we have for the unmigrated section,

$$\tan \phi_z = \frac{V \Delta t}{\Delta x} = V \delta_t, \tag{8.15b}$$

where $\delta_t = \Delta t/\Delta x$ is the dip of the reflector in the unmigrated time section. Substituting these into Equation (8.14) gives

$$\delta_t = \delta_t'/\sqrt{1 + V^2 \delta_t'^2}, \quad \text{or} \quad \delta_t' = \delta_t/\sqrt{1 - V^2 \delta_t^2}. \tag{8.16a}$$

These equations apply to a flat dipping reflector. They can be generalized to curved reflectors by replacing δ_t with dt/dx, which gives the dip along the curved reflection event on the unmigrated time section, i.e., the stack data; and by replacing δ_t' with $(dt/dx)'$, which gives the dip along the curved reflection event on the migrated time section (see Rothman, Levin, and Rocca, 1985).

Note also that $\delta_t = \Delta t / \Delta x = 1/c$ and that $\delta_t{}' = (\Delta t / \Delta x)' = 1/c'$, where c and c' are the apparent velocities associated with the dipping reflections on the unmigrated and migrated stack time sections respectively. Therefore, one may also make these replacements in the modified equations (8.16a). With a little algebra, one obtains the following:

$$c = c'\sqrt{1 + V^2/c'^2}, \qquad \text{or} \qquad c' = c\sqrt{1 - V^2/c^2}. \tag{8.16b}$$

These equations clearly show that $c' < c$, i.e., migration reduces the apparent velocity of a flat dipping reflector and increases the dip.

Note that either of the equations in (8.16b) can be expanded and simplified to yield the following:

$$c'^2 = c^2 - V^2 = c^2 - (v/2)^2. \tag{8.16c}$$

Note that the migration velocity v must be chosen so that $V \leq c$, i.e., $v \leq 2c$, so that $c' \geq 0$.

As a simple example, suppose that c for a flat dipping reflection on the unmigrated time section is measured to be $c = 2.5$ km/s, and the migration velocity is chosen to be $v = 2$ km/s. Then $V = 1$ km/s, and $c' = \sqrt{2.5^2 - 1^2} = 2.29$ km/s.

When computing the frequency-wavenumber spectrum, errors due to *aliasing* can occur. Given the samples $s_t = \{s_0, s_1, s_2, \ldots\}$ of a time signal whose frequency spectrum lies in the range $0 \leq f \leq f_{\max}$, the true continuous time signal $s(t)$ can be fully recovered from its samples as long as $f_{\max} \leq f_N$, where the Nyquist frequency $f_N = 1/(2\Delta t)$ with Δt being the sample interval. If $f_{\max} > f_N$, then $s(t)$ cannot be fully recovered because the spectrum of s_t is then distorted (not the same as the spectrum of $s(t)$) – components with frequencies above f_N "alias" as components with frequencies below f_N. For a sampled spatial signal, aliasing occurs if the continuous spatial signal contains wavenumbers k_x such that $k_x > k_{xN} = 1/(2\Delta x)$, where Δx is spatial sample interval. For example, on a seismic section, the sequence of samples lying along some line of constant time that crosses all the traces is a spatial signal, with Δx being the interval between traces. If small-scale structure of spatial extent less than Δx is present in the subsurface, then it may go undetected – the true wavenumber spectrum contains wavenumbers above k_{xN} (which are due to the small-scale structure). This is sometimes a problem in 3D surveys, or in making contour isochron maps from criss-crossing seismic lines.

Reflecting interfaces with dips larger than a certain maximum value $(\phi_z')_{\max}$ will be incorrectly migrated, due to spatial aliasing. Referring to Figure 8.7, which shows monofrequency reflections from dipping interfaces, the spatial signal of wavenumber $\frac{3}{2}k_{xN}$ aliases as a signal of wavenumber $\frac{1}{2}k_{xN}$, hence the steeply dipping reflector will not be correctly represented in the frequency-wavenumber

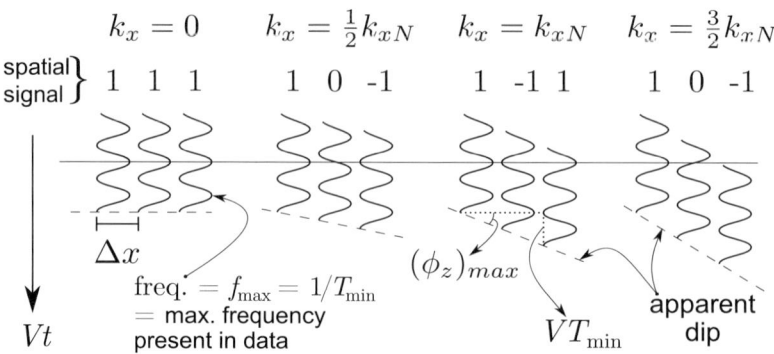

Figure 8.7 An illustration of spatial aliasing. The diagrams show the reflections of the highest-frequency component of a pulse from interfaces with increasing apparent dips. For the steepest dip, the component whose wavenumber is $\frac{3}{2}k_{xN}$ aliases as the component whose wavenumber is $\frac{1}{2}k_{xN}$. $(\phi_z)_{\max}$ is the steepest apparent dip that can be correctly represented in the f-k_x spectrum.

domain, and will be incorrectly migrated. From Figure 8.7, we see that the steepest dip $(\phi'_z)_{\max}$ that can be correctly migrated is given by

$$\sin(\phi'_z)_{\max} = \tan(\phi_z)_{\max} = \frac{VT_{\min}}{2\Delta x} = \frac{v}{4f_{\max}\Delta x}, \qquad (8.17)$$

where $f_{\max}$ is the highest frequency present in the data. Conversely, to ensure that all dips between $0°$ and some chosen value $(\phi'_z)_{\max}$ are correctly migrated, one must filter out from the data any frequencies above $f_{\max}$. For example, consider the migration of a CMP stack section with $\Delta x = 15\,\mathrm{m}$ (the geophone group interval is $30\,\mathrm{m}$) and $v = 3{,}000\,\mathrm{m/sec}$. To ensure that all dips up to $(\phi'_z)_{\max} = 30°$ are correctly migrated, one must filter out any frequencies above $f_{\max} = 100\,\mathrm{Hz}$. Note that the higher $f_{\max}$ is, the lower is $(\phi'_z)_{\max}$.

As mentioned previously, the better the velocity input, the more accurate the migration. In the preceding discussion of frequency-wavenumber migration, we assumed that the velocity v for the medium was constant. In fact, f-k_x migration via Equation (8.11) cannot handle the case of $v = v(z)$ (e.g., see Claerbout, 1985, p. 33, and Stolt and Benson, 1986, p. 207). A more accurate migration might be obtained by a scheme that was capable of treating the variable velocity case. As we will see in the analysis that follows, phase-shift migration can be generalized to handle this case.

Let us assume that the recorded wavefield (the data) approximately satisfies the wave equation (8.2) with $V = V(z)$ (see Equation 5.34). A 2D Fourier transform then leads to the following Helmholtz equation:

$$\frac{\partial^2 \tilde{\tilde{u}}}{\partial z^2} = -\kappa_z^2 \tilde{\tilde{u}} \qquad \text{where} \qquad \kappa_z = \pm \sqrt{\frac{\omega^2}{V^2(z)} - \kappa_x^2}. \qquad (8.18)$$

This can be solved approximately using the WKBJ method (see Equations 5.36–5.43), or by "factoring" the Helmholtz equation into first-order equations for upgoing and downgoing waves (see Equations 5.45 to 5.49). The equation for upgoing waves is

$$\frac{\partial \tilde{\tilde{u}}}{\partial z} = i \kappa_z \tilde{\tilde{u}}, \qquad \text{where} \qquad \kappa_z = -\text{sign}(\omega) \sqrt{\frac{\omega^2}{V^2(z)} - \kappa_x^2}, \qquad (8.19a)$$

and its solution is

$$\tilde{\tilde{u}}(\kappa_x, z, \omega) = \tilde{\tilde{u}}(\kappa_x, 0, \omega) \exp\left[i \int_0^z \kappa_z \, dz \right]. \qquad (8.19b)$$

Transforming back to the *x*-*t* domain and putting $t = 0$ to obtain the image (see Equations 8.8 and 8.9) leads to a formula for the migrated depth section $u(x, z, 0)$ via phase-shift migration when velocity varies with depth:

$$u(x, z, 0) = \frac{1}{4\pi^2} \int_{-\infty}^{\infty} \int_{-\infty}^{\infty} \tilde{\tilde{u}}(\kappa_x, 0, \omega) \exp\left[i\kappa_x x + i \int_0^z \kappa_z \, dz \right] d\kappa_x \, d\omega, \qquad (8.20)$$

where κ_z is given in Equation (8.19a).

For more details on frequency-wavenumber migration, see Chun and Jacewitz (1981), Claerbout (1985), Gazdag and Sguazzero (1984), Robinson (1983), Rothman, Levin, and Rocca (1985), Stolt (1978), and Stolt and Benson (1986), among others.

8.7 Finite Difference Migration

This technique involves solving a modified wave equation by the **method of finite differences**, so we first look at a very simple example of solving a differential equation by this method. Suppose we wish to solve the equation

$$\frac{dy}{dx} = y(x), \quad \text{with} \quad y(0) = 1. \qquad (8.21)$$

The exact solution is $y(x) = e^x$, which we can use to check the accuracy of the finite difference solution.

We now discretize the *x* axis into a 1D grid and let Δx be the spacing between adjacent grid points. If Δx is small, then using the definition of a derivative, dy/dx can be approximately written as

$$\frac{dy}{dx} \approx \frac{y(x + \Delta x) - y(x)}{\Delta x}, \qquad (8.22)$$

which is the **forward difference formula** for dy/dx. Equivalently, one may write

$$\frac{dy}{dx} \approx \frac{y(x) - y(x - \Delta x)}{\Delta x}, \qquad (8.23)$$

which is the **backward difference formula** for dy/dx. Using the forward difference formula, our differential equation becomes

$$\frac{y(x + \Delta x) - y(x)}{\Delta x} = y(x), \quad y(0) = 1. \qquad (8.24)$$

Solving for $y(x + \Delta x)$ gives

$$y(x + \Delta x) = (1 + \Delta x)y(x). \qquad (8.25)$$

This equation states that if we know the value of y at x, we can calculate the value of y at the next point $x + \Delta x$. In this way, we can then calculate y at all values of x, beginning at $x = 0$. The following table compares the results of the exact solution, $y = e^x$, and the approximate finite difference (FD) solution, for the first five points only, for $\Delta x = 0.1$.

x	Exact	FD	% Error
0.0	1.0000	1.0000	0.00
0.1	1.1052	1.1000	0.47
0.2	1.2214	1.2100	0.93
0.3	1.3499	1.3310	1.40
0.4	1.4918	1.4641	1.86

For $x = 0$, the FD solution is just the initial condition, $y(0) = 1$. With $1 + \Delta x = 1.1$, each entry in the FD column is $1.1 \times$ the previous entry (see Equation 8.25). The percent error is calculated as follows: for example, for $x = 0.4$, the percent error is $[(1.4918 - 1.4641)/1.4918] \times 100 = 1.86\,\%$.

Clearly, the error grows rapidly as x increases – the scheme is *unstable*. There are ways, though, to reduce the error. For instance, we can make the grid spacing Δx smaller. Also, it turns out that the error is significantly reduced if we use the more accurate **central difference formula** to approximate dy/dx, i.e.,

$$\frac{dy}{dx} \approx \frac{y(x + \Delta x) - y(x - \Delta x)}{2\Delta x}. \qquad (8.26a)$$

The reason the central difference formula is more accurate than the forward difference formula is because the slope of the line connecting the points $y(x + \Delta x)$ and $y(x - \Delta x)$, which is given in (8.26a), is closer to the true slope at x than is the slope of the line connecting the points $y(x + \Delta x)$ and $y(x)$, which is given in (8.22).

Using the central difference formula results in the differential equation (8.21) being approximated by

$$\frac{y(x + \Delta x) - y(x - \Delta x)}{2\Delta x} = y(x), \tag{8.26b}$$

which in turn results in the central difference scheme

$$y(x + \Delta x) = (2\Delta x)y(x) + y(x - \Delta x). \tag{8.26c}$$

Note that to compute $y(x + \Delta x)$ by this scheme, values of y at two previous levels, $y(x)$ and $y(x - \Delta x)$, are required, unlike the forward difference scheme in Equation (8.25), where the value of y at only one previous level, $y(x)$, is required. This means that to start off the calculation, one needs to know both $y(0)$, which is given by the known boundary condition in Equation (8.21), and $y(\Delta x)$, which could be computed by the forward difference scheme in Equation (8.25). Or better still, $y(\Delta x)$ can be computed by the forward difference scheme in Equation (8.25) using a much smaller value of the grid spacing Δx. For instance, with $\Delta x = 0.1$, a value of $\Delta x = 0.005$ could be used in equation (8.25) to compute $y(0.1)$, which in 20 steps gives $y(0.1) = 1.1049$, which in turn can be used in the central difference scheme for $y(\Delta x)$. The resulting central difference scheme is still unstable, but the errors for the same x values are much smaller compared to the forward difference scheme (e.g., at $x = 0.4$, the error is only 0.069%), and the error increases less rapidly.

Furthermore, higher-order difference formulas for dy/dx can be used to improve accuracy (the preceding difference formulas are first order), although this may be uneconomical from a computational point of view.

If the differential equation we want to solve contains higher-order derivatives, such as d^2y/dx^2, these can also be approximated with finite differences. For instance, using forward differences, and with $y'(x) \equiv dy/dx$, we have

$$\begin{aligned}
\frac{d^2y}{dx^2} &= \frac{d}{dx}y'(x) \approx \frac{y'(x + \Delta x) - y'(x)}{\Delta x} \\
&= \frac{1}{\Delta x}\left[\frac{y(x + 2\Delta x) - y(x + \Delta x)}{\Delta x} - \frac{y(x + \Delta x) - y(x)}{\Delta x}\right] \\
&= \frac{\left[y(x + 2\Delta x) - 2y(x + \Delta x) + y(x)\right]}{(\Delta x)^2}.
\end{aligned} \tag{8.27}$$

Once again, accuracy can be substantially improved by using central differences to approximate d^2y/dx^2, instead of forward or backward differences.

In seismic wave theory, the finite difference method is often used to numerically solve partial differential equations, such as the wave equation. Stable schemes, in which the error does not grow without bound, can normally be designed.

To discuss finite difference migration, we begin with the wave equation (8.2), and we assume a constant velocity medium. As we have seen, when the wave equation is Fourier-transformed into the frequency-wavenumber domain and "factored," it leads to an approximate differential equation for upgoing waves, i.e.,

$$\frac{\partial \overline{\overline{u}}}{\partial z} = i\kappa_z \overline{\overline{u}}, \qquad \kappa_z = -\frac{\omega}{V}\sqrt{1 - \left(\frac{V\kappa_x}{\omega}\right)^2}, \tag{8.28}$$

(Equation 8.19a, with V being constant, i.e., independent of z). This differential equation could be transformed back to the x-t domain and solved using the method of finite differences. Before doing so, however, we introduce a new time variable, namely, the **advanced time** t_a, given by

$$t_a = t + \frac{z}{V}. \tag{8.29}$$

Along with this, we define the **advanced wavefield** u_a as

$$u_a\left(x, z, t + \frac{z}{V}\right) = u(x, z, t). \tag{8.30}$$

From a computational viewpoint, it is more convenient and economical to compute u_a than u by the finite difference method (see Claerbout, 1985; Stolt and Benson, 1986).

From Fourier transform theory, we know that if $\overline{f}(\omega)$ is the transform of $f(t)$, then $e^{-i\omega\tau}\overline{f}(\omega)$ is the transform of $f(t + \tau)$. Applying this rule while taking the 2D Fourier x-t transform of Equation (8.30) gives

$$\overline{\overline{u}}(\kappa_x, z, \omega) = e^{-i\omega z/V}\,\overline{\overline{u}}_a(\kappa_x, z, \omega). \tag{8.31}$$

The partial derivative with respect to z of this is

$$\frac{\partial \overline{\overline{u}}}{\partial z} = e^{-i\omega z/V}\frac{\partial \overline{\overline{u}}_a}{\partial z} - \frac{i\omega}{V}e^{-i\omega z/V}\,\overline{\overline{u}}_a. \tag{8.32}$$

Substituting Equations (8.31) and (8.32) into the differential equation (8.28) yields

$$\frac{\partial \overline{\overline{u}}_a}{\partial z} = i\left(\kappa_z + \frac{\omega}{V}\right)\overline{\overline{u}}_a. \tag{8.33}$$

To Fourier-transform this equation back to the x-t domain, we first note from Equation (8.3) that doing a forward 2D Fourier transform amounts to replacing $(\partial/\partial t)$ with $-i\omega$ and $(\partial/\partial x)$ with $i\kappa_x$. Hence, to do an inverse 2D Fourier transform of Equation (8.33), we simply make these replacements in Equation (8.33) for ω and κ_x. There is one problem, though: the operator $\partial/\partial x$ then appears under the square root sign in the formula for κ_z (see Equation 8.28), and it is not clear what the meaning of the square root of a derivative operator is. However, the problem

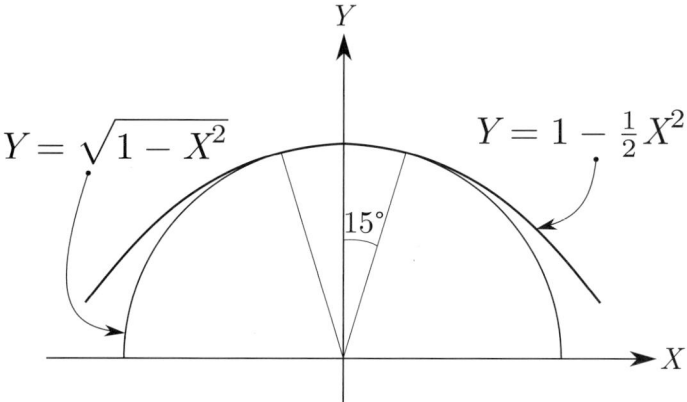

Figure 8.8 The parabolic approximation.

can eliminated by expanding κ_z as a power series. Using the rule that $\sqrt{1 + \epsilon} = 1 + \frac{1}{2}\epsilon + \ldots$, and assuming for now that $(V\kappa_x/\omega)$ is small compared to 1, we obtain

$$\kappa_z \approx -\frac{\omega}{V}\left[1 - \frac{1}{2}\left(\frac{V\kappa_x}{\omega}\right)^2\right]. \tag{8.34}$$

This is known as the **parabolic approximation** for the following reason. If we let $X \equiv V\kappa_x/\omega$ and $Y \equiv -V\kappa_z/\omega$, then the exact formula for κ_z in (8.28) can be written as $Y = \sqrt{1 - X^2}$, which is the equation of a circle in XY space, and the approximate formula for κ_z in (8.34) can be written as $Y = 1 - \frac{1}{2}X^2$, which is the equation of a parabola in XY space. In other words, (8.34) approximates a circle with a parabola (see Figure 8.8). The approximation is good only to within about $15°$ from the Y axis, i.e., for small values of κ_x (i.e., X). This means that the resulting migration will work correctly only for reflectors with small enough dips, in fact, for dips no greater than about $15°$. To prove this, consider a flat dipping reflector with a dip angle of θ. In the exploding reflector model, this reflector will generate an upgoing wave whose wavevector components are $\kappa_x = (\omega/V)\sin\theta$ and $\kappa_z = -(\omega/V)\cos\theta$, meaning that $X = \sin\theta$ and $Y = \cos\theta$. In these last two equations, θ is the angle measured from the Y axis. Since θ is also the reflector dip, the point is proven.

The parabolic approximation is also called the **$15°$ approximation**. Better approximations (for steeper dips) are also possible.

Substituting the parabolic approximation into Equation (8.33) gives

$$\frac{\partial \tilde{\bar{u}}_a}{\partial z} = \frac{iV}{2\omega}\kappa_x^2 \tilde{\bar{u}}_a. \tag{8.35}$$

This equation is now transformed back to the x-t domain by replacing ω and κ_x with the appropriate derivative operators, as mentioned previously. The result is

$$\frac{\partial^2 u_a}{\partial z \, \partial t} - \frac{V}{2} \frac{\partial^2 u_a}{\partial x^2} \qquad\qquad (8.36)$$

This equation is often called the **15° equation**. The method of finite differences is then used to solve this equation for u_a. Setting $t = 0$ gives the migrated data $u_a(x, z, z/V) = u(x, z, 0)$, which lie on the plane $t = z/V$ in x-z-t space, i.e., the solution space of u_a.

Better approximations for κ_z can be made. For instance, consider the iteration relation

$$Y_{n+1} = 1 - \frac{X^2}{1 + Y_n}, \qquad n = 0, 1, 2, \ldots, \qquad \text{where} \qquad Y_0 \equiv 1. \qquad (8.37)$$

It turns out that $Y_0, Y_1, Y_2, Y_3, \ldots$ are successively better approximations of $Y = \sqrt{1 - X^2}$. $Y \approx Y_1$ is the parabolic approximation. $Y \approx Y_2$, i.e.,

$$Y \approx 1 - \frac{X^2}{1 + 1 - \frac{1}{2}X^2} = \frac{1 - \frac{3}{4}X^2}{1 - \frac{1}{4}X^2}, \qquad\qquad (8.38)$$

leads to a partial differential equation for u_a that is more complicated than Equation (8.36), but dips up to about 45° can be correctly migrated with it.

Equation (8.36) is solved on a 3D grid with spacings Δx, Δz, and Δt in the x, z, and t directions. The finer the grid spacing, the better the migration, although the expense increases geometrically with fineness. There is typically also an "optimum" fineness, i.e., making the grid finer may actually increase the computational errors. Usually first-order approximations are used for the derivatives. Higher-order approximations to derivatives will generally improve the migration, although the computational expense increases with order.

The input seismic section $u(x, 0, t) = u_a(x, 0, t)$ is sampled at the grid points on the plane $z = 0$. These data are then downward-continued, via the finite difference solution of Equation (8.36), to depth z. If the grid spacing is too coarse, aliasing will occur – high spatial and temporal frequencies may alias as lower ones in the migrated section. This results in a "frequency dispersion" effect, i.e., waveforms on the migrated section may be broader than the ones on the unmigrated section.

For the case of a depth-dependent velocity, i.e., $V = V(z)$, the advanced time t_a becomes

$$t_a = t + \int_0^z \frac{dz}{V(z)}. \qquad\qquad (8.39)$$

The remainder of the analysis can be carried through in the same way. The 15° equation (8.36) is still obtained, except that $V = V(z)$. By switching to a new depth variable ζ, where $\zeta \equiv (1/V_0) \int_0^z V(z) \, dz$ and V_0 is a constant reference velocity, the 15° equation takes on a form identical to the constant velocity equation (8.36), except that z is replaced by ζ and $V(z)$ is replaced by V_0 (Stolt and Benson, 1986, p. 206). The migrated data are found on the surface $t = \int_0^z dz/V(z)$ in x-z-t space.

For more details, see Claerbout (1976, 1985) and Stolt and Benson (1986), among others.

8.8 Kirchoff Migration

This method is essentially an improved version of diffraction-summation migration. The theory is rather involved, so only a qualitative discussion is given here. The improvement involves the scaling and filtering of the digital signal consisting of the trace amplitude values along the diffraction curve being considered *before* summation along the curve. The technique is based on the **Kirchoff integral solution** to the wave equation. Given a finite volume bounded by a closed surface, Kirchoff's formula provides a way to calculate the wavefield at a point inside the volume if we know the values of the wavefield on the surface of the volume. For seismic migration, part of this closed surface consists of the ground surface, which contains the sources and receivers, and on which we know the wavefield. To obtain a closed surface, the source–receiver surface is allowed to extend to ∞, and then a hemisphere of infinite radius (in the subsurface) is attached to it. The values of the wavefield on the hemisphere are, of course, zero, and so the wavefield at an interior point (in the subsurface) can be calculated by doing a double integral over the source–receiver plane $z = 0$, where we know the wavefield.

For more details, see French (1975), Schneider (1978), and Stolt and Benson (1985, pp. 102–105), among others.

8.9 Depth Migration

The conventional $1\frac{1}{2}$D migration methods discussed in this chapter have been applied to data from regions in which the velocity varies both horizontally and vertically. They do a reasonably good job if the medium is not too structurally complex. However, they are, after all, based on the simplifying assumption that the velocity v for the medium is constant, or that $v = v(z)$. Hence, errors will always be present in conventionally migrated data from regions in which $v = v(x, z)$.

Figure 8.9 shows that a diffracting point below a dipping reflector will be *incorrectly* migrated by conventional methods. For instance, Kirchoff migration would sum the energy along the diffraction curve (after scaling and filtering) and place the result at the apex, whose horizontal location is at point A. However, point A is *not* the true horizontal position of the diffracting point, and so Kirchoff migration gives the wrong answer. The same is true for the conventional finite difference

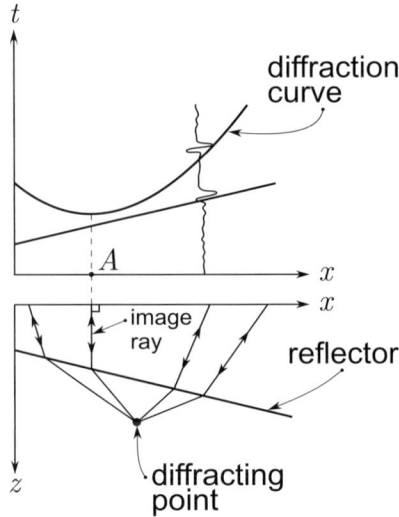

Figure 8.9 An illustration of the incorrect migration of a diffracting point by conventional $1\frac{1}{2}$D methods. Modified from Larner, Hatton, Gibson, and Hsu (1981), with permission from the Society of Exploration Geophysicists.

and frequency-wavenumber migration methods, since they are based on the same assumptions about velocity.

How could we migrate the data correctly? First note that the zero-offset ray from the diffracting point with the shortest travel time (corresponding to the apex of the diffraction curve) is the one that emerges *vertically* at the surface. It is known as the **image ray** (Hubral, 1977). Suppose then that we estimate the 2D velocity function $v(x, z)$ from the data. We can use the conventionally migrated section for this (it is presumably still better than the unmigrated section). Knowing the time t at the apex, we can use $v(x, z)$ to trace the image ray back to the diffracting point. Among all the zero-offset rays, the image ray is the easiest to trace back because we know its angle of emergence (90° to the vertical). We then have the correct horizontal position of the diffracting point, depending on how accurate our guess for $v(x, z)$ is, and we place the diffracted energy there. These steps can be iterated, and the velocity function $v(x, z)$ can be refined at each iteration.

The preceding discussion is a brief qualitative introduction to **depth migration**. For more details, see Hatton, Larner, and Gibson (1981) and Larner, Hatton, Gibson, and Hsu (1981), and references found therein, among others. See also the texts by Claerbout (1985), Ikelle and Amundsen (2018), Liner (2016), Robinson (1983), and Stolt and Benson (1986).

Seismic migration is about producing a more accurate image of the subsurface from seismic data. Modern approaches to this involve the subject of **seismic inversion**, on which there is a large body of literature. See, for example, the text by Wang (2017).

Exercises

1. Finite differences:

   (a) Use the forward difference formula to compute the finite difference solution of the equation $dy/dx = y(x)$, with $y(0) = 1$ (Equation 8.21), using $\Delta x = 0.01$ and $\Delta x = 0.001$, and compare with the exact solution. (Hint: to avoid doing many repetitive computations, use Equation 8.25 to derive a general formula for $y(n\Delta x), n = 0, 1, 2, ...$).

   (b) Use the central difference formula to compute the finite difference solution of the equation $dy/dx = y(x)$, $y(0) = 1$, using $\Delta x = 0.1$. Note that you will need $y(0.1)$ as input, as well as $y(0)$. Obtain $y(0.1)$ from part (a). Compare your results with the exact solution, with the forward difference solution for $\Delta x = 0.1$ (in the text), and with the forward difference solution of part (a).

2. Equation (8.36) is the "15° equation," which is used in finite difference migration. Derive the "45° equation."

3. Consider a medium consisting of a single flat dipping reflector. A constant velocity 1.5D migration is performed on the CMP stack section using a migration velocity v_1. The output time section, i.e., the migrated time section, is undermigrated, so a residual migration is performed using a migration velocity v_2. In other words, the output of the first migration is used as the input for the second migration. The output time section of the second migration is correctly migrated. If the first migration had been done with the correct migration velocity v_c, the residual migration would not have been necessary. Derive a formula for v_c in terms of v_1 and v_2. If $v_1 = 5.2$ km/s and $v_2 = 2.1$ km/s, what is v_c?

4. Consider a medium containing a single flat dipping reflector (e.g., Figure 8.2). The medium velocity above the reflector is 4 km/s. On the unmigrated time section, i.e., the CMP stack section, the reflector has a dip of 20 ms/trace, with the trace interval being 50 m. A certain point on the reflector (on the stack section) has a two-way time of 1.6 s. What is the true depth of this point, i.e., the depth after migration? How far does the migration process move the point horizontally, and in which direction (left or right)?

9

Plane Waves in Anisotropic Media

9.1 Anisotropy

Anisotropy generally means that the physical parameters of a medium vary with direction. Anisotropic media can be either homogeneous or heterogeneous. In a seismically anisotropic medium, the velocity of a seismic wave depends on the direction in which the wave is traveling. This means that a point source in a homogeneous anisotropic medium will generate a wave with a *nonspherical* wavefront.

Consider a medium consisting of a stack of thin flat horizontal isotropic homogeneous layers (where "thin" generally means thinner than one-quarter of the dominant wavelength of the seismic waves passing through the medium). Body waves propagating in a direction normal to the layers will generally travel slower than body waves traveling parallel to the layers. Such a medium can often be treated as a single *anisotropic* medium. The existence of fractures in a medium can also induce seismic anisotropy.

9.2 3D Plane Wave Solutions

Consider a generally anisotropic homogeneous medium. The equation of motion is

$$\sum_{j=1}^{3} \frac{\partial \sigma_{ij}}{\partial x_j} + f_i = \rho \frac{\partial^2 u_i}{\partial t^2}, \quad \text{with} \quad \sigma_{ij} = \sum_{k=1}^{3} \sum_{l=1}^{3} c_{ijkl} \frac{\partial u_k}{\partial x_l}, \quad i,j = 1,2,3. \quad (9.1)$$

(see Equation 2.27). Now substitute into (9.1) a trial plane wave, $u_k = U_k \exp[i\omega(\mathbf{s} \cdot \mathbf{x} - t)]$, where $\mathbf{s}$ is the slowness vector of the wave, which can be written as $\mathbf{s} = (1/v)\mathbf{n}$, where $\mathbf{n}$ is a unit vector in the direction of $\mathbf{s}$, i.e., the direction of wave propagation, and v is the wave speed. The derivatives can be easily worked out, e.g., $\partial u_k / \partial x_l = i\omega s_l u_k = i\omega n_l u_k / v$, etc. One obtains the following equation:

$$\frac{1}{\rho} \sum_{j=1}^{3} \sum_{l=1}^{3} \sum_{k=1}^{3} c_{ijkl} n_j n_l u_k = v^2 u_i, \quad i = 1,2,3. \quad (9.2)$$

294

If we define **A** as the 3×3 matrix with elements A_{ik} given by

$$A_{ik} \equiv \frac{1}{\rho} \sum_{j=1}^{3} \sum_{l=1}^{3} c_{ijkl} n_j n_l, \tag{9.3a}$$

then (9.2) can be written as

$$\sum_{k=1}^{3} A_{ik} u_k = v^2 u_i, \quad \text{i.e.,} \quad \mathbf{Au} = v^2 \mathbf{u}, \tag{9.3b}$$

where **u** is the column vector containing the components of **u**. Note that this is an **eigenvalue equation**, where v^2 is the eigenvalue.

For isotropic media, the elements A_{ik} of the matrix **A** can be obtained by substituting the formula for the stiffness tensor c_{ijkl} for an isotropic medium (Equation 2.29) into the preceding formula for A_{ik}. Solving the eigenvalue equation (which can be done for the simplified case $n_2 = 0$, i.e., a wave propagating in the x-z plane, without loss of generality) then yields the three eigenvalues $v^2 = \{\alpha^2, \beta^2, \beta^2\}$, i.e., the P, SV, and SH wave speeds in an isotropic medium, along with the three mutually orthogonal eigenvectors giving the direction of particle motion for P, SV, and SH waves (parallel, perpendicular-vertical, and perpendicular-horizonal, respectively, to the direction of wave propagation). The fact that two of the eigenvalues are the same (which happens for isotropy) is referred to as **degeneracy** in the parlance of eigenvalue theory. For a generally anisotropic medium, the three eigenvalues are all different.

As a memory refresher, here is a brief review on how to solve the eigenvalue equation **Ax**=λ**x**. In this equation, **A** is an $m \times m$ matrix, **x** is a $m \times 1$ column vector, and λ is the (scalar) eigenvalue (or "characteristic value"). First, write the equation as $(\mathbf{A}-\lambda\mathbf{I})\mathbf{x}=\mathbf{0}$, where **I** is the identity matrix. This is a system of linear equations. Since the right-hand side is zero, this system has the trivial solution **x**=**0** *unless* $\det[\mathbf{A}-\lambda\mathbf{I}]=0$ (which leads to the desired nontrivial solution). Solving the determinant equation, often called the **characteristic equation**, gives the eigenvalues λ of the matrix **A**. The eigenvectors **x** are then obtained by substituting the eigenvalues into $(\mathbf{A}-\lambda\mathbf{I})\mathbf{x}=\mathbf{0}$ and solving the system for **x**. Here is a simple example for $m = 2$:

$$\mathbf{A} = \begin{bmatrix} 1 & 2 \\ 2 & 1 \end{bmatrix}, \quad \mathbf{x} = \begin{bmatrix} x \\ y \end{bmatrix} \implies \det \begin{bmatrix} 1-\lambda & 2 \\ 2 & 1-\lambda \end{bmatrix} = (\lambda-3)(\lambda+1) = 0. \tag{9.4}$$

Consequently, the eigenvalues are $\lambda = \{\lambda_1, \lambda_2\} = \{-1, 3\}$. Substituting $\lambda = \lambda_1 = -1$ into $(\mathbf{A}-\lambda\mathbf{I})\mathbf{x}=\mathbf{0}$ gives $2x+2y = 0$, whose solution is $y = -x$. So the eigenvector is $\mathbf{x} = [x, -x]^T$. Usually, eigenvectors are normalized so that they have unit length. Normalizing this eigenvector, we have $x^2 + (-x)^2 = 1$, leading to $x = 1/\sqrt{2}$.

So the normalized eigenvector for λ_1 is $\mathbf{x}_1 = (1/\sqrt{2})[1, -1]^T$. Similarly, the normalized eigenvector for λ_2 works out to be $\mathbf{x}_2 = (1/\sqrt{2})[1, 1]^T$. Summarizing, the eigenvalues and normalized eigenvectors are

$$\lambda_1 = -1, \quad \mathbf{x}_1 = \frac{1}{\sqrt{2}}\begin{bmatrix} 1 \\ -1 \end{bmatrix}, \quad \text{and} \quad \lambda_2 = 3, \quad \mathbf{x}_2 = \frac{1}{\sqrt{2}}\begin{bmatrix} 1 \\ 1 \end{bmatrix}. \tag{9.5}$$

Note that the two eigenvectors are orthogonal (perpendicular) to each other, because $\mathbf{x}_1^T \mathbf{x}_2 = 0$ (or in vector notation, $\mathbf{x}_1 \cdot \mathbf{x}_2 = 0$).

Now consider a medium that is anisotropic and homogeneous. In such a medium, the wave speeds vary with direction. In particular, we look at the case of a **transversely isotropic (TI) medium**, in which the speeds of waves traveling in the *x-z* plane (the offset-depth plane) vary with direction, but the speeds of waves traveling in the *x-y* plane (the transverse direction) do not. More precisely, such a medium is often called a VTI medium, i.e., a TI medium with a vertical symmetry axis (the *z* axis). For example, a sequence of horizontal thin shale layers can be modeled as a VTI medium, because vertically propagating waves in this medium travel slower than horizontally propagating waves. Also, anisotropy can be induced by the formation of cracks in a medium.

For isotropic media, the stress–strain relation $\sigma_{ij} = \lambda(\sum_k e_{kk})\delta_{ij} + 2\mu e_{ij}$ can also be written in matrix form as follows:

$$\begin{bmatrix} \sigma_{11} \\ \sigma_{22} \\ \sigma_{33} \\ \sigma_{23} \\ \sigma_{31} \\ \sigma_{12} \end{bmatrix} = \begin{bmatrix} \ell & \lambda & \lambda & 0 & 0 & 0 \\ \lambda & \ell & \lambda & 0 & 0 & 0 \\ \lambda & \lambda & \ell & 0 & 0 & 0 \\ 0 & 0 & 0 & \mu & 0 & 0 \\ 0 & 0 & 0 & 0 & \mu & 0 \\ 0 & 0 & 0 & 0 & 0 & \mu \end{bmatrix} \begin{bmatrix} e_{11} \\ e_{22} \\ e_{33} \\ 2e_{23} \\ 2e_{31} \\ 2e_{12} \end{bmatrix}, \quad \text{where} \quad \ell \equiv \lambda + 2\mu. \tag{9.6}$$

As we can see, there are only two independent elastic constants, λ and μ, in an isotropic medium. However, in a VTI medium, there are five independent elastic constants, e.g., λ, μ, λ', λ'', and μ' (other parameter definitions and symbols have been used as well, as discussed later in this chapter). The stress–strain relation for a VTI medium is

$$\begin{bmatrix} \sigma_{11} \\ \sigma_{22} \\ \sigma_{33} \\ \sigma_{23} \\ \sigma_{31} \\ \sigma_{12} \end{bmatrix} = \begin{bmatrix} \ell & \lambda & \lambda' & 0 & 0 & 0 \\ \lambda & \ell & \lambda' & 0 & 0 & 0 \\ \lambda' & \lambda' & \ell'' & 0 & 0 & 0 \\ 0 & 0 & 0 & \mu' & 0 & 0 \\ 0 & 0 & 0 & 0 & \mu' & 0 \\ 0 & 0 & 0 & 0 & 0 & \mu \end{bmatrix} \begin{bmatrix} e_{11} \\ e_{22} \\ e_{33} \\ 2e_{23} \\ 2e_{31} \\ 2e_{12} \end{bmatrix}, \quad \text{where} \quad \ell'' \equiv \lambda'' + 2\mu. \tag{9.7a}$$

For a general anisotropic medium, the stress strain relation, $\sigma_{ij} = \sum_k \sum_l c_{ijkl} e_{kl}$, i.e., Hooke's law, can also be written in the same matrix form: $\boldsymbol{\sigma} = \mathbf{ce}$, where $\boldsymbol{\sigma}$ and $\mathbf{e}$ are the column vectors in (9.7a) on the left and right, respectively, and where $\mathbf{c}$ is a 6×6 symmetric matrix whose elements c_{ij}, $i, j = 1, \ldots, 6$, contain the elastic constants for the anisotropic medium (see Equation 2.26). For example, for a VTI medium, there are five independent constants, namely, c_{11}, c_{13}, c_{33}, c_{44}, and c_{66}, with $c_{11} = \lambda + 2\mu$, $c_{13} = \lambda'$, $c_{33} = \lambda'' + 2\mu$, $c_{44} = \mu'$, and $c_{66} = \mu$. The stress–strain relation for a VTI medium can then also be written as

$$
\begin{bmatrix} \sigma_{11} \\ \sigma_{22} \\ \sigma_{33} \\ \sigma_{23} \\ \sigma_{31} \\ \sigma_{12} \end{bmatrix} = \begin{bmatrix} c_{11} & c_{12} & c_{13} & 0 & 0 & 0 \\ c_{12} & c_{11} & c_{13} & 0 & 0 & 0 \\ c_{13} & c_{13} & c_{33} & 0 & 0 & 0 \\ 0 & 0 & 0 & c_{44} & 0 & 0 \\ 0 & 0 & 0 & 0 & c_{44} & 0 \\ 0 & 0 & 0 & 0 & 0 & c_{66} \end{bmatrix} \begin{bmatrix} e_{11} \\ e_{22} \\ e_{33} \\ 2e_{23} \\ 2e_{31} \\ 2e_{12} \end{bmatrix}, \quad c_{12} = c_{11} - 2c_{66}. \quad (9.7b)
$$

For a perfectly isotropic medium, in terms of the $\mathbf{c}$ matrix, there are two independent constants, c_{11} and c_{66}, because the other three can be expressed in terms of c_{11} and c_{66} as follows:

$$
c_{13} = c_{12} = c_{11} - 2c_{66}, \ c_{33} = c_{11}, \ c_{44} = c_{66} \quad \text{(isotropy)}. \quad (9.7c)
$$

The matrix $\mathbf{c}$ is symmetric, implying that only 21 of its 36 elements c_{ij} are independent. This means that 21 independent elastic constants would be required to specify the most general anisotropic medium.

The elements c_{ijkl} of the stiffness tensor and the elements c_{ij} of the 6×6 matrix $\mathbf{c}$ can be related to each other by comparing the equations for the elements σ_{ij} obtained from the matrix equation $\boldsymbol{\sigma} = \mathbf{ce}$ with those from Hooke's law, $\sigma_{ij} = \sum_k \sum_l c_{ijkl} e_{kl}$. For example, from the matrix equation, we have

$$
\sigma_{11} = c_{11} e_{11} + c_{12} e_{22} + c_{13} e_{33} + 2c_{14} e_{23} + 2c_{15} e_{31} + 2c_{16} e_{12}, \quad (9.8)
$$

and from Hooke's law, we have, using the symmetry relations $c_{ijkl} = c_{ijlk}$ and $e_{ij} = e_{ji}$,

$$
\sigma_{11} = c_{1111} e_{11} + c_{1122} e_{22} + c_{1133} e_{33} + 2c_{1123} e_{23} + 2c_{1131} e_{31} + 2c_{1112} e_{12}. \quad (9.9)
$$

Comparing, we have $c_{11} = c_{1111}$, $c_{12} = c_{1122}$, etc. By comparing the equations for the other σ_{ij}, it is easy to see that the following rule applies: $c_{ij} = c_{klmn}$ where the subscripts kl and mn are the same as those of the ith and jth element, respectively, of the 6×1 stress or strain column vector (see Equation 9.6). Here are some more examples: $c_{25} = c_{2231}$, $c_{43} = c_{2333}$.

To solve the eigenvalue problem $\mathbf{Au} = v^2 \mathbf{u}$, the elements A_{ik} are first computed from (9.3a). For example, for a VTI medium, consider for simplicity, without loss of generality, a wave propagating in the *x-z* plane (the medium is isotropic in the transverse direction, i.e., any vertical plane will do). Setting $n_2 = 0$, one obtains, using the rule stated at the end of the previous paragraph,

$$\rho A_{11} = c_{1111}n_1^2 + c_{1113}n_1n_3 + c_{1311}n_3n_1 + c_{1313}n_3^2 = c_{11}n_1^2 + c_{55}n_3^2$$
$$= c_{11}n_1^2 + c_{44}n_3^2 = (\lambda + 2\mu)n_1^2 + \mu'n_3^2, \tag{9.10a}$$

where $c_{55} = c_{44}$ (for a VTI medium). The other A_{ik} can be obtained in the same way. They are as follows:

$$\rho A_{13} = \rho A_{31} = (c_{13} + c_{44})n_1n_3, \quad \rho A_{22} = c_{66}n_1^2 + c_{44}n_3^2, \tag{9.10b}$$
$$\rho A_{33} = c_{44}n_1^2 + c_{33}n_3^2, \quad A_{12} = A_{21} = A_{23} = A_{32} = 0. \tag{9.10c}$$

In these equations,

$$n_1 = \sin\theta \quad \text{and} \quad n_3 = \cos\theta, \tag{9.10d}$$

where θ is the angle in the *x-z* plane that the slowness vector **s** of the wave, or **n**, makes with the *z* axis. Once all the A_{ik} are known, the eigenvalue problem can be solved. We apply this procedure to obtain the eigenvalues and eigenvectors for a VTI medium. The characteristic equation (the one analogous to Equation 9.4) is

$$H(AB \quad C) = 0, \qquad \text{where} \tag{9.10e}$$
$$H \equiv c_{66}n_1^2 + c_{44}n_3^2 - \rho v^2, \qquad A \equiv c_{11}n_1^2 + c_{44}n_3^2 - \rho v^2, \tag{9.10f}$$
$$B \equiv c_{44}n_1^2 + c_{33}n_3^2 - \rho v^2, \qquad C \equiv (c_{13} + c_{44})^2 n_1^2 n_3^2. \tag{9.10g}$$

Equation (9.10e) results in two equations: $H = 0$ and $AB - C = 0$. We will see later that the former is for *SH* waves and the latter for *P-SV* waves. Solving these two equations results in three sets of eigenvalues (the values of v^2) and eigenvectors, corresponding to the *P*, *SV*, and *SH* waves. Note in the following that the VTI eigenvectors, which give the direction of particle displacement, are no longer parallel and perpendicular to the direction of wave propagation for *P* and *SV* waves, respectively. Consequently, these waves are not "pure" *P* and *SV* waves, and are called *quasi-P* (*qP*) and *quasi-SV* (*qSV*) waves. Some details of the derivation of the formulas for the eigenvalues v^2 are given in Appendix 9A. For a VTI medium, we obtain

$$2\rho v^2 = \ell_1 + \ell_2 \sin^2\theta \pm D(\theta), \tag{9.11a}$$

where the + sign is chosen for *qP* waves and the − sign for *qSV* waves, where

$$D(\theta) \equiv \left\{ \ell_3^2 + 2[2\ell_4^2 - \ell_3\ell_5]\sin^2\theta + [\ell_5^2 - 4\ell_4^2]\sin^4\theta \right\}^{1/2} \tag{9.11b}$$

with

$$\ell_1 \equiv \lambda'' + 2\mu + \mu' = c_{33} + c_{44}, \tag{9.12}$$

$$\ell_2 \equiv \lambda - \lambda'' = c_{11} - c_{33}, \tag{9.13}$$

$$\ell_3 \equiv \lambda'' + 2\mu - \mu' = c_{33} - c_{44}, \tag{9.14}$$

$$\ell_4 \equiv \lambda' + \mu' = c_{13} + c_{44}, \tag{9.15}$$

$$\ell_5 \equiv \lambda + \lambda'' + 4\mu - 2\mu' = c_{11} + c_{33} - 2c_{44}. \tag{9.16}$$

and where, as mentioned previously, θ is the angle in the *x-z* plane that the slowness vector **s** of the wave, or **n**, makes with the *z* axis. The angle θ is usually called the *phase angle*, because it is the direction in which the planes of constant phase are traveling. In anisotropic media, the phase angle differs from the ray angle (discussed later in this section). The preceding formulas give the phase speeds v_{qP} and v_{qSV} as a function of θ. For example, for waves traveling in the vertical (*z*) and horizontal (*x*) directions ($\theta = 0°$ and $90°$, respectively), they reduce to

$$v_{qP}(0°) \equiv \alpha_0 = \sqrt{\frac{c_{33}}{\rho}} = \sqrt{\frac{\lambda'' + 2\mu}{\rho}}, \tag{9.17a}$$

$$v_{qSV}(0°) \equiv \beta_0 = \sqrt{\frac{c_{44}}{\rho}} = \sqrt{\frac{\mu'}{\rho}}, \tag{9.17b}$$

$$v_{qP}(90°) \equiv \alpha_{90} = \sqrt{\frac{c_{11}}{\rho}} = \sqrt{\frac{\lambda + 2\mu}{\rho}}, \tag{9.18}$$

$$v_{qSV}(90°) \equiv \beta_{90} = \sqrt{\frac{c_{44}}{\rho}} = \sqrt{\frac{\mu'}{\rho}}. \tag{9.19}$$

For *SH* waves, we easily obtain, from $H = 0$,

$$v^2 = \beta_h^2 \sin^2 \theta + \beta_v^2 \cos^2 \theta, \tag{9.20}$$

$$\beta_h^2 \equiv \frac{\mu}{\rho} = \frac{c_{66}}{\rho}, \qquad \beta_v^2 \equiv \frac{\mu'}{\rho} = \frac{c_{44}}{\rho} \tag{9.21}$$

with β_h and β_v being the *SH* wave speeds in the horizontal (*x*) and vertical (*z*) directions.

We also need to keep in mind that the angle θ is the phase angle for *SH* waves in (9.20), for *qP* waves in (9.11a) with the $+$ sign, and for *qSV* waves in (9.11a) with the $-$ sign. These three angles are generally different.

The normalized eigenvectors $\bar{\mathbf{u}}$ (the polarization vectors) for either *qP* and *qSV* waves in a VTI medium are given by $\mathbf{d} = \bar{\mathbf{u}} = [\bar{u}_1, 0, \bar{u}_3]^T$, where

$$\bar{u}_1 = \sqrt{\frac{c_{44}\sin^2\theta + c_{33}\cos^2\theta - \rho v^2}{c_1\sin^2\theta + c_3\cos^2\theta - 2\rho v^2}}, \tag{9.22}$$

$$\bar{u}_3 = \sqrt{\frac{c_{11}\sin^2\theta + c_{44}\cos^2\theta - \rho v^2}{c_1\sin^2\theta + c_3\cos^2\theta - 2\rho v^2}}, \tag{9.23}$$

where $\quad c_1 \equiv c_{11} + c_{44}, \quad c_3 \equiv c_{33} + c_{44}.$ (9.24)

For qP (qSV) waves, v in $\bar{u}$ is replaced by the formula giving v for qP (qSV) waves in (9.11). The eigenvector for SH waves is the same as the isotropic one, i.e., $\bar{u} = [0, 1, 0]^T$, meaning no prefix "q" is needed.

Other symbols have also been used for the five independent elastic constants in a VTI medium. For example, since α_{90} in Equation (9.18) is the qP wave speed parallel to the transverse (x-y) plane, i.e., the plane of isotropy, the symbols $\lambda_{\parallel} \equiv \lambda$ and $\mu_{\parallel} \equiv \mu$ have been used, so that $\alpha_{90} = \sqrt{(\lambda_{\parallel} + 2\mu_{\parallel})/\rho}$ and $\beta_h = \sqrt{\mu_{\parallel}/\rho}$, and since α_0 in Equation (9.17a) is the qP wave speed perpendicular to the transverse (x-y) plane, the symbols $\lambda_{\perp} \equiv \lambda'$ and $2\mu_{\perp} \equiv \lambda'' + 2\mu - \lambda'$ have been used, so that $\alpha_0 = \sqrt{(\lambda_{\perp} + 2\mu_{\perp})/\rho}$ (see, e.g., Grant and West, 1965; Sheriff and Geldart, 1995). Note, though, that $\beta_0 = \beta_v = \beta_{90} = \sqrt{\mu'/\rho}$, not $\sqrt{\mu_{\perp}/\rho}$ or $\sqrt{\mu_{\parallel}/\rho}$.

The form of Equation (9.20) (the equation of an ellipse in polar coordinates) tells us that SH waves in a homogeneous VTI medium have ellipsoidal wavefronts whose

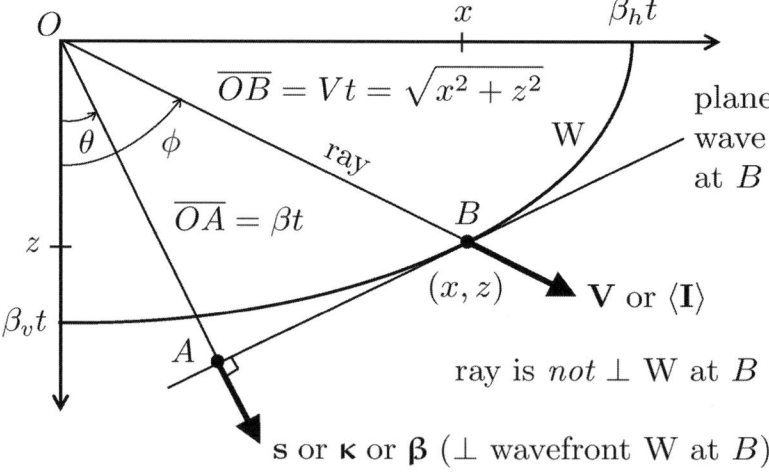

Figure 9.1 An SH ellipsoidal wavefront W at time t in a VTI medium. $\beta \equiv v_{SH}$. The wavefront emanates from the origin O. The plane wave has $n_2 = 0$. Note that the ray is *not* perpendicular to the wavefront in a VTI medium. The straight line tangential to the wavefront at the point B (located at (x, z)) is the local plane wave associated with the wavefront at point B. Also, $\overline{OA} = \beta t$ and $\overline{OB} = Vt = \sqrt{x^2 + z^2}$.

intersections with the x-z plane are elliptical. Figure 9.1 shows such a wavefront. It also shows that in anisotropic elastic media, the direction of mean energy flow $\langle \mathbf{I} \rangle$, which is along the ray angle ϕ, is in general different from the direction of phase propagation θ. Because the group velocity $\mathbf{V} = (\partial \omega / \partial \kappa_x, \partial \omega / \partial \kappa_y, \partial \omega / \partial \kappa_z)$ is along the ray, ϕ is also called the **group** angle. Note that

$$\beta = V \cos(\phi - \theta), \tag{9.25}$$

where $\beta \equiv v$ in (9.20). Substituting $\sin \theta = \kappa_x / \kappa$, $\cos \theta = \kappa_z / \kappa$, and $\beta = v = \omega / \kappa$ into (9.20) gives the **SH dispersion relation** for a VTI medium,

$$\omega^2 = \beta_h{}^2 \kappa_x{}^2 + \beta_v{}^2 \kappa_z{}^2. \tag{9.26}$$

Other equations that can be derived are

$$\tan \phi = \frac{V_x}{V_z} = \left(\frac{\beta_h}{\beta_v} \right)^2 \tan \theta, \tag{9.27}$$

$$\frac{1}{V^2} = \frac{\cos^2 \phi}{\beta_v{}^2} + \frac{\sin^2 \phi}{\beta_h{}^2}, \tag{9.28}$$

$$V^2 = \beta^2 + \left(\frac{d\beta}{d\theta} \right)^2. \tag{9.29}$$

Equations (9.26) through (9.28) apply only to *SH* waves in a VTI medium, whereas (9.29) applies also to *qP* and *qSV* waves (with β replaced by α).

From Figure 9.1 we have $\cos \phi = z/(Vt)$ and $\sin \phi = x/(Vt)$. Substituting these into the $1/V^2$ equation gives the equation for the *SH* wavefront:

$$\frac{x^2}{(\beta_h t)^2} + \frac{z^2}{(\beta_v t)^2} = 1, \tag{9.30}$$

which is an ellipse. Substituting $\kappa_x = \omega s_x$ and $\kappa_z = \omega s_z$ into the dispersion relation gives the **slowness ellipse**

$$\beta_h{}^2 s_x{}^2 + \beta_v{}^2 s_z{}^2 = 1, \tag{9.31}$$

which is the intersection of the **slowness surface** with the s_x-s_z plane.

More details on dispersion relations and slowness surfaces, including how they apply to tilted TI media, can be found in Appendix 9A.

9.3 Weak VTI-Anisotropy

In this case, the formulas for v in Section 9.2 can be simplified using the parameters ε, γ, and δ, which are measures of anisotropy, are small for many rocks, and are defined by (Thomsen, 1986)

$$\varepsilon = \frac{c_{11} - c_{33}}{2c_{33}} - \frac{\lambda - \lambda''}{2(\lambda'' + 2\mu)}, \tag{9.32}$$

$$\gamma \equiv \frac{c_{66} - c_{44}}{2c_{44}} = \frac{\mu - \mu'}{2\mu'} \tag{9.33}$$

$$\delta \equiv \frac{(c_{13} + c_{44})^2 - (c_{33} - c_{44})^2}{2c_{33}(c_{33} - c_{44})} = \frac{(\lambda' + \mu')^2 - (\lambda'' + 2\mu - \mu')^2}{2(\lambda'' + 2\mu)(\lambda'' + 2\mu - \mu')}. \tag{9.34}$$

For weak VTI, i.e., for $|\varepsilon|$, $|\gamma|$, and $|\delta| \ll 1$, a Taylor expansion of v in these small parameters yields

$$v_{qP}(\theta) \approx \alpha_0(1 + \delta \sin^2 \theta \cos^2 \theta + \varepsilon \sin^4 \theta), \tag{9.35}$$

$$v_{qSV}(\theta) \approx \beta_0 \left[1 + (\alpha_0/\beta_0)^2 (\varepsilon - \delta) \sin^2 \theta \cos^2 \theta \right], \tag{9.36}$$

$$v_{SH}(\theta) \approx \beta_0(1 + \gamma \sin^2 \theta). \tag{9.37}$$

For weak VTI, the following easy-to-prove relations can be used to experimentally determine ε, γ, and δ:

$$\varepsilon = \left[v_{qP}(90°) - \alpha_0 \right] / \alpha_0, \tag{9.38}$$

$$\gamma = \left[v_{SH}(90°) - \beta_0 \right] / \beta_0, \tag{9.39}$$

$$\delta = 4 \left\{ \left[v_{qP}(45°)/\alpha_0 \right] - 1 \right\} - \varepsilon. \tag{9.40}$$

The special case $\varepsilon = \delta$ is known as **elliptical** VTI. For weak elliptical VTI, the preceding equations reduce to

$$v_{qP}(\theta) \approx \alpha_0(1 + \delta \sin^2 \theta), \qquad v_{qSV}(\theta) \approx \beta_0. \tag{9.41a,b}$$

Equation (9.41a) has the form of an ellipse in polar coordinates (like Equation 9.20, with $\cos^2 \theta$ replaced by $1 - \sin^2 \theta$). It states that qP wavefronts in weak elliptical VTI media are ellipsoidal, whereas Equation (9.41b) states that qSV wavefronts are spherical. This is also true for general VTI.

For general VTI, substituting $\varepsilon = \delta$ into (9.11) and doing the math also shows that (a) $v_{qSV}(\theta) = \beta_0$, i.e., v_{qSV} is independent of θ, meaning the qSV wavefronts are spherical (the qSV waves behave isotropically), and that (b) the qP wavefronts are elliptical in the x-z plane (hence the name elliptical VTI). Rocks containing dry gas-filled cracks are approximately elliptical VTI media.

More details on the theory for elliptical VTI can be found in the following section and in Appendix 9A.

9.4 Additional Comments

It can also be shown that the matrix **A** in Equation (9.3) can be written as

$$\mathbf{A} = \frac{1}{\rho} \mathbf{BcB}^T, \quad \text{where} \tag{9.42}$$

$$\mathbf{B} = \begin{bmatrix} n_1 & 0 & 0 & 0 & n_3 & n_2 \\ 0 & n_2 & 0 & n_3 & 0 & n_1 \\ 0 & 0 & n_3 & n_2 & n_1 & 0 \end{bmatrix}. \tag{9.43}$$

Using this rule, the matrix **A** can be computed directly from the matrix **c**.

Note from (9.17b) and (9.19) through (9.21) that in a VTI medium, the speed of horizontally traveling qSV waves is different from the speed of horizontally traveling SH waves. In other words, the two polarizations of a shear wave (horizontal and vertical) travel with different speeds, unlike the isotropic case. This means that a single shear wave polarized at an acute angle to the x-z plane cannot propagate in the horizontal direction (or in any other direction for which $v_{qSV} \neq v_{SH}$). Such a wave, if set in motion, would break up into two waves with different speeds, one with a vertical polarization (qSV) and one with a horizontal polarization (SH). This is called **shear wave splitting**. In crystal optics, it is called **birefringence**.

We should note, however, that the term "shear wave splitting" is usually associated with waves propagating in an **HTI medium**, i.e., a TI medium with a *horizontal* symmetry axis (e.g., a VTI medium rotated about the x axis or y axis by $90°$), or more generally, with waves in an **azimuthally anisotropic** medium (see, e.g., Thomsen, 1988). A simple example of an HTI medium would be a medium containing a distribution of plane vertical cracks or fractures all with the same strike. Shear waves traveling in the vertical direction in such a medium can have only two polarizations – one being parallel to the planes of the cracks and the other perpendicular – and these two differently polarized shear waves would travel vertically with different speeds, exhibiting shear wave splitting.

To understand why the two polarizations travel with different wave speeds, and which polarization has the greater speed, note that the shear waves whose polarization (i.e., particle motion) is perpendicular to the fractures can displace or deform the medium by a larger amount (they can close the gaps in the fractures) than can the shear waves whose polarization is parallel to the fractures. In other words, the waves with perpendicular polarization experience a lower effective rigidity, and therefore have a lower wave speed.

Shear wave splitting can be detected relatively easily, and can be used to infer the presence of azimuthal anisotropy in a medium.

What about reflection and transmission coefficients in anisotropic media? The Zoeppritz equations for anisotropy can be derived in the same way as for isotropic media (see Chapter 3), and are usually solved numerically. In some cases, though, analytical formulas can be obtained. For example, for SH waves incident upon an interface between two VTI media, the coefficients are

$$R_{SH} = \frac{W_1 - W_2}{W_1 + W_2}, \qquad T_{SH} = \frac{2W_1}{W_1 + W_2}, \tag{9.44a}$$

where

$$W_n \equiv \frac{\rho_n (\beta_v)_n^2}{\beta_n} \cos \theta_n, \qquad n = 1, 2, \tag{9.44b}$$

and where β_n is the phase velocity $\beta = \beta(\theta)$ for medium n (see Equation 9.25). For other examples, see Daley and Hron (1977, 1979) and Thomsen (1993).

Note that Snell's law is also more complicated, because phase velocities now depend on angle: for *SH* waves, it is

$$\frac{\sin \theta_1}{\beta_1 (\theta_1)} = \frac{\sin \theta_2}{\beta_2 (\theta_2)}. \tag{9.45}$$

Equations (9.11) for the phase velocities of *qP* and *qSV* waves in a VTI medium can also be expressed in the following useful form (Gassmann, 1964):

$$2\rho v_{qP}^2 = 2 \left(c_{11} \sin^2 \theta + c_{33} \cos^2 \theta \right) + c_\alpha \left(\sqrt{1 + 4\kappa_D} - 1 \right), \tag{9.46}$$

and

$$2\rho v_{qSV}^2 = 2c_{44} - c_\alpha \left(\sqrt{1 + 4\kappa_D} - 1 \right), \tag{9.47}$$

where

$$c_\alpha = (c_{11} - c_{44}) \sin^2 \theta + (c_{33} - c_{44}) \cos^2 \theta = c_{11} \sin^2 \theta + c_{33} \cos^2 \theta - c_{44}, \tag{9.48}$$

$$\kappa_D = \left(c_D \sin^2 \theta \cos^2 \theta \right) / c_\alpha^2, \tag{9.49}$$

$$c_D = (c_{13} + c_{44})^2 - (c_{11} - c_{44})(c_{33} - c_{44}). \tag{9.50}$$

These equations can be derived by algebraically rearranging the terms in (9.11), or perhaps more easily by re-solving the equation $AB - C = 0$ from (9.10e), with the aim of obtaining the velocities in the form of (9.46) and (9.47).

It can be easily shown that the elliptical VTI condition, i.e., $\varepsilon = \delta$, is equivalent to the condition $c_D = 0$. Putting $c_D = 0$ then in (9.46) and (9.47) yields the phase velocities for elliptical VTI:

$$\rho v_{qP}^2 = c_{11} \sin^2 \theta + c_{33} \cos^2 \theta \tag{9.51}$$

which has the form of the equation of an ellipse in polar coordinates (hence the name elliptical VTI), and

$$\rho v_{qSV}^2 = c_{44}, \tag{9.52}$$

which states that the *qSV* phase velocity is constant, i.e., independent of θ, meaning *qSV* waves in the elliptical VTI case have spherical wavefronts, i.e., they behave as if they were propagating in an isotropic medium.

We can now see that the first term on the right-hand side of (9.46) is the elliptical VTI term and the second term on the right-hand side of (9.46) gives the deviation from the elliptical VTI case. The same is true of (9.47).

As mentioned previously, more details on the theory for elliptical VTI can be found in Appendix 9A.

Although only waves in TI media are discussed in this chapter, waves in other types of anisotropic media, such as orthorhombic media, have also been used in the seismic wave theory literature to effectively simulate seismic wave propagation in various types of subsurface media, e.g., media with complex fracture sets.

9.5 Appendix 9A: Velocities, Elliptical Anisotropy, Slowness

This appendix deals with velocity formulas, elliptical anisotropy, slowness surfaces and group (ray) velocity, slowness diagrams, and tilted TI media.

9.5.1 Velocity Formulas

As indicated earlier, the qP and qSV wave velocities for the VTI case can be obtained by solving the equation $AB - C = 0$ in (9.10e) in the text. The solution was given in (9.11) without derivation. Here, some details are provided.

Note that $AB - C = 0$ is a quadratic equation in ρv^2 when it is expanded. The generic quadratic equation and its solutions are

$$ax^2 + bx + c = 0, \qquad \Longrightarrow \qquad x = \left[-b \pm \sqrt{b^2 - 4ac} \,\right]/2a. \qquad (9.53)$$

Applying this to $AB - C = 0$ results in

$$\rho v^2 = \tfrac{1}{2}\left[G \pm \sqrt{G^2 - 4J} \,\right], \qquad (9.54)$$

where the $+$ sign is for qP waves and the $-$ sign is for qSV waves, and where

$$G \equiv (c_{11} + c_{44}){n_1}^2 + (c_{33} + c_{44}){n_3}^2 = c_{11}{n_1}^2 + c_{33}{n_3}^2 + c_{44}, \qquad (9.55)$$

$$J \equiv \left(c_{11}{n_1}^2 + c_{44}{n_3}^2\right)\left(c_{44}{n_1}^2 + c_{33}{n_3}^2\right) - (c_{13} + c_{44})^2 {n_1}^2 {n_3}^2, \qquad (9.56)$$

$$G^2 - 4J = \left[(c_{11} - c_{44}){n_1}^2 - (c_{33} - c_{44}){n_3}^2\right]^2 + 4(c_{13} + c_{44})^2 {n_1}^2 {n_3}^2. \qquad (9.57)$$

It is straightforward to confirm that the two equations for ρv^2 reduce to the correct formulas for perfect isotropy, i.e., $\rho v^2 = \lambda + 2\mu$ for P waves and μ for SV waves.

v^2 in (9.11) can be obtained by using ${n_1}^2 = \sin^2 \theta$ and replacing ${n_3}^2 = \cos^2 \theta$ with $1 - \sin^2 \theta$ in the preceding formulas and doing some algebra.

9.5.2 Elliptical Anisotropy

Expansion of (9.57) shows that if we set

$$(c_{13} + c_{44})^2 = (c_{11} - c_{44})(c_{33} - c_{44}), \qquad (9.58)$$

then $G^2 - 4J$ becomes a perfect square, i.e.,

$$G^2 - 4J = \left[(c_{11} - c_{44})n_1^2 + (c_{33} - c_{44})n_3^2\right]^2, \qquad (9.59)$$

meaning that (9.54) reduces to

$$\rho v^2 = \begin{cases} c_{11}n_1^2 + c_{33}n_3^2, & qP, \\ c_{44}, & qSV. \end{cases} \qquad (9.60)$$

This is the case of *elliptical anisotropy* – see (9.51) and (9.52).

9.5.3 Slowness Surface and Ray (Group) Velocity

In the VTI case, the dispersion relation is given by (9.26) and the slowness ellipse is given by (9.31), i.e.,

$$\omega^2 = \beta_h^2 \kappa_x^2 + \beta_v^2 \kappa_z^2 \qquad \text{and} \qquad \beta_h^2 s_x^2 + \beta_v^2 s_z^2 = 1. \qquad (9.61)$$

For general anisotropy, the dispersion relation can be written formally as

$$\omega^2 = f(\kappa_x, \kappa_y, \kappa_z) \equiv f(\kappa_j) \equiv f(\boldsymbol{\kappa}). \qquad (9.62)$$

The **slowness surface** is then obtained by substituting $\kappa_j = \omega s_j$ into this:

$$f(\omega s_j)/\omega^2 \equiv \psi(s_j) = 1. \qquad (9.63)$$

This is a surface in 3D slowness space ($s_x s_y s_z$ space). Noting that the jth component of the ray (group) velocity is given by $V_j = \partial\omega/\partial\kappa_j$, differentiation of (9.62) leads to

$$\frac{\partial\omega^2}{\partial\kappa_j} = 2\omega\frac{\partial\omega}{\partial\kappa_j} = 2\omega V_j = \frac{\partial f}{\partial\kappa_j} \quad\Rightarrow\quad V_j = \frac{1}{2\omega}\frac{\partial f}{\partial\kappa_j} \quad\Rightarrow\quad \mathbf{V} = \frac{1}{2\omega}\nabla_\kappa f, \quad (9.64)$$

where ∇_κ is the gradient in angular wavenumber space ($\kappa_x \kappa_y \kappa_z$ space). From vector analysis, we know that ∇f is everywhere perpendicular to (or normal to) the surface $f = \text{const}$. Hence, we see that in wavenumber space, $\mathbf{V}$ is everywhere normal to the wavenumber surface $f(\boldsymbol{\kappa}) = \omega^2$ (the dispersion relation).

Alternatively, we have

$$2\omega V_j = \frac{\partial f}{\partial\kappa_j} = \frac{\partial f}{\partial(\omega s_j)} \quad\Rightarrow\quad V_j = \frac{1}{2}\frac{\partial(f/\omega^2)}{\partial s_j} \quad\Rightarrow\quad V_j = \frac{1}{2}\frac{\partial\psi}{\partial s_j}$$

$$\Rightarrow\quad \mathbf{V} = \frac{1}{2}\nabla_s\psi, \qquad (9.65)$$

where ∇_s is the gradient in slowness space. In slowness space, the ray (group) velocity **V** is everywhere normal to the slowness surface $\psi(\mathbf{s}) = 1$.

9.5.4 Slowness Diagrams

These are plots of the slowness surfaces. They can be used to gain insight into anisotropic wave propagation effects. For simplicity, we consider the 2D case in which waves are propagating parallel to the xz plane (i.e., waves for which $s_y = 0$), with the z axis pointing upward. Slowness surfaces in the 2D case are slowness curves. Let the slowness vector $\mathbf{s} = (s_x, s_y, s_z) = (p, 0, \eta)$.

Figure 9.2 shows the slowness curves for a generic isotropic case. In this case, the slowness curves in media 1 and 2 are circles. The outer circle is the slowness curve for either the incident wave or reflected wave in medium 1, and the inner one is for the transmitted wave in medium 2. A vertical line for a given p value intersects the slowness curves in two places. For the outer circle, this gives the η values (vertical component of slowness) for the incident ($\eta^{\text{inc}} < 0$) and reflected ($\eta^{\text{ref}} > 0$) waves, and for the inner circle it gives the η value for the transmitted wave ($\eta^{\text{tra}} < 0$). For a supercritical p value, e.g., p_{sc}, the vertical line intersects

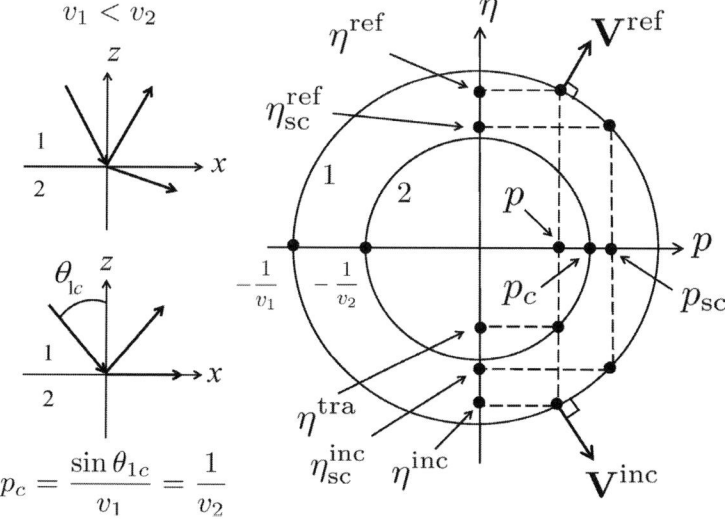

Figure 9.2 Schematic diagrams of rays (left) at the precritical and critical angles for waves of one type only (either P or SV or SH) in the isotropic case for $v_1 < v_2$, and the corresponding slowness curves (the concentric circles). The outer circle is for the incident and reflected waves in medium 1 and it satisfies $p^2 + \eta^2 = 1/v_1^2$. The inner circle is for the transmitted waves in medium 2 and it satisfies $p^2 + \eta^2 = 1/v_2^2$. The subscript "sc" stands for "supercritical."

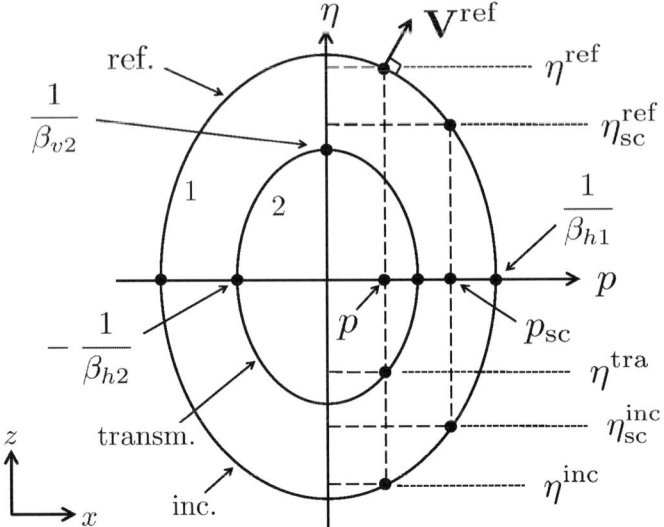

Figure 9.3 A schematic diagram of the slowness curves (ellipses) for *SH* waves in the VTI case, for the case $\beta_{v1} < \beta_{h1} < \beta_{v2} < \beta_{h2}$. The outer ellipse is for the incident and reflected waves in medium 1, and it satisfies $\beta_{h1}^2 p^2 + \beta_{v1}^2 \eta^2 = 1$. The inner ellipse is for the transmitted waves in medium 2, and it satisfies $\beta_{h2}^2 p^2 + \beta_{v2}^2 \eta^2 = 1$. The subscript "sc" stands for "supercritical."

only the slowness curve for medium 1, giving η_{sc}^{ref} (> 0) – there is no downgoing transmitted wave (it is evanescent). The ray velocity **V** is normal to the slowness curve. Note that $|\eta^{inc}| = |\eta^{ref}|$, meaning that the angle of incidence equals the angle of reflection.

Figure 9.3 shows an analogous diagram for *SH* waves in the VTI case. Here, the slowness curves are ellipses. Again, $|\eta^{inc}| = |\eta^{ref}|$ shows that the angle of incidence equals the angle of reflection.

Figure 9.4 shows that some interesting wave propagation effects occur when a VTI medium is tilted. The energy of the wave flows in the direction of the ray velocity **V**, which is everywhere normal to the slowness curve. Suppose that the slowness curve in Figure 9.4 is for the incidence medium. We then see that only for p values in the range $[p_2, p_4]$ does the incident wave energy travel downward and to the right (in the $+x$ and $-z$ direction) because it is only for p values in this range that **V** goes from vertical downward to horizontal and to the right (i.e., $0 \leq \phi \leq 90°$). $p = 0$ (which corresponds to a vertically downward incident ray in Figures 9.2 and 9.3) would correspond to an incident ray going down and to the left, as is also the case for $p = p_1$. Note also that for $p = p_2$, we see that a vertically incident ray generates a reflected ray that is *not* vertical ($|\eta^{inc}| \neq |\eta^{ref}|$), i.e., *the angle of incidence does not equal the angle of reflection*. $p = p_3$ also

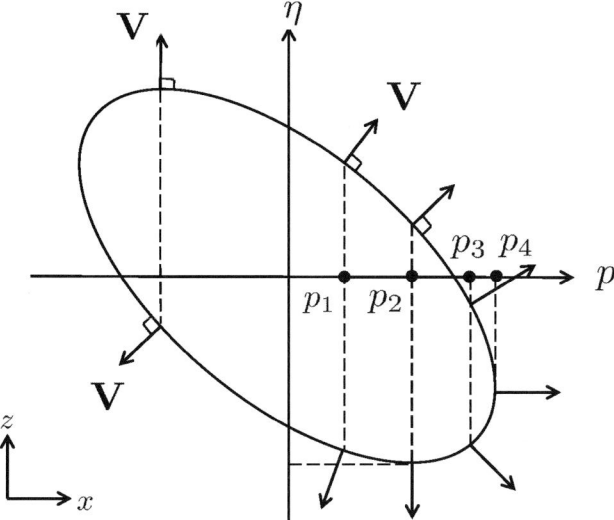

Figure 9.4 A schematic diagram of the slowness curve (ellipse) for *SH* waves in the case of a tilted VTI medium, for the case $\beta_v < \beta_h$.

shows this. $p = p_3$ also shows that **V** for the reflected wave, i.e. $\mathbf{V}^{\mathrm{ref}}$, points upward but $\eta^{\mathrm{ref}} < 0$, i.e., the wavevector or slowness vector for the reflected wave points downward, i.e., $\theta^{\mathrm{ref}} < 90°$ but $\phi^{\mathrm{ref}} > 90°$. Another way to look at it is as follows: if the slowness curve is for the incident wave, then at $p = p_3$, there appear to be two possibilities for the incident wave, because both η values are negative, but the upper η value corresponds to upgoing energy (an upgoing ray), so the lower one is the correct choice for the incident wave (and the upper one for the reflected wave). At $p = p_1$, an incident ray traveling downward and to the left produces a reflected ray traveling upward and to the right. At the negative p value shown on the left, an incident ray traveling downward and to the left produces a reflected ray traveling vertically upward.

Suppose now that the slowness ellipse in Figure 9.4 corresponds to the transmitted wave. We then see that only for p values in the range $[p_2, p_4]$ does the transmitted wave energy travel downward and to the right (in the $+x$ and $-z$ direction). At $p = p_3$, both values of η are negative, meaning that there are two possible choices for downward-pointing slowness vectors or wavevectors, i.e., there are apparently two possible choices for the transmitted wave. But the upper η value is rejected because it corresponds to an upgoing transmitted ray, which is unphysical. Hence the lower η value is correct one for the transmitted wave.

By superimposing the slowness curves for the incidence and transmission media on one graph, as in Figures 9.2 and 9.3 (and they could have different tilts), all the

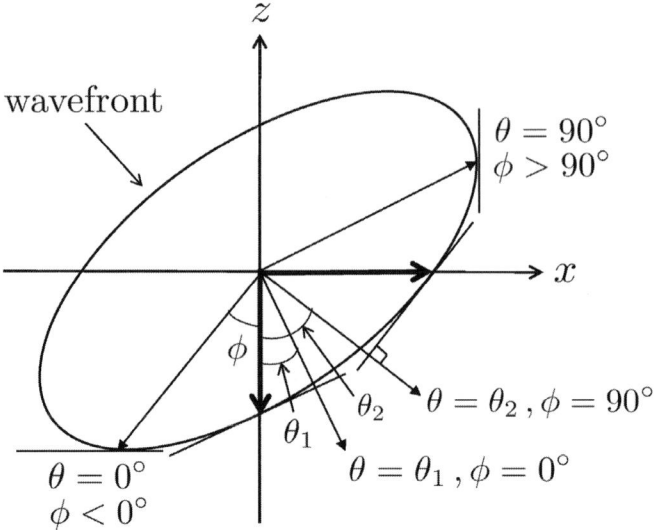

Figure 9.5 A schematic diagram of a wavefront in the tilted VTI medium of Figure 9.4.

waves (incident, reflected, and transmitted) and possible p values, and the various interesting effects, can be seen on one diagram.

Figure 9.5 shows a corresponding wavefront in a tilted VTI medium, along with some phase angles θ and their corresponding ray angles ϕ. The arrows ending on the wavefront are rays (along which the wave energy flows). The boldface arrows are rays with ray angles of $\phi = 0°$ and $90°$. The two arrows with phase angles θ_1 and θ_2 are in the direction of the corresponding wavevectors or slowness vectors, i.e., the directions of phase propagation. Note that only phase angles θ in the range $[\theta_1, \theta_2]$ correspond to rays traveling downward and to the right. A phase angle slightly greater than θ_2 could not be a transmitted wave phase angle, because it corresponds to a ray with a ray angle slightly greater than $90°$, i.e., a ray traveling upward. When the VTI medium is not tilted, then transmitted wave phase angles θ can take values between $0°$ and $90°$, but in the tilted case, they can take values only between θ_1 and θ_2.

Exercises

1. The plane wave eigenvalue problem for a perfectly isotropic medium:

    (a) Compute the matrix **A** by substituting Equation (2.29) into Equation (9.3a).

(b) Solve the eigenvalue equation for the eigenvalues and eigenvectors. For mathematical simplicity, assume the plane wave is propagating in the z direction in the x-z plane (no generality is lost by this assumption, because in an isotropic medium, wave propagation properties are the same in all directions). Does your solution produce the three wave types, P, SV, and SH?

2. Derive Equations (9.27) and (9.28).

3. For a VTI medium, derive a formula for the SH group velocity V as a function of the phase angle θ.

4. For a VTI medium, derive a formula for the SH phase velocity β as a function of the group angle ϕ.

5. Matrix form of the equation of motion:

   (a) Write the 6×1 strain column vector **e** in Equation (9.6a) in the form **e=Eu**, where **E** is a 6×3 matrix of partial derivative operators.

   (b) We have seen how the stress–strain relation (Hooke's law) can be written in the matrix form **σ=ce** (see Equations 9.6 and 9.7), where **σ** is the 6×1 stress column vector. Write the equation of motion for a general anisotropic medium in matrix form. Use part (a) to express this equation in terms of **u**.

   (c) Substitute a time harmonic wave $u_i = U_i \exp[-i\omega t]$ into the matrix equation of motion you obtained in part (b), and carry out the time derivatives only. What kind of equation do you get as a result?

6. Solve the plane wave eigenvalue problem for a VTI medium.

7. Is the matrix **A** in Equation (9.3) symmetric? Prove your answer.

8. Do the elements of the matrix **A** in Equation (9.3) form the components of a rank-2 tensor? Prove your answer.

9. Derive (9.44).

10. A VTI medium has the following parameter values: $\rho = 2$ g/cm^3, $\alpha_0 = 2$ km/s, $\beta_0 = 1$ km/s, $\epsilon = \gamma = \delta = 0.3$. Assume that this is not a "weak" VTI medium. Compute the elements c_{ij} of the **c**-matrix. To which components c_{ijkl} of the stiffness tensor are they equivalent? In this medium, what is the speed of a horizontally propagating qP wave? A horizontally propagating qSV wave? Horizontally and vertically propagating SH waves?

10

Plane Waves in Anelastic Media

This chapter provides an introduction to the theory of plane wave propagation in absorbing or anelastic media. More detailed and thorough accounts can be found in the books by Achenbach (1973), Aki and Richards (1980, 2002), Ben-Menahem and Singh (2000), Bland (1960), Borcherdt (2009), Carcione (2001, 2007), Červený (2001), Christensen (1971), Fung (1965), and Hudson (1980).

10.1 Absorption of Seismic Waves

In a perfectly elastic medium, there is no absorption of seismic wave energy. But in real materials, which are **anelastic**, wave energy is absorbed due to **internal friction**. Anelastic media are sometimes called **dissipative** media, or **viscoelastic** media if the anelasticity is modeled by the theory of viscoelasticity. Absorption is frequency dependent, i.e., different frequencies are absorbed by different amounts. One consequence of this is that the waveform changes with distance traveled.

We will see in this chapter that 1D waves in anelastic media can be conveniently studied by allowing the angular wavenumber κ and the wave speed v to be complex numbers (defined later in this section). To distinguish these from their real counterparts in the perfectly elastic case, we will use K and V for the real angular wavenumber and real wave speed, with $K = \mathrm{Re}(\kappa)$ and $V = \mathrm{Re}(v)$.

We have already studied the absorption of seismic waves to some extent in Section 7.5, where we saw that absorption can be described via the **quality factor** Q or the **loss factor** $1/Q$. In addition, we saw that if $-\Delta A$ (> 0) is the amplitude lost per cycle of oscillation of a sinusoidal wave of angular frequency ω traveling with speed V in the x direction in a dissipative material, where x is the distance traveled by the wave, and A is the amplitude of the wave, then assuming $Q \gg 1$ (which is typically the case), we have

$$\frac{1}{Q} \equiv -\frac{\Delta A}{\pi A} \quad \Longrightarrow \quad A(x) = A_0 \exp\left(-\frac{\omega x}{2VQ}\right) \tag{10.1}$$

(see Equation 7.7c), i.e., the amplitude decays with distance traveled. For a perfectly elastic medium, $Q = \infty$ and $1/Q = 0$, meaning $A(x) = A_0$, i.e., the amplitude does not decay with x.

Consider a 1D sinusoidal wave $u = A\exp[i(Kx - \omega t)]$, $K = \omega/V$, traveling in the positive x direction in an absorbing medium. Replacing A with $A(x)$ gives

$$u = A_0\exp[-ax]\exp[i(Kx - \omega t)] = A_0\exp[i(\kappa x - \omega t)], \quad \text{where} \quad (10.2)$$

$$a = \frac{\omega}{2VQ}, \quad \kappa = K + ia = \frac{\omega}{V}\left(1 + \frac{i}{2Q}\right) = \frac{\omega}{v}, \quad \frac{1}{v} = \frac{1}{V}\left(1 + \frac{i}{2Q}\right), \quad (10.3)$$

where κ is called the **complex angular wave number**, and where $1/v$ is called the **complex slowness**, with v being the **complex wave speed** (K and V are real numbers, as mentioned previously). Note that u can still be written in the standard form for a harmonic wave, i.e., $u = A_0\exp[i(\kappa x - \omega t)]$, except that κ is complex. This shows that absorption can be included in wave motion by making the wavenumber complex (with the frequency ω being real). For $Q \gg 1$, we can also derive simple equations for v and v^2 as follows:

$$(1+z)^\epsilon \approx 1 + \epsilon z, \quad |\epsilon z| \ll 1 \quad \Rightarrow \quad v = V\left(1 - \frac{i}{2Q}\right) \quad \Rightarrow \quad v^2 = V^2\left(1 - \frac{i}{Q}\right),$$
$$(10.4)$$

where the $1/Q^2$ term has been dropped in v^2 because $Q \gg 1$. Formulas like $\beta^2 = \mu/\rho$ from the elastic case suggest that we can define a **complex modulus** M for the medium by setting $v^2 = M/\rho$, which gives

$$M \equiv \rho v^2 = \rho V^2\left(1 - \frac{i}{Q}\right) \quad \Longrightarrow \quad V^2 = \frac{\text{Re}(M)}{\rho}, \quad \frac{1}{Q} = -\frac{\text{Im}(M)}{\text{Re}(M)}. \quad (10.5)$$

The last equation is often used as a definition for Q. Note that the two definitions that we now have for Q are slightly different. If the first equation in (10.1) is used as the definition of Q, then $1/v$ in (10.3) is exact, but v and v^2 in (10.4) are approximate (holding for $Q \gg 1$). But if the last equation in (10.5) is used as the definition of Q, then v^2 in (10.4) is exact, but v and $1/v$ are approximate. However, for $Q \gg 1$ the two definitions give essentially the same results.

Generally, an absorbing medium is **dispersive**, i.e., $V = V(\omega)$ and $Q = Q(\omega)$, implying $M = M(\omega)$. But for seismic body waves, Q is nearly independent of frequency.

10.2 Waveforms and Dispersion

Consider a plane wave traveling in the x direction in a dissipative medium. Let $u(x, t)$ be the waveform at position x and time t in the medium. Assume $u(x, t)$

is a real number (the sample values on seismograms are real numbers). Let an overhead bar, as well as the symbol $\mathcal{F}_t$, represent the Fourier time transform, i.e., $\bar{u}(x,\omega) = \mathcal{F}_t\{u(x,t)\}$. $u(x,t)$ can be represented as a sum of harmonic waves of different frequencies:

$$u(x,t) = \frac{1}{2\pi}\int_{-\infty}^{\infty} \bar{s}(\omega)e^{i(\kappa x - \omega t)}\,d\omega = \mathcal{F}_t^{-1}\{\bar{s}(\omega)e^{i\kappa x}\}$$

$$= \frac{1}{2\pi}\int_{-\infty}^{\infty} \bar{s}(\omega)e^{-ax}e^{-i\omega(t-x/V)}\,d\omega, \tag{10.6}$$

$u(0,t) = \mathcal{F}_t^{-1}\{\bar{s}(\omega)\} = s(t) = $ the waveform at the source point $x = 0$,

i.e., the source pulse,

$\bar{u}(x,\omega) = \bar{s}(\omega)e^{i\kappa x} = $ the frequency spectrum of the waveform at x,

$\bar{u}(0,\omega) = \bar{s}(\omega) = $ the frequency spectrum of the waveform at $x = 0$,

i.e., the frequency spectrum of the source pulse,

$\bar{s}(\omega) = \mathcal{A}(\omega)e^{i\psi(\omega)}, \quad \mathcal{A}(\omega)$ and $\psi(\omega) = $ the amplitude and phase spectra

of the source pulse.

Note that if the absorption factor a is defined as in (10.3), then the integral over the negative frequencies in (10.6) diverges (if $V, Q > 0$). One way to overcome this difficulty is to redefine a as $a \equiv |\omega|/(2VQ)$, and to assume that $V(-\omega) = V(\omega)$ and $Q(-\omega) = Q(\omega)$, i.e., that V and Q are even functions of frequency (meaning $a(\omega)$ is even). This results in a decaying exponential for both negative and positive frequencies, and the integral converges. If a is not redefined, i.e., if the definition of a in (10.3) is used, then Q would have to be defined as an odd function of ω.

In addition, if we assume that $\mathcal{A}(-\omega) = \mathcal{A}(\omega)$ and $\psi(-\omega) = -\psi(\omega)$ (meaning $\bar{s}(-\omega) = \bar{s}(\omega)^*$), where * denotes the complex conjugate, then it is easy to show that $u(x,t)$ is real, as required. If $u(x,t)$ is real, one can show that u can also be written in a form that involves only positive frequencies (see exercise 28 in Chapter 1):

$$u(x,t) = \frac{1}{\pi}\mathrm{Re}\left[\int_0^{\infty} \bar{s}(\omega)e^{-i\omega(t-x/v)}\,d\omega\right]. \tag{10.7}$$

Example: $s(t) = \delta(t) \Rightarrow \bar{s}(\omega) = 1$. First, assume there is **no** dispersion, i.e., V, Q are independent of ω.

Case (a), no absorption: $Q = \infty \Rightarrow$ using (1.137a), (10.6) gives $u(x,t) = \delta(t - x/V) \Rightarrow$ there is no change in the waveform – it is a δ function at both the source and the receiver.

Case (b), with absorption:

$$u(x,t) = \frac{1}{2\pi} \int_{-\infty}^{\infty} e^{-|\omega|x/2VQ} e^{-i\omega(t-x/V)} \, d\omega$$

$$= \frac{1}{2\pi} \left[\int_{-\infty}^{0} e^{\omega x/2VQ} e^{-i\omega(t-x/V)} \, d\omega + \int_{0}^{\infty} e^{-\omega x/2VQ} e^{-i\omega(t-x/V)} \, d\omega \right]$$

$$= \frac{1}{\pi} \frac{(x/2VQ)}{(x/2VQ)^2 + (t-x/V)^2}. \tag{10.8}$$

One can also get this with (10.7) or with Fourier sine or cosine transforms (to avoid negative frequencies). Note that $u(x,0) \neq 0$, i.e., the wave arrives before the source is activated! This is a **noncausal** or **acausal** response. To avoid acausality, and to keep linearity (so that Fourier analysis can be used), one must include **dispersion**, i.e., $V = V(\omega)$ and $Q = Q(\omega)$. In Figure 10.1a, the expression in (10.8) is curve 1, and curve 2 is what u would typically look like if dispersion is included.

Theory and experimental measurements suggest, for seismic body waves,

$$V(\omega) = V(\omega_0)\left[1 + \frac{1}{\pi Q} \ln\left(\frac{\omega}{\omega_0}\right)\right], \quad \text{and} \quad Q(\omega) \sim \text{constant (independent of } \omega\text{)}, \tag{10.9}$$

where ω_0 is some reference frequency. $V(\omega)$ is graphed in Figure 10.1b. Absorption implies dispersion, but dispersion does not imply absorption (e.g., surface waves in perfectly elastic media are dispersive).

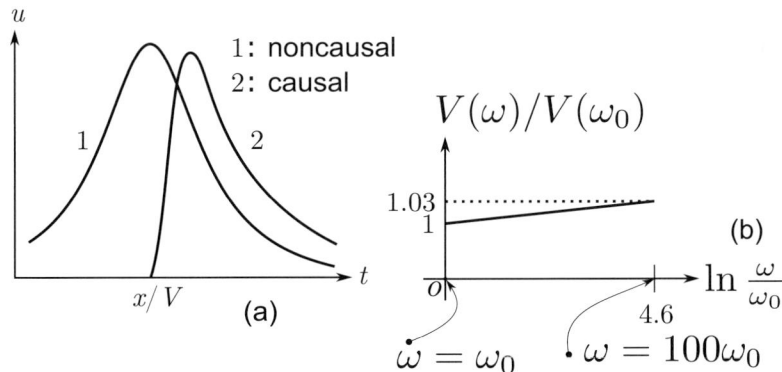

Figure 10.1 (a) The noncausal pulse (curve 1) in Equation 10.8, where dispersion was not included, and a schematic diagram of a causal version of the pulse (curve 2), i.e., what it might look like if dispersion was included, and (b) Equation (10.9) for $Q = 50$.

10.3 Anelasticity Models and Stress–Strain Relations

Anelastic media can be described with spring-dashpot models. The spring is the elastic element, and obeys the stress–strain relation $\sigma = \mu e = \mu(\partial u/\partial x)$, where σ and e are the stress and strain, resp. (see Figure 10.2). μ is the spring constant. The dashpot is the viscous element, and obeys $\sigma = \eta\dot{e}$, where η is the viscosity constant, and where the overhead dot again represents a time derivative. The strain energy density for the spring can be obtained from $\mathcal{W} = \frac{1}{2}\sigma_{ij}e_{ij}$, which reduces to $\mathcal{W} = \frac{1}{2}\sigma e = \sigma^2/2\mu$. The rate of energy density dissipation for the dashpot can be obtained from $\dot{\mathcal{W}} = \sigma_{ij}\dot{e}_{ij}$, which reduces to $\dot{\mathcal{W}} = \sigma\dot{e} = \sigma^2/\eta$. Because the stress–strain relations (derived in this section) for the spring-dashpot models are linear, and because the dashpot brings viscosity into the model, materials modeled with springs and dashpots are often called **linear viscoelastic** materials.

For a **Maxwell Solid** (see Figure 10.2), the stress is the same across each element, but $e = e_\eta + e_\mu$, so

$$\dot{e} = \dot{e}_\eta + \dot{e}_\mu = (\sigma/\eta) + (\dot{\sigma}/\mu) \quad \Rightarrow \quad \dot{\sigma} + (\mu/\eta)\sigma = \mu\dot{e} \quad \text{or} \quad \sigma + (\eta/\mu)\dot{\sigma} = \eta\dot{e}, \tag{10.10a}$$

where as before, $e = \partial u/\partial x$.

For a **Voigt Solid** (see Figure 10.2), the strain is the same across each element, but $\sigma = \sigma_\mu + \sigma_\eta$, so

$$\sigma = \mu e + \eta\dot{e}. \tag{10.10b}$$

For perfect elasticity, $\eta = \infty$ in the Maxwell model, and $\eta = 0$ in the Voigt model.

A **Standard Linear Solid (SLS)** (see Figure 10.3) is either a Voigt element in series with an elastic element, or a Maxwell element in parallel with an elastic element.

One can show that both SLS-I and SLS-II in Figure 10.3 give the stress–strain relation

$$\sigma + \tau_\sigma\dot{\sigma} = M_r(e + \tau_e\dot{e}), \tag{10.10c}$$

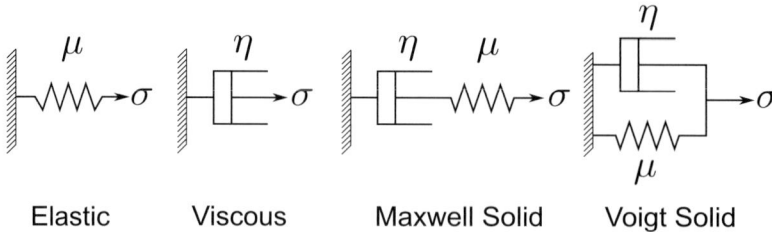

Elastic   Viscous   Maxwell Solid   Voigt Solid

Figure 10.2 Spring-dashpot models for the elastic, viscous, Maxwell and Voigt solids.

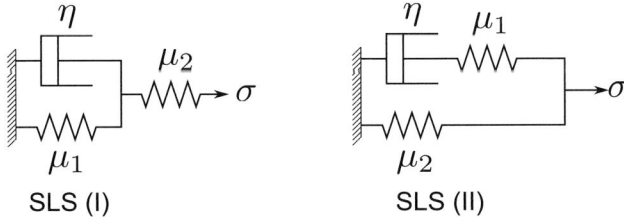

Figure 10.3 Spring-dashpot models for two versions of the standard linear solid (SLS).

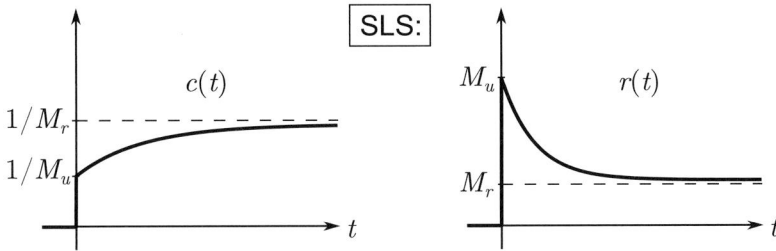

Figure 10.4 The creep function $c(t)$ and the relaxation function $r(t)$ for the standard linear solid.

where τ_σ, τ_e, and M_r are functions of μ_1, μ_2, and η, and the functions for SLS-I are different from those for SLS-II. Also, $\tau_\sigma < \tau_e$. The SLS has been used in seismology. Unlike the Maxwell and Voigt models, the SLS contains all four quantities σ, $\dot\sigma$, e, and $\dot e$.

The **creep function** $c(t)$ is the strain e produced by a unit step function $H(t)$ in stress ($\sigma = H(t)$).

The **relaxation function** $r(t)$ is the stress σ produced by a unit step function $H(t)$ in strain ($e = H(t)$).

For example, for a standard linear solid, using $\dot H(t) = \delta(t)$ gives the ordinary differential equations

$$H(t) + \tau_\sigma \delta(t) = M_r\left[c + \tau_e \dot c\right] \qquad \text{and} \qquad r + \tau_\sigma \dot r = M_r\left[H(t) + \tau_e \delta(t)\right], \quad (10.10d)$$

whose solutions (which can be obtained, for instance, by the method of Laplace transforms) are

$$c(t) = \frac{1}{M_r}\left[1 - \left(1 - \frac{\tau_\sigma}{\tau_e}\right)\exp\left(-\frac{t}{\tau_e}\right)\right]H(t), \qquad (10.10e)$$

$$r(t) = M_r\left[1 + \left(\frac{\tau_e}{\tau_\sigma} - 1\right)\exp\left(-\frac{t}{\tau_\sigma}\right)\right]H(t), \qquad (10.10f)$$

which are graphed in Figure 10.4.

τ_e is the strain relaxation time, τ_σ is the stress relaxation time, M_r is the **relaxed modulus**, and $M_u \equiv (\tau_e/\tau_\sigma)M_r$ is the **unrelaxed modulus**.

Lab measurements of creep (Lomnitz, 1956, 1957) indicate that $c(t) \sim 1 + q\ln(1+at)$, where q, a are constants – this is similar to the $c(t)$ for the SLS, except that it grows with time t and doesn't level off. This experimental $c(t)$ also leads to (10.9) (but $c(t)$ in Equation 10.10e does not, meaning that it is not a good enough model for seismic body waves).

Equations of motion can also be obtained for the preceding viscoelastic models of solids. For example, for the Voigt solid, substituting (10.10b) with $e = \partial u/\partial x$ into the 1D equation of motion $\partial\sigma/\partial x = \rho\ddot{u}$ leads to

$$\mu\frac{\partial^2 u}{\partial x^2} + \eta\frac{\partial^3 u}{\partial x^2 \partial t} = \rho\frac{\partial^2 u}{\partial t^2}, \tag{10.11}$$

which looks like the 1D wave equation except for the additional third-order derivative term. Substituting a trial solution $u = A\exp[i(\kappa x - \omega t)]$ with $\kappa = \omega/v$ and $v = \sqrt{M/\rho}$ and doing the math is one way to derive a formula for $Q^{-1}(\omega)$ for the Voigt solid (although there is an easier way, as we will see later).

10.4 General Linear Viscoelasticity

Spring-dashpot models have specific stress–strain relations that depend on the structure of the model. However, it is desirable to derive a general stress–strain relation that does not depend on the structure of the model. To do this, consider a 1D anelastic medium under stress. Using a formulation due to L. Boltzmann, the strain $e(t)$ at the present time t is due to the stress history up to present time t (not just the stress *at* time t), i.e., the medium has a "memory." In a time interval $d\tau$ centered on a past time τ, the increment of stress is $d\sigma(\tau) = [d\sigma(\tau)/d\tau]d\tau = \dot{\sigma}(\tau)\,d\tau$. It contributes an amount $de(t)$ to the strain at the present time t, with the proportionality constant c being a function of $t - \tau$, i.e., $de(t) = c(t - \tau)\,d\sigma(\tau)$. Similarly, $d\sigma(t) = r(t - \tau)\,de(\tau)$. Hence, we obtain the stress–strain relations

$$e(t) = \int_{-\infty}^{t} c(t-\tau)\dot{\sigma}(\tau)\,d\tau = c * \dot{\sigma} \quad \text{and} \quad \sigma(t) = \int_{-\infty}^{t} r(t-\tau)\dot{e}(\tau)\,d\tau = r * \dot{e}. \tag{10.12}$$

These are also sometimes written as $e = c * d\sigma$ and $\sigma = r * de$. Regarding the lower limit, for the steady-state response (after all transients have died out), it is convenient to assume stress begins at $\tau = -\infty$ (for other applications, $\tau = 0$ is used). Regarding the upper limit, one can formally also use ∞ as the upper limit, because future times $\tau > t$ do not contribute to $e(t)$ or $\sigma(t)$, so $c(t - \tau) = 0$ and $r(t - \tau) = 0$ for $\tau > t$, i.e., the functions $c(t)$ and $r(t)$ are zero for $t < 0$.

We have already used c and r for the creep and relaxation functions. So why did we use c and r as the proportionality "constants" in the preceding functions? Because they *are* in fact the creep and relaxation functions. The proof follows. Using the preceding definitions of the creep and relaxation functions, we have

$$\sigma(t) = H(t) \quad \Rightarrow \quad e(t) = \int_{-\infty}^{t} c(t - \tau)\delta(\tau)\, d\tau = c(t), \tag{10.13a}$$

$$e(t) = H(t) \quad \Rightarrow \quad \sigma(t) = \int_{-\infty}^{t} r(t - \tau)\delta(\tau)\, d\tau = r(t), \tag{10.13b}$$

indicating that proportionality "constants" c and r are indeed the creep and relaxation functions. Equations (10.12) are the basis of the **linear theory of viscoelasticity**.

10.5 Complex Modulus and Q

How does the more general approach in the preceding section relate to Q? As before, let $\bar{g}(\omega) \equiv \mathcal{F}_t\{g(t)\}$, and let $\sigma(t) = \mu(t) * \dot{e}(t)$ be the stress–strain relation, where $\mu(t)$ is now the relaxation function. $\sigma(t) = \mu(t) * \dot{e}(t)$ is generalization to viscoelasticity of the elastic relation $\sigma(t) = \mu e(t)$. Taking the Fourier time transform $\mathcal{F}_t$ of $\sigma(t) = \mu(t) * \dot{e}(t)$, and using (1.145) and the rule in (1.150), which allows us to make the replacement $\partial/\partial t \to -i\omega$ to find the Fourier transform, we get

$$\sigma(t) = \mu(t) * \dot{e}(t) \quad \Rightarrow \quad \bar{\sigma}(\omega) = \bar{\mu}(\omega)\bar{\dot{e}}(\omega) = \bar{\mu}(\omega)(-i\omega)\bar{e}(\omega). \tag{10.14a}$$

Hence,

$$\bar{\sigma}(\omega) = M(\omega)\bar{e}(\omega), \quad M(\omega) \equiv -i\omega\bar{\mu}(\omega) = -i\omega \int_0^{\infty} \mu(t)e^{i\omega t}\, dt \tag{10.14b}$$

$$\text{and} \quad \frac{1}{Q} \equiv -\frac{\text{Im}[M(\omega)]}{\text{Re}[M(\omega)]}, \tag{10.14c}$$

where $M(\omega)$ is the complex modulus. This is the same complex modulus that was defined in (10.5), which can be shown by calculating the complex wave speed v for this stress–strain relation, and showing that it is given by $v^2 = M/\rho$ (as in Equation 10.5), which is done in (10.23) later in this chapter. The lower limit in the preceding integral, $-\infty$, has been replaced by 0 because $\mu(t) = 0$ for $t < 0$. One can also prove $\bar{\sigma} = M\bar{e}$ by evaluating the double integral that arises in $\mathcal{F}_t\{\sigma = \mu * \dot{e}\}$, or by substituting $\sigma = \sigma_0 \exp[-i\omega t]$ and $e = e_0 \exp[-i\omega t]$ into the stress–strain relation $\sigma = \mu * \dot{e}$ and doing the math, which gives $M(\omega) = \sigma_0/e_0 = \sigma/e$.

Note that the 1D anelastic stress–strain relation in the frequency domain, $\bar{\sigma}(\omega) = M(\omega)\bar{e}(\omega)$, has the same form as the one for perfectly elastic 1D media, $\bar{\sigma}(\omega) = \mu\bar{e}(\omega)$, where μ is a real constant, i.e., the stress is a product of the modulus and the strain, except that the modulus M is complex and frequency dependent. This suggests that problems in viscoelasticity can be treated by transforming the equations into the frequency domain, solving them as if they were for elastic media (but with complex moduli), and transforming back into the time domain. Equivalently, any elastic solution can be made into a viscoelastic one by transforming it into the frequency domain, replacing the elastic moduli with complex frequency-dependent viscoelastic ones, and transforming back into the time domain (although care must be taken in applying this to certain specific cases). This is called the **elastic-viscoelastic correspondence principle**.

The fact that the modulus $M(\omega)$ is complex means that the stress and strain are out of phase in a viscoelastic medium.

The complex modulus $M(\omega)$ can be computed from the integral in (10.14b) given $\mu(t)$, but if the stress–strain relation is known, e.g., for a spring-dashpot model, it is generally easier to Fourier transform the stress–strain relation in time, and use $M(\omega) = \bar{\sigma}(\omega)/\bar{e}(\omega)$.

For example, for a standard linear solid,

$$\sigma + \tau_\sigma \dot{\sigma} = M_r(e + \tau_e \dot{e}) \quad \Rightarrow \quad \bar{\sigma} - i\omega\tau_\sigma\bar{\sigma} = M_r(\bar{e} - i\omega\tau_e\bar{e}) \quad \Rightarrow \quad (10.15)$$

$$M(\omega) = \frac{\bar{\sigma}}{\bar{e}} = \frac{M_r(1 - i\omega\tau_e)}{1 - i\omega\tau_\sigma} = \frac{M_r(1 - i\omega\tau_e)(1 + i\omega\tau_\sigma)}{(1 - i\omega\tau_\sigma)(1 + i\omega\tau_\sigma)}$$

$$= \frac{M_r\big[(1 + \omega^2\tau_e\tau_\sigma) + i\omega(\tau_\sigma - \tau_e)\big]}{1 + \omega^2\tau_\sigma{}^2} \quad (10.16)$$

$$\Longrightarrow \quad |M(\omega)| = M_r\sqrt{\frac{1 + \omega^2\tau_e{}^2}{1 + \omega^2\tau_\sigma{}^2}}, \quad \frac{1}{Q} = \frac{\omega(\tau_e - \tau_\sigma)}{1 + \omega^2\tau_e\tau_\sigma}. \quad (10.17)$$

$|M|$ and Q^{-1} are shown in Figure 10.5. $M(\omega)$ could also have been obtained by substituting $\sigma = \sigma_0\exp[-i\omega t]$ and $e = e_0\exp[-i\omega t]$ into the stress–strain relation

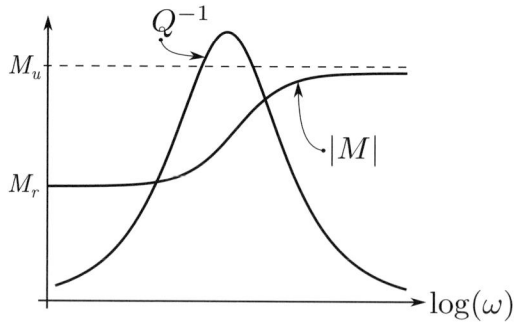

Figure 10.5 The loss factor Q^{-1} and the magnitude $|M|$ of the complex modulus for the standard linear solid.

(the leftmost equation in Equation 10.15) and using $M(\omega) = \sigma_0/e_0$. Note that Q is not constant, i.e., Q varies with frequency. However, a network of SLS (or Maxwell or Voigt) solids can give a Q that is almost independent of frequency, which is generally the case for seismic body waves.

10.6 Generalized Standard Linear Solid

This is a model consisting of a network of N SLSs in parallel, as shown in Figure 10.6. The Nth element is described by the constants τ_{eN}, $\tau_{\sigma N}$, and M_{rN}. The strain e is the same across each element, but the stress is the sum of all the stresses σ_n, $n = 1, \ldots, N$ in the individual elements.

Letting $D \equiv \partial/\partial t$, we have, for one SLS, from (10.10c) or (10.15),

$$(1 + \tau_\sigma D)\sigma = M_r(1 + \tau_e D)e \implies \sigma = M_r(1 + \tau_\sigma D)^{-1}(1 + \tau_e D)e, \quad (10.18)$$

and the relaxation function $\mu(t)$ is given by $r(t)$ in (10.10f). Note also the use of differential operator algebra. For the generalized SLS, $\mu(t)$ (the stress for $e = H(t)$) is the sum of the μ_n for the individual elements, and the stress σ is the sum of the σ_n:

$$\mu(t) = \sum_{n=1}^{N} \mu_n(t) = \left[\sum_{n=1}^{N} M_{rn}\big(1 + \gamma_n \exp(-t/\tau_{\sigma n})\big) \right] H(t), \quad (10.19)$$

$$\text{where} \quad \gamma_n \equiv (\tau_{en}/\tau_{\sigma n}) - 1 \quad (\gamma_n > 0), \quad \text{and}$$

$$\sigma = \sum_{n=1}^{N} \sigma_n = \left[\sum_{n=1}^{N} M_{rn}(1 + \tau_{\sigma n}D)^{-1}(1 + \tau_{en}D) \right] e. \quad (10.20)$$

Performing a Fourier time transform on the preceding expression for σ by replacing D with $-i\omega$ gives $\bar{\sigma}(\omega) = M(\omega)\bar{e}$, where

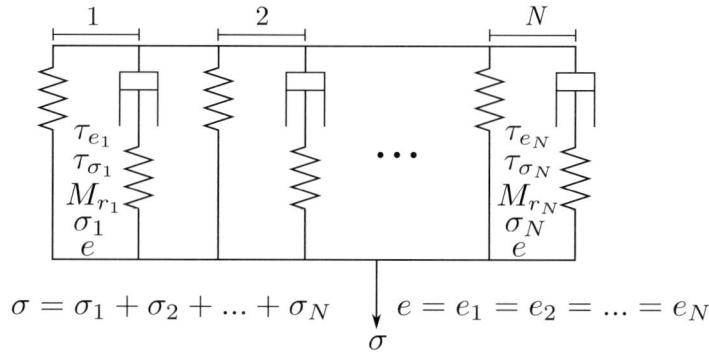

Figure 10.6 The spring-dashpot model for the generalized standard linear solid.

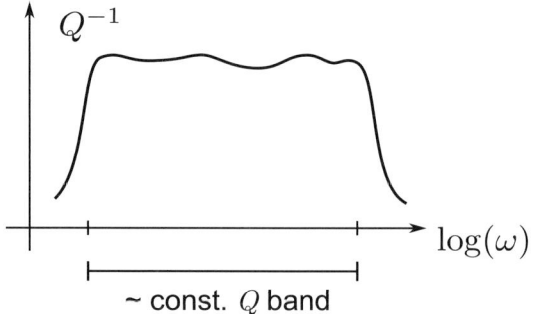

Figure 10.7 A schematic diagram of the loss factor Q^{-1} for the generalized standard linear solid for a specific choice of model parameters that makes Q nearly constant in a certain frequency band.

$$
M(\omega) = \sum_{n=1}^{N} M_{rn}\left(\frac{1 - i\omega\tau_{en}}{1 - i\omega\tau_{\sigma n}}\right)
$$

$$
= \left[\sum_{n=1}^{N} M_{rn}\left(\frac{1 + \omega^2\tau_{en}\tau_{\sigma n}}{1 + \omega^2\tau_{\sigma n}^2}\right)\right] + i\left[\omega\sum_{n=1}^{N} M_{rn}\left(\frac{\tau_{\sigma n} - \tau_{en}}{1 + \omega^2\tau_{\sigma n}^2}\right)\right]. \quad (10.21)
$$

The loss factor is then given by $Q^{-1}(\omega) = -\mathrm{Im}(M)/\mathrm{Re}(M)$. Choosing the parameters M_{rn}, τ_{en}, and $\tau_{\sigma n}$ judiciously can give a nearly constant-Q function, i.e., a function Q that is nearly independent of frequency within a certain frequency band – see Figure 10.7. Equation (10.21) also leads to the formula for $V(\omega)$ in (10.9).

In physical terms, the generalized standard linear solid represents a discrete distribution of different physical relaxation mechanisms in a rock as a superposition of relaxation functions. The different mechanisms combine to give a nearly constant Q. Kanamori and Anderson (1977) and Liu, Anderson, and Kanamori (1976) have shown that a continuous distribution of relaxation mechanisms can also lead to a nearly constant-Q function that looks like the one in Figure 10.7 but is smooth in the constant-Q frequency band.

The generalized Maxwell solid also consists of a network of elementary Maxwell solids (see Exercise 5), and as mentioned previously, it can yield a nearly constant Q through a judicious choice of network parameters. The same can be said for a generalized Voigt solid.

The generalized standard linear solid and the generalized Maxwell solid have been combined with the finite difference method to produce synthetic seismograms for nearly constant-Q media. See, e.g., Carcione (2007), Emmerich and Korn (1987), and Krebes and Quiroga-Goode (1994).

In general, the stress–strain relation for a spring-dashpot network model of a viscoelastic medium can be written as $P_\sigma(D)\sigma = P_e(D)e$, where $P_\sigma(D)$ and $P_e(D)$ are polynomials in D. See, e.g., Fung (1965).

10.7 The 1D Anelastic Equation of Motion

Substitution of the stress–strain relation $\sigma(t) = \mu(t) * \dot{e}(t)$, where the strain $e = (\partial u / \partial x)$, into the general 1D equation of motion (without the body force), $(\partial \sigma / \partial x) = \rho(\partial^2 u / \partial t^2)$, where $u = u(x, t)$, leads to the 1D anelastic equation of motion for a homogeneous isotropic viscoelastic medium. The derivation goes as follows:

$$\sigma(t) = \mu(t) * \dot{e}(t) = \mu(t) * \frac{\partial}{\partial t}\left(\frac{\partial u}{\partial x}\right) \quad \Rightarrow \quad \frac{\partial \sigma}{\partial x} = \mu(t) * \frac{\partial}{\partial t}\left(\frac{\partial^2 u}{\partial x^2}\right) = \rho \frac{\partial^2 u}{\partial t^2}. \tag{10.22a}$$

Taking the Fourier time transform of this equation of motion produces

$$-i\omega\bar{\mu}(\omega)\frac{\partial^2 \bar{u}(x, \omega)}{\partial x^2} = \rho(-i\omega)^2 \bar{u}(x, \omega) \quad \Rightarrow \quad \frac{\partial^2 \bar{u}}{\partial x^2} + \kappa^2 \bar{u} = 0, \tag{10.22b}$$

$$\text{where} \quad \kappa^2 = \frac{\rho\omega^2}{M(\omega)} = \frac{\omega^2}{v^2}, \quad M(\omega) = -i\omega\bar{\mu}(\omega), \quad v^2 = \frac{M(\omega)}{\rho}, \tag{10.23}$$

where κ and v are the complex angular wavenumber and complex wave speed, respectively. The rightmost equation in (10.22b) is the anelastic 1D Helmholtz equation, whose solution is $\bar{u} = \bar{s}(\omega)\exp(i\kappa x)$ (which is also suggested by the elastic-viscoelastic correspondence principle). Hence, taking the inverse Fourier transform,

$$u(x, t) = \frac{1}{2\pi}\int_{-\infty}^{\infty} \bar{s}(\omega)e^{i(\kappa x - \omega t)}\, d\omega = \frac{1}{2\pi}\int_{-\infty}^{\infty} \bar{s}(\omega)e^{-ax}e^{i(Kx - \omega t)}\, d\omega, \tag{10.24}$$

where $\kappa = K + ia$, is a solution, representing a superposition of plane waves. It can also be obtained by substituting a trial solution $u = \bar{s}(\omega)\exp[i(\kappa x - \omega t)]$ into the equation of motion in (10.22a) and doing the math – one finds that the trial wave u is a solution if $\kappa^2 = \rho\omega^2/M(\omega)$. Summing over all frequencies then gives (10.24). $K = \omega/V$, where $V = V(\omega)$ is the (real) phase velocity. Note that the leftmost equation of (10.24) has the same form as the elastic solution, except κ is complex and frequency dependent.

A formula for the absorption factor $a(\omega)$ as a function of Q can be derived as follows. First, it can be shown that

$$\frac{1}{Q} = -\frac{\text{Im}(M)}{\text{Re}(M)} = \frac{\text{Im}(\kappa^2)}{\text{Re}(\kappa^2)}, \tag{10.25}$$

which is left as an exercise for the reader. Hence, $\kappa = K + ia$, where $K = \omega/V$, implies

$$\kappa^2 = (K^2 - a^2) + i(2Ka) \quad \Rightarrow \quad (1/Q) = \left[2Ka/(K^2 - a^2)\right] \quad \Rightarrow \tag{10.26a}$$

$$a^2 + (2KQ)a - K^2 = 0 \quad \Rightarrow \quad a = \left(\frac{\omega}{2VQ}\right)\left[2Q^2\left(\sqrt{1 + Q^{-2}} - 1\right)\right], \tag{10.26b}$$

where the standard formula for the solution to a quadratic equation (the equation in a) has been applied, with the formula for a in (10.26b) being obtained after some algebraic manipulation. Note that this "a" is different from the one we have been using so far, i.e., $a = \omega/(2VQ)$. That is because the latter "a" is derived from the energy-based definition in (10.1), which is based on the assumption $Q \gg 1$. If this assumption is applied to the formula for a in (10.26b), then the term in square brackets becomes $[\cdots] \approx 2Q^2(1 + \frac{1}{2}Q^{-2} - 1) = 1$, resulting in the same formula for "a" as before. Actually, a more careful calculation of "a," based on (10.1), which does not assume $Q \gg 1$, also gives a in (10.26b) (this calculation involves deriving formulas for the mean strain energy density, etc., for a plane 1D anelastic wave). Consequently, in 1D, the energy- and modulus-based definitions of Q give the same "a" for all Q. In 3D, however, they give the same "a" for all Q only for "homogeneous" plane waves (defined later). In 3D, different formulas for "a" are obtained for general plane waves even for $Q \gg 1$.

Note that, even for Q as low as 5, the quantity $\left[2Q^2(\sqrt{1 + Q^{-2}} - 1)\right] = 0.9902 \approx 1$, so $a = (\omega/2VQ)$ is still accurate enough for most practical purposes even for such low Q.

A little algebra shows that $a(\omega)$ in (10.26b) can also be written as

$$a(\omega) = \frac{\omega}{V}\left[\frac{\sqrt{1 + Q^{-2}} - 1}{\sqrt{1 + Q^{-2}} + 1}\right]^{1/2}. \tag{10.27}$$

Useful formulas for κ^2, $M(\omega)$, and v^2 can also be derived in terms of Q and V.

10.8 The 3D Anelastic Equation of Motion

In this section, we use the summation convention. The general 3D equation of motion, without the body force, is

$$(\partial\sigma_{ij}/\partial x_j) = \rho(\partial^2 u_i/\partial t^2), \quad i = 1, 2, 3, \quad \text{where} \quad u_i = u_i(\mathbf{x}, t), \tag{10.28a}$$

(see Equation 2.14a). For a homogeneous isotropic viscoelastic medium, the stress–strain relation in Equation (2.30) can be generalized to

$$\sigma_{ij}(\mathbf{x}, t) = \delta_{ij}\lambda(t) * \dot{e}_{kk}(\mathbf{x}, t) + 2\mu(t) * \dot{e}_{ij}(\mathbf{x}, t), \tag{10.28b}$$

which when substituted into the 3D equation of motion in Equation (10.28a) leads to

$$(\lambda + \mu) * \partial_t\left[\nabla(\nabla \cdot \mathbf{u})\right] + \mu * \partial_t(\nabla^2\mathbf{u}) = \rho\partial_t^2\mathbf{u}, \tag{10.28c}$$

where "∂_t" means "$\partial/\partial t$." The derivation is similar to the one that produced Equation (2.41), and (10.28c) could have also been directly obtained from Equation (2.41). The Fourier time transform of (10.28c) is

$$(\Lambda + M)\nabla(\nabla \cdot \bar{\mathbf{u}}) + M\nabla^2\bar{\mathbf{u}} = -\rho\omega^2\bar{\mathbf{u}}, \quad \Lambda \equiv -i\omega \int_0^\infty \lambda(t)e^{i\omega t}\, dt, \quad (10.29)$$

where M is defined as before (see Equation 10.14b). This looks like the Fourier time transform of the equation of motion in the elastic case, except that Λ and M are complex and frequency dependent. Hence, the elastic-viscoelastic correspondence principle implies that the solutions (in the frequency domain) have the same form as the elastic ones, i.e.,

$$\bar{\mathbf{u}} = U\exp[i\kappa \cdot \mathbf{x}]\mathbf{d} = U\exp[i\kappa_n x_n]\mathbf{d}, \quad U = U(\omega), \quad (10.30)$$

where κ is complex. The polarization vector $\mathbf{d}$ is, in general, complex too. For example, for P waves, $\mathbf{d} = \kappa/|\kappa| = \kappa/\sqrt{\kappa \cdot \kappa}$, hence, if κ is complex, then so is $\mathbf{d}$. Also,

$$\kappa_\alpha^2 = \omega^2/\alpha^2, \quad \kappa_\beta^2 = \omega^2/\beta^2, \quad \alpha^2 = (\Lambda + 2M)/\rho, \quad \beta^2 = M/\rho, \quad (10.31)$$

where α and β are the complex velocities for P and S waves. Since κ is complex, we could write $\kappa = \mathbf{K} + i\mathbf{a}$, which is a generalization of $\kappa = K + ia$ in (10.3) or (10.24) to 3D. But instead, we use a notation that has often been used in the literature, i.e., $\kappa = \mathbf{P} + i\mathbf{A}$. Hence, we have

$$\kappa = \mathbf{P} + i\mathbf{A} \quad \Rightarrow \quad Ue^{i\kappa \cdot \mathbf{x}} = Ue^{-\mathbf{A}\cdot\mathbf{x}}e^{i\mathbf{P}\cdot\mathbf{x}}, \quad V = \omega\big/|\mathbf{P}| \quad (10.32)$$

(for either P or S waves) where $\mathbf{P}$ is the **real angular wavevector** and it can also be called the **propagation vector**, and where $\mathbf{A}$ is the **attenuation vector** or **absorption vector**, V is the (real) phase velocity, and $Ue^{-\mathbf{A}\cdot\mathbf{x}}$ is the amplitude of the plane wave. $\mathbf{P}$ is in the direction of wave propagation and $\mathbf{A}$ is in the direction of maximum attenuation. The plane wavefronts are described mathematically by the **planes of constant phase**, $\mathbf{P} \cdot \mathbf{x} = $ const. The amplitude of the plane wave is constant on the planes $\mathbf{A} \cdot \mathbf{x} = $ const. These are the **planes of constant amplitude**. In general, $\mathbf{P}$ and $\mathbf{A}$ are not parallel, which means that the planes of constant amplitude do not coincide with the wavefronts, i.e., the amplitude *varies* along a plane wavefront. We will refer to these types of waves as **general plane waves** — see Figure 10.8.

In Figure 10.8, θ is the **phase propagation angle** and γ is the **attenuation angle** or **absorption angle**. If $\gamma \neq 0$, the general plane wave is an **inhomogeneous** or **nonuniform** plane wave. If $\gamma = 0$, it is a **homogeneous** or **uniform** plane wave. The phase of the wave increases most rapidly in the $\mathbf{P}$ direction, and the wave amplitude attenuates most rapidly in the $\mathbf{A}$ direction. Also,

$$\kappa^2 = \kappa \cdot \kappa = (\mathbf{P} + i\mathbf{A}) \cdot (\mathbf{P} + i\mathbf{A}) = P^2 - A^2 + 2i\mathbf{P} \cdot \mathbf{A} = \omega^2/v^2 \quad (10.33a)$$

$$\text{with} \quad P \equiv |\mathbf{P}|, \quad A \equiv |\mathbf{A}|, \quad \mathbf{P} \cdot \mathbf{A} = PA\cos\gamma, \quad (10.33b)$$

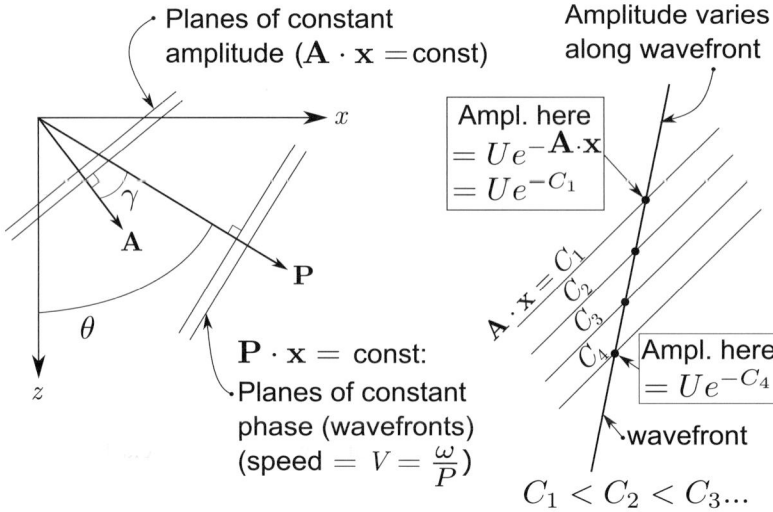

Figure 10.8 General plane waves. C_1, C_2, etc., are constants.

where v is the complex wave speed, and where A is not to be confused with the amplitude A in (10.1). For a perfectly elastic medium, $\text{Im}(\kappa^2) = 0$, (because v, and hence κ, is real), i.e., $\mathbf{P} \cdot \mathbf{A} = 0$. This implies that for a perfectly elastic medium, *either* $\{A = 0 \ (\Rightarrow \gamma$ is undefined)$\}$ *or* $\{A \neq 0$ and $\gamma = \pm 90°\}$ – these describe either regular body waves or evanescent waves, respectively, in elastic media. On the other hand, for a viscoelastic medium, $\text{Im}(\kappa^2) \neq 0$ (because v is complex), hence, γ can never be $\pm 90°$. Normally, $-90° < \gamma < +90°$ (otherwise, the wave amplitude would *increase* in the direction of phase propagation).

10.9 Reflection and Transmission of General Plane Waves

Just like in the perfectly elastic case, Snell's law in the viscoelastic case states that κ_x is continuous across the interface. This condition can be derived in the same way as in the perfectly elastic case – by application of the welded contact boundary conditions (continuity of stress and displacement). Equivalently, p is continuous across the interface, where $p = \kappa_x/\omega$ is the horizontal component of the slowness. Since $\kappa_x = P_x + iA_x$, Snell's law states that both P_x and A_x are continuous across the interface. From Figure 10.8, this means that *both* $(\sin\theta)/V$ *and* $A\sin(\theta - \gamma)$ must be continuous across the interface (in a perfectly elastic medium, only the first of these is continuous). Another way to look at it is as follows: define a complex angle χ such that $\kappa_x = (\omega/v)\sin\chi$, where v is the complex wave speed. Then, Snell's law states that $(\sin\chi)/v$ is continuous.

As an example, consider an elastic-viscoelastic interface, with a wave incident from the elastic side, as in Figure 10.9.

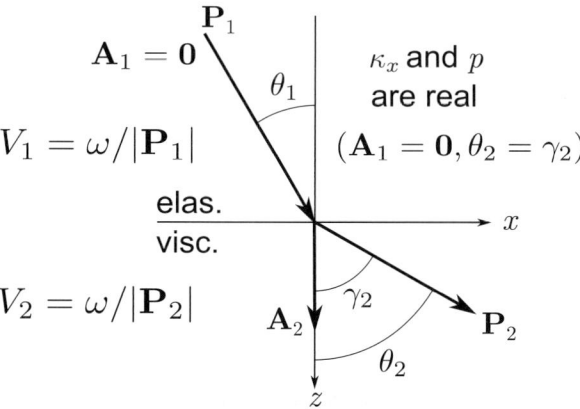

Figure 10.9 An example of general plane waves at an elastic-viscoelastic interface. The continuity of P_x implies $(\sin\theta_1)/V_1 = (\sin\theta_2)/V_2$, and the continuity of A_x implies $A_{x2} = A_{x1} = 0$, meaning $\theta_2 = \gamma_2$. In this example, κ_x and p are real because $\mathbf{A_1} = \mathbf{0}$.

We see in Figure 10.9 that $\mathbf{A}$ for the transmitted wave must always point in the z direction in this case, i.e., for all incidence angles θ_1. But in general (for a viscoelastic–viscoelastic interface), it can point in other directions. Note also that κ_x and p are real in this example, because $\theta_2 = \gamma_2$. But in general, they are complex.

As for plane wave reflection and transmission coefficients, the Zoeppritz equations have the same form as those for a perfectly elastic medium, except that p, the velocities α and β, $\sin\theta$, $\cos\theta$, etc., are replaced by their complex and frequency dependent analogs ($\sin\theta \rightarrow \sin\chi$, etc.). Hence, the same is true for the coefficients: in the formulas for the coefficients, one simply replaces the the velocities, moduli, angles, etc., with their complex frequency-dependent counterparts. For example, for normally incident P waves, we have $R = (Z_2 - Z_1)/(Z_2 + Z_1)$, where $Z_j = \rho\alpha_j, j = 1, 2$, where α_j is the complex velocity (see Equation 10.31) for the jth medium. For nonnormal incidence, the reflection and transmission coefficients can be computed in a straightforward way once θ_{inc} and γ_{inc}, the phase propagation angle and the absorption angle for the incident wave, are determined (which is not always straightforward). There are specific cases, though, in which anti-intuitive concepts and computational difficulties can arise (see, e.g., Krebes and Daley, 2007; Richards, 1984; Ruud, 2006; Sidler, Carcione and Holliger, 2008; Ursin, Carcione, Gei, 2017). Another interesting facet of the reflection and transmission of waves from an interface separating anelastic media is that reflections can occur even for an interface contrast in Q only (i.e., the densities and velocities are the same on both sides of the interface, with only Q differing). See Lines, Wong, Innanen, Vasheghani, Sondergeld, Treitel, and Ulrych (2014) for a discussion of physical modeling experiments that demonstrate this effect.

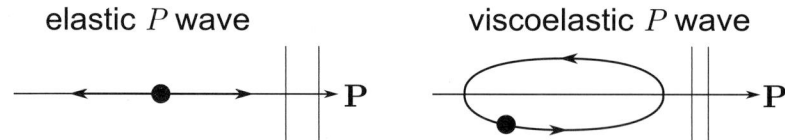

Figure 10.10 A schematic diagram of the particle motion for a *P* wave. The vertical lines are plane wavefronts.

The particle motion for a plane wave in a perfectly elastic medium is linear, i.e., particles move back and forth along straight lines – for a *P* wave, they move parallel to the direction of wave propagation, and for an *S* wave, perpendicular – see Figure 10.10. But in a viscoelastic medium, it can be shown that the particle motion is no longer perfectly linear, but elliptical (Borcherdt, 1973; Buchen, 1971).

Loss factors for *P* and *S* waves can be defined as follows:

$$Q_\alpha^{-1} = -\frac{\text{Im}[\Lambda + 2M]}{\text{Re}[\Lambda + 2M]} \quad \text{and} \quad Q_\beta^{-1} = -\frac{\text{Im}[M]}{\text{Re}[M]}. \tag{10.34}$$

Normally, $Q_\alpha > Q_\beta$. If the amplitude-based definition of Q in (10.1), i.e., $1/Q = -\Delta A/\pi A$ (where A is not to be confused with the magnitude of the absorption vector **A**), or the energy-based definition $1/Q = -\Delta E/2\pi E$ (see the text that follows Equation 7.7a), is used to derive Q-formulas for any Q (not just $Q \gg 1$), then one obtains the preceding formulas only for homogeneous waves – for inhomogeneous waves, the amplitude- or energy-based definitions of Q give a result that is slightly different from the modulus-based definition of Q in (10.25) (see Borcherdt, 1973). In spite of this, the formulas in (10.34) are often used in the literature as definitions of Q. However, these definitions do not work as well in a medium that is both anelastic and anisotropic – in that case, the energy-based definition works better (Červený and Pšenčík, 2008; Krebes and Le, 1994).

From (10.34), we can write

$$v^2 = \frac{N}{\rho} = \frac{\text{Re}[N]}{\rho}\left[1 - \frac{i}{Q}\right], \quad \begin{cases} N = \Lambda + 2M, \ v = \alpha, \ Q = Q_\alpha & \text{for } P \text{ waves} \\ N = M, \ v = \beta, \ Q = Q_\beta & \text{for } S \text{ waves} \end{cases} \tag{10.35}$$

$$\Rightarrow \ \kappa^2 = \frac{\omega^2}{v^2} = P^2 - A^2 + 2iPA\cos\gamma = \frac{\rho\omega^2}{\text{Re}[N]\,(1 - i/Q)} = \frac{\rho\omega^2(1 + i/Q)}{\text{Re}[N]\,(1 + Q^{-2})}. \tag{10.36}$$

Solving for *P* and *A* gives

$$P^2 = \frac{\rho\omega^2}{\text{Re}[N]}\left[\frac{\sqrt{1 + (Q\cos\gamma)^{-2}} + 1}{2(1 + Q^{-2})}\right], \quad A^2 = \frac{\rho\omega^2}{\text{Re}[N]}\left[\frac{\sqrt{1 + (Q\cos\gamma)^{-2}} - 1}{2(1 + Q^{-2})}\right]. \tag{10.37}$$

The (real) phase velocity of general plane waves is $V = \omega/P$. For 1D waves, $\gamma = 0$, and it is not difficult to show that V and A reduce to V and a in (10.5) and (10.3), respectively, for $Q \gg 1$.

A formula for V_H, the phase velocity of homogeneous plane waves, can be obtained from $V_H = \omega/P_H$ by putting $\gamma = 0$ into P^2:

$$V_H{}^2 = \frac{\text{Re}[N]}{\rho}\left[\frac{2(1+Q^{-2})}{1+\sqrt{1+Q^{-2}}}\right]. \tag{10.38}$$

This formula can then be used to eliminate $\text{Re}[N]$ in (10.36) and (10.37) to obtain useful formulas for P, A, V, v, κ, etc., in terms of V_H and Q only. For example,

$$v^2 = V_H{}^2\left[\frac{1+\sqrt{1+Q^{-2}}}{2(1+Q^{-2})}\right]\left[1-\frac{i}{Q}\right], \quad V^2 = V_H{}^2\left[\frac{1+\sqrt{1+Q^{-2}}}{1+\sqrt{1+(Q\cos\gamma)^{-2}}}\right]. \tag{10.39a}$$

Note that the phase speed V for an inhomogeneous general plane wave decreases as γ, the "degree of inhomogeneity," increases.

Note also that the formula for v does not involve γ. In practice, it is usually easier to work with the complex parameters, i.e., the complex velocity v and the complex horizontal slowness $p = \kappa_x/\omega$ (or the complex angle χ), rather than P, A, θ, and γ. This is especially true for numerical computations of ray travel times and amplitudes in viscoelastic media.

In addition, note that for large Q, the formulas for $V_H{}^2$ and v^2 can be approximated as

$$V_H{}^2 \approx \frac{\text{Re}[N]}{\rho} \quad \text{and} \quad v^2 \approx V_H{}^2\left[1-\frac{i}{Q}\right], \quad Q \gg 1, \tag{10.39b}$$

which resemble the 1D equations in (10.4) and (10.5).

10.10 An Alternative Form of the Stress–Strain Relation

Note from (10.14a) and (10.14b) that

$$\sigma(t) = \mu(t) * \dot{e}(t) \quad \Rightarrow \quad \bar{\sigma}(\omega) = M(\omega)\bar{e}(\omega) = -i\omega\bar{\mu}(\omega)\,\bar{e}(\omega)$$
$$\Rightarrow \quad \sigma(t) = \dot{\mu}(t) * e(t) \equiv m(t) * e(t), \tag{10.40}$$

where (1.150) has been used. In words, the overhead dot, i.e., the time derivative, can be switched from the e to the μ in the stress–strain relation. In addition, with $m(t) \equiv \dot{\mu} = d\mu/dt$, we can write the stress–strain relation as $\sigma = m * e$, which is a more familiar form of the convolution (in that e, rather than $\dot{e}$, appears). Also, the complex modulus $M(\omega)$ is just the Fourier transform of $m(t)$, i.e., $M(\omega) = \bar{m}(\omega)$. The same approach can be used in 3D as well, e.g., in the analysis of Section 10.8.

The stress–strain relation in the form $\sigma(t) = m(t) * e(t)$ is sometimes used in the literature.

10.11 Measuring Absorption

Q is often estimated from seismic data using some version of the method of **spectral ratios**. This involves determining the frequency spectrum of a waveform at different receiver locations. The ratios of these spectra can be used to estimate Q. As a simple example, consider a wave traveling in a 1D medium from the source at $x = 0$ to the receiver at x. For a perfectly elastic medium, the spectrum of the source pulse is $\bar{s}(\omega)$, whose inverse Fourier transform is $s(t)$ (the source pulse), and the spectrum of the waveform arriving at the receiver is $\bar{s}(\omega) \exp(i\omega x/V)$, whose inverse Fourier transform is $s(t - x/V)$. In other words, the waveform has not changed. However, for an absorbing medium, the spectrum at the receiver is $\bar{s}(\omega) \exp[-a(\omega)x] \exp(i\omega x/V)$, where $a = \omega/2VQ$, resulting in a different waveform. The amplitude spectrum is simply the magnitude of the spectrum, i.e., $\mathcal{A}(\omega) = |\bar{s}(\omega)| \exp[-a(\omega)x]$. A formula for Q can be easily derived in terms of the ratio of the amplitude spectra at two receiver points:

$$\mathcal{A}_j(\omega) = |\bar{s}(\omega)| \exp\left[-a(\omega)x_j\right], \quad j = 1, 2 \quad \Rightarrow$$

$$Q(\omega) = \frac{\omega(x_2 - x_1)}{2V(\omega) \ln\left[\mathcal{A}_1(\omega)/\mathcal{A}_2(\omega)\right]}. \tag{10.41}$$

Often, V is assumed independent of ω so that an estimate of Q can be obtained more easily. In the 3D case involving heterogeneous media, other factors need to be included in the calculation, such as the product of reflection and transmission coefficients, and the geometrical spreading factor, and Q is often determined in a sequential fashion, i.e., layer by layer.

When there are layers, the absorption term in the spectrum, $\exp(-ax)$, is determined by summing up the quantities $x/2VQ$ for each ray segment, where x is the length of the ray segment. If the sum is denoted by b, then the absorption term is $\exp(-b\omega)$. When velocity varies smoothly with position, then an integration over x, rather than a sum, must be performed to determine b. Often, the quantity t^* is used, where $t^* \equiv x/VQ$. In terms of t^*, we have $\exp(-ax) = \exp(-\pi f t^*)$, where f is the actual frequency. For a curved ray path in a smoothly varying medium, $t^* = \int [V(s)Q(s)]^{-1} ds$, where s is the arc length along the ray path and the limits of the integral are the s-values of the source and receiver points.

It should be kept in mind that the attenuation of waveforms is caused not only by absorption, but also by scattering due to small heterogeneities in a medium, and by geometrical spreading, and reflection and transmission losses. Geometrical spreading is the dominant effect in the attenuation of amplitudes. Sometimes it is difficult to say whether attenuation in a given data set is due to intrinsic absorption or to scattering. One possible way to distinguish the two is to use the fact that scattering produces a **coda**, which is a long noisy "tail" attached to the seismogram, after the *P* and *S* arrivals.

10.12 Exactly Constant *Q*

We have seen above that *Q* is nearly constant, i.e., nearly independent of frequency for seismic body waves. In theory, it is possible for *Q* to be exactly constant, and still have causality preserved, as long as there is velocity dispersion. In the standard nearly constant-*Q* theory, based on a spring-dashpot network, *Q* is nearly constant over a certain frequency band only. This means that $Q^{-1} \to 0$ as $\omega \to \infty$, i.e., high frequencies behave elastically. High frequencies arrive first at the receiver (because *V* is large) – in practice, one gets an elastic "precursor" on an arriving waveform. This is known as the "pedestal effect" (after Strick, 1970). In constant-*Q* theory, there is no pedestal effect, because there is no high-frequency elastic response (see Figure 10.11). This is physically unrealistic, but it can be argued that in practice, this is a small, unimportant effect – it does not significantly affect computed waveforms.

The formulas for *M*, *V*, etc., in constant-*Q* theory are (Kjartansson, 1979):

$$M(\omega) = M_0 \left[\frac{i\omega}{\omega_0}\right]^{2\gamma}, \quad \gamma \equiv \frac{1}{\pi}\tan^{-1}\left[\frac{1}{Q}\right], \quad \gamma \approx \frac{1}{\pi Q} \ (Q \gg 1), \quad V(\omega) = V_0 \left|\frac{\omega}{\omega_0}\right|^{\gamma},$$
(10.42)

$$V_0 = \frac{\sqrt{M_0/\rho}}{\cos(\pi\gamma/2)}, \quad u = e^{-ax}e^{-i\omega[t-x/V]}, \quad a(\omega) = \tan\left[\frac{\pi\gamma}{2}\right]\mathrm{sgn}(\omega)\frac{\omega}{V}. \quad (10.43)$$

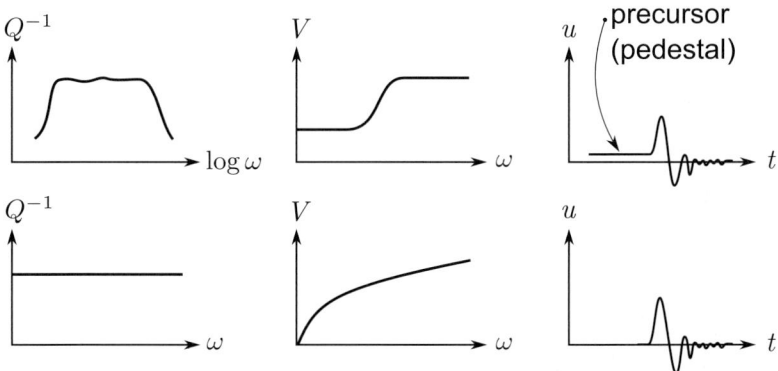

Figure 10.11 Schematic diagrams comparing the standard (upper row) and constant-*Q* (lower row) theories.

Exercises

1. Derive formulas for M_r, τ_σ, and τ_e as functions of μ_1, μ_2, and η for standard linear solids I and II. Is $\tau_\sigma < \tau_e$? Also calculate M_u and τ_e/τ_σ in both cases.

2. Derive the formulas for the creep and relaxation functions for the elastic element (a spring), the viscous element (a dashpot), the Maxwell solid, and the Voigt solid. Sketch graphs of the functions and discuss them, explaining how they make physical sense.

3. Derive the formulas for the creep and relaxation functions for the standard linear solid. Sketch graphs of the functions, and discuss them, explaining how they make physical sense.

4. Prove that the Fourier time transform of $\mu(t) * \dot{e}(t)$ is $M(\omega)\bar{e}(\omega)$, where

$$M(\omega) = -i\omega \int_0^\infty \mu(t)e^{i\omega t}\, dt \qquad \text{and} \qquad \bar{e}(\omega) = \int_{-\infty}^\infty e(t)e^{i\omega t}\, dt,$$

by doing the convolution integral.

5. Derive $M(\omega)$ for a generalized Maxwell solid (a network of N Maxwell elements in parallel with themselves and with a spring element).

6. Derive formulas for the complex modulus $M(\omega)$ and the loss factor $Q^{-1}(\omega)$ for the Maxwell and Voigt models of viscoelasticity. Sketch graphs of $Q^{-1}(\omega)$ and $|M(\omega)|$ vs. ω (superimpose the two graphs for $|M(\omega)|$ on one graph). Briefly discuss their physical meanings. Are they realistic (for rocks)? Also, explain their variation with ω in terms of the behavior of the springs and dashpots.

7. By substituting the time-harmonic expressions $\sigma(t) = \sigma_0 e^{-i\omega t}$ and $e(t) = e_0 e^{-i\omega t}$ into the 1D anelastic stress–strain relation

$$\sigma(t) = \int_{-\infty}^t \mu(t - \tau)\dot{e}(\tau)\, d\tau,$$

show that $\sigma(t) = M(\omega)e(t)$, where $M(\omega)$ is the complex modulus. Note that since $M(\omega)$ is complex, this implies that in an anelastic medium, the stress is out of phase with the strain for harmonic waves.

8. Derive $\kappa^2 = \rho\omega^2/M(\omega)$ by substituting a plane wave solution for u into the 1D anelastic equation of motion in the time domain (given in the following equation) and doing the integral. The 1D anelastic EOM is

$$\mu(t) * d\left(\frac{\partial^2 u}{\partial x^2}\right) = \rho\frac{\partial^2 u}{\partial t^2}, \quad \text{where} \quad f(t) * dg(t) \equiv \int_{-\infty}^t f(t-\tau)\frac{\partial g(\tau)}{\partial \tau}\, d\tau.$$

9. Show that the anelastic loss factor Q^{-1} ($= -\mathrm{Im}(M)/\mathrm{Re}(M)$) is also given by $Q^{-1} = \mathrm{Im}(\kappa^2)/\mathrm{Re}(\kappa^2)$, where $\kappa = \omega/v$, with v being the complex wave speed.

10. Derive (10.37). Express κ^2 in (10.36) in terms of ω, V_H, and Q.

11. Derive expressions for (a) the mean kinetic energy density, and (b) the mean energy flux, for a plane harmonic P, SV, and SH wave in an infinite homogeneous isotropic viscoelastic medium.

Answers to Selected Exercises

For exercises requiring graphs and/or discussions along with the formulas and/or numerical answers, only the latter are generally given here.

Chapter 1.

1. $\mathbf{A} \cdot \mathbf{B} = 4$. $\mathbf{A} \times \mathbf{B} = (23, 2, -9)$. Angle $= 80.83°$.

2. $\nabla \psi = (y^2 z - 2x, 2xyz, xy^2)$. $\nabla \cdot \mathbf{A} = 3yz^2 + 12xy^2 - x^2 y$. $\nabla \times \mathbf{A} = (-x^2 z, 8xyz, 4y^3 - 3xz^2)$.

5. $\int_V \nabla \cdot \mathbf{A} \, dV = \int_S \mathbf{A} \cdot \mathbf{n} \, dS = -1/4$.

9. Approximate surface area of Saskatchewan $= 632{,}256$ km^2.

11. Mass of dome $= 2\pi R^2 (\rho_0 + a)$.

12. **(d)** Oblate spheroid: volume $= 1.083 \times 10^{12}$ km^3, surface area $= 5.101 \times 10^8$ km^2. Sphere: volume $= 1.087 \times 10^{12}$ km^3, surface area $= 5.112 \times 10^8$ km^2.

22. Rotate the xz frame clockwise until the z' axis lines up with $\mathbf{u}$. The corresponding angle of rotation is $\delta = 26.565°$. Row 1, 2, and 3 of the rotation matrix are $(c, 0, s)$, $(0, 1, 0)$, and $(-s, 0, c)$, respectively, where $c = \cos \delta = 0.8944$ and $s = \sin \delta = 0.4472$. Amplitude $|\mathbf{u}| = 2.236$.

23. **(c)** $\bar{g}(\omega) = 2i\omega/(a^2 + \omega^2)$. Amplitude spectrum $= |\bar{g}(\omega)| = 2|\omega|/(a^2 + \omega^2)$. Phase spectrum $= \psi(\omega) = \{+\pi/2$ for $\omega > 0, -\pi/2$ for $\omega < 0,$ undefined for $\omega = 0\}$.

29. **(a)** $|\bar{g}(\omega_0/2)| = 4\sqrt{2}/3\omega_0^2$. $\psi(\omega_0/2) = -135°$.

Chapter 2.

4. Dilatation $= -6.174 \times 10^{-4}$. There is a 0.1 mm decrease in the radius. The relative pressure difference between the top and bottom is $\approx 10^{-4}$ and so can be neglected.

6. The summation convention and the "comma" derivative convention is used in the following answers.

(a) $\mathcal{W} = \frac{1}{2}\lambda\mathcal{D}^2 + \mu e_{ij}e_{ij}$.

(b) $\mathcal{W} = \frac{1}{2}\lambda(\nabla\cdot\mathbf{u})^2 + \frac{1}{2}\mu(u_{i,j}u_{i,j} + u_{i,j}u_{j,i})$.

(c) $\mathcal{W} = \frac{1}{2}k_B\mathcal{D}^2 + \mu e'_{ij}e'_{ij}$.

(d) $\mathcal{W} = \frac{1}{2}k_B\mathcal{D}^2 = \frac{1}{2}\lambda\mathcal{D}^2$.

(e) $\mathcal{W} = \frac{1}{2}(\lambda + 2\mu)(\partial u/\partial x)^2$.

11. Assume A is real.

(a) *SV*.

(b) $\sigma_{13} = \sigma_{31}(= \sigma_{xz} = \sigma_{zx}) = -\rho\omega\beta A \sin\{\omega[(z/\beta) - t]\}$ with all other $\sigma_{ij} = 0$.
   $\mathbf{I} = \rho\beta\omega^2 A^2 \sin^2\{\omega[(z/\beta) - t]\}\mathbf{e}_z$.
   $\mathcal{W} = \mathcal{K} = \frac{1}{2}\rho\omega^2 A^2 \sin^2\{\omega[(z/\beta) - t]\}$.

(c) The stress σ_{xz} and the displacement u_x are $90°$ out of phase with each other. Also, the magnitude of the stress varies linearly with frequency, whereas the displacement does not. The magnitude of the stress also varies linearly with impedance $\rho\beta$.

(d) $\partial(K + W)/\partial t = -\nabla\cdot\mathbf{I} = -\rho\omega^3 A^2 \sin\{2\omega[(z/\beta) - t]\}$.

(e) $\langle\sigma_{ij}\rangle = 0$, all i,j . $\langle\mathbf{I}\rangle = \frac{1}{2}\rho\beta\omega^2 A^2 \mathbf{e}_z$. $\langle\mathcal{W}\rangle = \langle\mathcal{K}\rangle = \frac{1}{4}\rho\omega^2 A^2$.

(f) 2.37×10^{10} ergs. 39.5 seconds.

14. $\langle\mathcal{B}\rangle = \rho\omega^2|U|^2$.

21. (c) $a \approx 10$ m.

22. (a) $\tau_1 = 3 + \sqrt{10} = 6.1623$, $\tau_2 = 3 - \sqrt{10} = -0.1623$, $\tau_3 = 0$. $\mathbf{n}_1 = [0.81124, 0.58471, 0]^T$, $\mathbf{n}_2 = [-0.58471, 0.81124, 0]^T$, $\mathbf{n}_3 = [0, 0, 1]^T$.

(b) Rows 1, 2, and 3 of the rotation matrix are $(c, s, 0)$, $(-s, c, 0)$ and $(0, 0, 1)$, respectively, where $c = \cos\phi = 0.81124$ and $s = \sin\phi = 0.58471$, with $\phi = 35.783°$.

25. (a) $u_y = 0$, $u_z = 0$. $4\pi\rho u_x = \{[2At/r^3] + [A\delta(t - r/\alpha)/\alpha^2 r]$, for $(r/\alpha) \leq t \leq (r/\beta)$; 0 for $t < (r/\alpha)$ and $t > (r/\beta)\}$.

Chapter 3.

3. $R = (\rho_1\beta_1 - \rho_2\beta_2)/(\rho_1\beta_1 + \rho_2\beta_2)$, $T = 2\rho_1\beta_1/(\rho_1\beta_1 + \rho_2\beta_2)$.

5. $\phi_1 = 34.70°$, $t_0 = 0.183$ s, $h = 108.33$ m.

9. $\phi_1 = 20°$: amplitude $= 0.119$, phase $= \pm180°$. $\phi_1 = 30°$: amplitude $= 1.0$, phase $= -139.89°$.

13. (a) Let $c_j \equiv \cos\theta_j$, $j = 1, 2$. Then $R = (Z_2c_1 - Z_1c_2)/(Z_2c_1 + Z_1c_2)$, $T = 2Z_1c_1/(Z_2c_1 + Z_1c_2)$.

(b) Same result as in part (a).

(c) $\bar{R} = R$, $\bar{T} = (Z_2/Z_1)T$, where R and T are given in (a).

(d) Opposite.

14. (a) $R_{PP} = \cos(\theta + \phi)/\cos(\theta - \phi)$, $R_{PS} = -(\sin 2\theta)/\cos(\theta - \phi)$.
    (b) u_z: polarity reversal for $\theta \leq 63.43°$. u_x: polarity reversal for $\theta \geq 63.43°$.
    (c) The *P-SV* conversion angle is $\theta_0 = 63.43°$.

Chapter 4.

1. $z_0 = 18.0$ m, which is 18.1% of the Rayleigh wavelength.
3. $c_g = \beta\sqrt{1 - [\beta(n + \frac{1}{2})\pi/\omega h]^2}$.
5. $c(f) = 60\pi f/(0.63f - 1.65)$. $c_g = 299.20$ m/s (c_g is independent of f).

Chapter 5.

2. $p(x) = 1/[\,v_0\sqrt{1 + (ax/2v_0)^2}\,]$. $t(x) = (2/a)\sinh^{-1}(ax/2v_0)$.
3. (a) $v_0 = 0.909$ km/s.
    (b) The integrand closely approximates an ellipse, so the formula for the area of an ellipse can be used. $z_1 = 210$ km, $v(z_1) = 1.25$ km/s.
    (c) From Simpson's rule, the exact result is $z_1 = 212$ km, $v(z_1) = 1.250$ km/s. The result in part (b) agrees well with this.

Chapter 6.

2. $\mathcal{H}[f(t)] = (1/\pi)\ln|(t - T)/(t + T)|$.

Chapter 7.

1. (a) Relative amplitude $= 4.86 \times 10^{-4}$ m^{-1}. Phase $= -180°$.
    (b) The relative amplitude is reduced by a factor of 0.66 to 3.19×10^{-4} m^{-1}.
6. (a) $x = 148.40$ m. $t = 0.381$ s. Relative amplitude $= 2.119$ km^{-1}. Phase $= -150.9°$.
    (b) Relative amplitude $= 0.672$ km^{-1}. Phase $= -150.9°$. In general, reflection and transmission coefficients also depend on frequency and Q, so a more careful treatment would result in slightly different numbers.

Chapter 8.

2. $(\partial^3 u_a/\partial z\partial t^2) - (V^2/4)(\partial^3 u_a/\partial z\partial x^2) + (V/2)(\partial^3 u_a/\partial t\partial x^2) = 0$.
4. True depth $= 1.92$ km. Migration moves the point 2.56 km horizontally to the right.

Chapter 9.

3. $V^2 = (\beta_h^4 \sin^2\theta + \beta_v^4 \cos^2\theta)/(\beta_h^2 \sin^2\theta + \beta_v^2 \cos^2\theta)$.
7. Yes.
8. Yes.

10. $c_{11} = c_{1111} = 12.8$ GPa, $c_{13} = c_{1133} = 6.05$ GPa, $c_{33} = c_{3333} = 8$ GPa, $c_{44} = c_{2323} = 2$ GPa, $c_{66} = c_{1212} = 3.2$ GPa. $\alpha_{90} = 2530$ m/s, $\beta_{90} - 1{,}000$ m/s. *SH*: $\beta_h = 1{,}265$ m/s, $\beta_v = 1{,}000$ m/s.

Chapter 10.

1. SLS I: $M_r = \mu_1\mu_2/(\mu_1+\mu_2)$, $\tau_\sigma = \eta/(\mu_1+\mu_2)$, $\tau_e = \eta/\mu_1$. Clearly, $\tau_\sigma < \tau_e$. $M_u = M_r(\tau_e/\tau_\sigma) = \mu_2 \cdot (\tau_e/\tau_\sigma) = 1 + (\mu_2/\mu_1)$.

6. Voigt solid: $M(\omega) = \mu - i\omega\eta$, $Q^{-1} = \eta\omega/\mu$. Not realistic for rocks.

References

Achenbach, J.D. (1973). *Wave Propagation in Elastic Solids*. New York: North-Holland Publishing Co.

Aki, K. and Richards, P.G. (1980). *Quantitative Seismology: Theory and Methods* (vols. I and II). San Francisco: W.H. Freeman and Co.

Aki, K. and Richards, P.G. (2002). *Quantitative Seismology*, 2nd edn. Sausalito: University Science Books.

Arfken, G. (1985). *Mathematical Methods for Physicists*, 3rd edn. New York: Academic Press.

Båth, M. and Berkhout, A.J. (1984). *Mathematical Aspects of Seismology*. London: Geophysical Press.

Ben-Menahem, A. and Singh, S.J. (2000). *Seismic Waves and Sources*, 2nd edn. (corrected). New York: Dover.

Bland, D.R. (1960). *The Theory of Linear Viscoelasticity*. New York: Pergamon Press.

Brekhovskikh, L.M. (1980). *Waves in Layered Media*, 2nd edn. New York: Academic Press.

Brown, R.J., Stewart, R.R., and Lawton, D.C. (2002). A proposed polarity standard for multicomponent seismic data. *Geophysics*, **67**, 1028–1037.

Borcherdt, R.D. (1973). Energy and plane waves in viscoelastic media. *J. Geophys. Res.*, **78**, 2442–2453.

Borcherdt, R.D. (1977). Reflection and refraction of type-II *S* waves in elastic and anelastic media. *Bull. Seism. Soc. Am.*, **67**, 43–67.

Borcherdt, R.D. (2009). *Viscoelastic Waves in Layered Media*. Cambridge: Cambridge University Press.

Bouzidi, Y. and Schmitt, D.R. (2012). Incidence-angle-dependent acoustic reflections from liquid-saturated porous solids. *Geophys. J. Int.*, **191**, 1427–1440.

Buchen, P.W. (1971). Plane waves in linear viscoelastic media. *Geophys. J. R. Astr. Soc.*, **23**, 531–542.

Bullen, K.E. and Bolt, B.A. (1985). *An Introduction to the Theory of Seismology*, 4th edn. Cambridge: Cambridge University Press.

Carcione, J.M. (2001). *Wave Fields in Real Media: Wave Propagation in Anisotropic, Anelastic and Porous Media*. Amsterdam: Pergamon Press, Elsevier Science Ltd.

Carcione, J.M. (2007). *Wave Fields in Real Media: Wave Propagation in Anisotropic, Anelastic, Porous and Electromagnetic Media*. Amsterdam: Pergamon Press, Elsevier Science Ltd.

Červený, V., 2001. *Seismic Ray Theory*. Cambridge: Cambridge University Press.

Červený, V. and Pšenčík, I. (2008). Quality factor Q in dissipative anisotropic media. *Geophysics*, **73**, T63–T75.

Červený, V. and Ravindra, R. (1971). *Theory of Seismic Head Waves*. Toronto: University of Toronto Press.

Červený, V., Molotkov, I.A., and Pšenčík, I. (1977). *Ray Method in Seismology*. Prague: Charles University.

Chabot, L. (2000). Supplemental visual aids for the Seismic Theory and Methods course, Department of Geoscience, University of Calgary. Unpublished.

Chabot, L., Henley, D.C., Brown, R.J., and Bancroft, J.C. (2001). Single-well seismic imaging using the full waveform of an acoustic sonic. *CREWES Research Report*, **13**, 583–600. Department of Geoscience, University of Calgary.

Chaisri, S. and Krebes, E.S. (2000). Exact and approximate formulas for *P-SV* reflection and transmission coefficients for a non-welded contact interface. *J. Geophys. Res.*, **105**, 28045–28054.

Chapman, C.H. (2004). *Fundamentals of Seismic Wave Propagation*. Cambridge: Cambridge University Press.

Christensen, R.M. (1971). *Theory of Viscoelasticity: An Introduction*. New York: Academic Press.

Chun, J.H. and Jacewitz, C.A. (1981). Fundamentals of frequency domain migration. *Geophysics*, **46**, 717–733.

Claerbout, J.F. (1976). *Fundamentals of Geophysical Data Processing*. New York: McGraw-Hill.

Claerbout, J.F. (1985). *Imaging the Earth's Interior*. Oxford: Blackwell Scientific Publications.

Cui, X., Lines, L., Krebes, E.S., and Peng, S. (2018). *Seismic Forward Modeling of Fractures and Fractured Medium Inversion*. Singapore: Springer Nature.

Dahlen, F.A. and Tromp, J. (1998). *Theoretical Global Seismology*. Princeton: Princeton University Press.

Dai, N., Kanasewich, E., and Vafidis, A. (1993). Finite-difference modeling of viscoelastic waves. *Presented at the national convention of the Canadian Society of Exploration Geophysicists.*

Daley, P.F. and Hron, F. (1977). Reflection and transmission coefficients for transversely isotropic media. *Bull. Seism. Soc. Am.*, **67**, 661–675.

Daley, P.F. and Hron, F. (1979). Reflection and transmission coefficients for seismic waves in ellipsoidally anisotropic media. *Geophysics*, **44**, 27–38.

Dettmer, J., Dosso, S.E., and Holland, C.W. (2007). Full wave-field reflection coefficient inversion. *J. Acoust. Soc. Am.*, **122**, 3327–3337.

Diebold, J.B. and Stoffa, P.L. (1981). The traveltime equation, tau-p mapping and inversion of common midpoint data. *Geophysics*, **46**, 238–254.

Dix, C.H. (1955). Seismic velocities from surface measurements. *Geophysics*, **20**, 68–86.

Dobrin, M.B. (1976). *Introduction to Geophysical Prospecting*, 3rd edn. New York: McGraw-Hill.

Dobrin, M.B., Lawrence, P.L., and Sengbush, R.L. (1954). Surface and near-surface waves in the Delaware Basin. *Geophysics*, **19**, 695–715.

Dobrin, M.B., Simon, R.F., and Lawrence, P.L. (1951). Rayleigh waves from small explosions. *Trans. Am. Geophys. Un.*, **32**, 822–832.

Eaton, D.W.S. (1993). Finite-difference traveltime calculation for anisotropic media. *Geophys. J. Int.*, **114**, 273–280.

Elmore, W.C. and Heald, M.A. (1969). *Physics of Waves*. New York: McGraw-Hill.

Emmerich, H. and Korn, M. (1987). Incorporation of attenuation into time-domain computation of seismic wavefields. *Geophysics*, **52**, 1252–1264.

Ewing, W.M., Jardetzky, W.S., and Press, F. (1957). *Elastic Waves in Layered Media*. New York: McGraw-Hill.

Feng, R. and McEvilly, T.V. (1983). Interpretation of seismic reflection profiling data for the structure of the San Andreas Fault zone. *Bull. Seism. Soc. Am.*, **73**, 1701–1720.

French, A.P. (1971). *Vibrations and Waves*. New York: W.W. Norton.

French, W.S. (1975). Computer migration of oblique seismic reflection profiles. *Geophysics*, **40**, 961–980.

Fung, Y.C. (1965). *Foundations of Solid Mechanics*. Englewood Cliffs: Prentice Hall.

Gassmann, F. (1964). Introduction to seismic travel time methods in anisotropic media. *Pure and Appl. Geophys.*, **58**, 63–112.

Gazdag, J. and Sguazzero, P. (1984). Migration of seismic data by phase shift plus interpolation. *Geophysics*, **49**, 124–131.

Graebner, M. (1992). Plane-wave reflection and transmission coefficients for a transversely isotropic solid. *Geophysics*, **57**, 1512–1519.

Grant, F.S. and West, G.F. (1965). *Interpretation Theory in Applied Geophysics*. New York: McGraw-Hill.

Graul, J.M. and Hilterman, F.J. (1979). Unpublished notes on seismic methods.

Guevara, S.E. (2001). Analysis and filtering of near-surface effects in land multicomponent seismic data. M.Sc. thesis, Department of Geoscience, University of Calgary.

Hagedoorn, J.G. (1954). A process of seismic reflection interpretation. *Geophys. Prosp.*, **2**, 85–127.

Hatton, L., Larner, K.L., and Gibson, B.S. (1981). Migration of seismic data from inhomogeneous media. *Geophysics*, **46**, 751–767.

Hilterman, F.J. (1970). Three-dimensional seismic modeling. *Geophysics*, **35**, 1020–1037.

Hilterman, F.J. (1975). Amplitudes of seismic waves – a quick look. *Geophysics*, **40**, 745–762.

Hubral, P. (1977). Time migration – some ray theoretical aspects. *Geophys. Prosp.*, **25**, 738–745.

Hudson, J.A. (1980). *The Excitation and Propagation of Elastic Waves*. Cambridge: Cambridge University Press.

Ikelle, L.T. and Amundsen, L. (2018). *Introduction to Petroleum Seismology*, 2nd edn. Tulsa: Society of Exploration Geophysicists.

Innanen, K.A. (2011). Inversion of the seismic AVF/AVA signatures of highly attenuative targets. *Geophysics*, **76**, R1–R14.

Innanen, K.A. (2012a). Anelastic P-wave, S-wave and converted-wave AVO approximations. *74th Annual International Conference and Exhibition, EAGE, Extended Abstracts*, P197. Netherlands: EAGE (European Association of Geoscientists and Engineers)

Innanen, K.A. (2012b). Exact and approximate anelastic reflection coefficients. *CREWES Research Report*, **24**, Department of Geoscience, University of Calgary.

Kanamori, H. and Anderson, D.L. (1977). Importance of physical dispersion in surface wave and free oscillation problems: review. *Reviews of Geophysics and Space Physics*, **15**, 105–112.

Kay, I. and Krebes, E.S. (1999). Applying finite element analysis to the memory variable formulation of wave propagation in anelastic media. *Geophysics*, **64**, 300–307.

Kennett, B.L.N (2001). *The Seismic Wavefield – Volume I: Introduction and Theoretical Development*. Cambridge: Cambridge University Press.

Kennett, B.L.N. (2002). *The Seismic Wavefield – Volume II: Interpretation of Seismograms on Regional and Global Scales*. Cambridge: Cambridge University Press.

Kennett, B.L.N. and Bunge, H.-P. (2018). *Geophysical Continua*. Cambridge: Cambridge University Press.

Kjartansson, E. (1979). Constant Q – wave propagation and attenuation. *J. Geophys. Res.*, **84**, 4737–4748.

Klem-Musatov, K., Hoeber, H.C., Moser, T.J., and Pelissier, M.A. (2016a). *Classical and Modern Diffraction Theory*. Geophysics Reprint Series No. 29. Tulsa: Society of Exploration Geophysicists.

Klem-Musatov, K., Hoeber, H.C., Moser, T.J., and Pelissier, M.A. (2016b). *Seismic Diffraction*. Geophysics Reprint Series No. 30. Tulsa: Society of Exploration Geophysicists.

Krebes, E.S. (1983). The viscoelastic reflection/transmission problem: two special cases. *Bull. Seism. Soc. Am.*, **73**, 1673–1683.

Krebes, E.S. (2004). Seismic forward modeling. *CSEG Recorder*, **29** (April), 28–39.

Krebes, E.S. and Daley, P.F. (2007). Difficulties with computing anelastic plane wave reflection and transmission coefficients. *Geophys. J. Int.*, **170**, 205–216.

Krebes, E.S. and Hron, F. (1980). Ray-synthetic seismograms for *SH* waves in anelastic media. *Bull. Seism. Soc. Am.*, **70**, 29–46.

Krebes, E.S. and Le, L.H.T. (1994). Inhomogeneous plane waves and cylindrical waves in anisotropic anelastic media. *J. Geophys. Res.*, **99**, 23899–23919.

Krebes, E.S. and Quiroga-Goode, G. (1994). A standard finite difference scheme for the time domain computation of anelastic wavefields. *Geophysics*, **59**, 290–296.

Larner, K.L., Hatton, L., Gibson, B.S., and Hsu, I-C. (1981). Depth migration of imaged time sections. *Geophysics*, **46**, 734–750.

Lay, T. and Wallace, T.C. (1995). *Modern Global Seismology*. New York: Academic Press.

Levander, A.R. (1988). Fourth-order finite-difference *P-SV* seismograms. *Geophysics*, **53**, 1425–1436.

Liner, C. (2016). *Elements of 3D Seismology*. Tulsa: Society of Exploration Geophysicists.

Lines, L., Wong, J., Innanen, K., Vasheghani, F., Sondergeld, C., Treitel, S., Ulrych, T. (2014). Research note: experimental measurements of *Q*-contrast reflections. *Geophys. Prosp.*, **62**, 190–195.

Liu, H-P., Anderson, D.L., and Kanamori, H. (1976). Velocity dispersion due to anelasticity; implications for seismology and mantle composition. *Geophys. J. R. Astr. Soc.*, **47**, 41–58.

Loewenthal, D., Lu, L., Roberson, R., and Sherwood, J.W.C. (1976). The wave equation applied to migration. *Geophys. Prosp.*, **24**, 380–399.

Lomnitz, C. (1956). Creep measurements in igneous rocks. *J. Geology*, **64**, 473–479.

Lomnitz, C. (1957). Linear dissipation in solids. *J. Appl. Phys.*, **28**, 201–205.

Mari, J.L. (1984). Estimation of static corrections for shear-wave profiling using the dispersion properties of Love waves. *Geophysics*, **49**, 1169–1179.

Martinez Fernandez, P.E. (2014). Application of a finite-difference scheme for the time-domain computation of 1D anelastic wavefields to fractured media. M.Sc. thesis, Department of Geoscience, University of Calgary.

Mathews, J. and Walker, R.L. (1970). *Mathematical Methods of Physics*, 2nd edn. New York: W.A. Benjamin, Inc.

May, B.T. and Covey, J.D. (1981). An inverse ray method for computing geologic structures from seismic reflections: zero-offset case. *Geophysics*, **46**, 268–287.

May, B.T. and Hron, F. (1978). Synthetic seismic sections of typical petroleum traps. *Geophysics*, **43**, 1119–1147.

Moczo, P., Bystrický, E., Kristek, J., Carcione, J.M., and Bouchon, M. (1997). Hybrid modeling of *P-SV* seismic motion at inhomogeneous viscoelastic topographic structures. *Bull. Seism. Soc. Am.*, **87**, 1305–1323.

Moczo, P., Kristek, J., and Galis, M. (2014). *The Finite-Difference Modelling of Earthquake Motions*. Cambridge: Cambridge University Press.

Moradi, S. and Innanen, K.A. (2016). Viscoelastic amplitude variation with offset equations with account taken of jumps in attenuation angle. *Geophysics*, **81**, N17–N29.

Narod, B.B. and Yedlin, M.J. (1986). A basic acoustic diffraction experiment for demonstrating the geometrical theory of diffraction. *Am. J. Phys.*, **54**, 1121–1126.

Officer, C.B. (1974). *Introduction to Theoretical Geophysics*. Berlin: Springer-Verlag.

Pekeris, C.L. (1948). Theory of propagation of explosive sound in shallow water. *Geol. Soc. Amer. Mem.*, **27**.

Petten, C.C. and Margrave, G.F. (2012). Using the Sharpe Hollow Cavity Model to investigate power and frequency content of explosive pressure sources. *CREWES Research Report*, **24**, Department of Geoscience, University of Calgary.

Pujol, J. (2003). *Elastic Wave Propagation and Generation in Seismology*. Cambridge: Cambridge University Press.

Richards, P.G. (1984). On wavefronts and interfaces in anelastic media. *Bull. Seism. Soc. Am.*, **74**, 2157–2165.

Ricker, N. (1953). The form and laws of propagation of seismic wavelets. *Geophysics*, **18**, 10–40.

Robinson, E.A. (1983). *Migration of Geophysical Data*. Boston: I. H. R. D. C.

Robinson, E.A. and Treitel, S. (1980). *Geophysical Signal Analysis*. Englewood Cliffs: Prentice Hall, Inc.

Rothman, D.H., Levin, S.A., and Rocca, F. (1985). Residual migration: applications and limitations. *Geophysics*, **50**, 110–126.

Ruud, B.O. (2006). Ambiguous reflection coefficients for anelastic media. *Stud. Geophys. Geod.*, **50**, 479–498.

Schneider, W.A. (1978). Integral formulation for migration in two and three dimensions. *Geophysics*, **43**, 49–76.

Schultz, P.S. and Claerbout, J.F. (1978). Velocity estimation and downward continuation by wavefront synthesis. *Geophysics*, **43**, 691–714.

Sengbush, R.L. (1961). Interpretation of synthetic seismograms. *Geophysics*, **26**, 138–157.

Sengbush, R.L. (1983). *Seismic Exploration Methods*. Boston: I. H. R. D. C.

Sharpe, J.A. (1942a). The production of elastic waves by explosion pressures. I. Theory and empirical field observations. *Geophysics*, **7**, 144–154.

Sharpe, J.A. (1942b). The production of elastic waves by explosion pressures. II. Results of observations near an exploding charge. *Geophysics*, **7**, 311–321.

Shearer, P.M. (2009). *Introduction to Seismology*, 2nd edn. Cambridge: Cambridge University Press.

Sheriff, R.E. and Geldart, L.P. (1982, 1995). *Exploration Seismology*. Cambridge: Cambridge University Press.

Shuey, R.T. (1985). A simplification of the Zoeppritz equations. *Geophysics*, **50**, 609–614.

Sidler, R., Carcione, J.M., and Holliger, K. (2008). On the evaluation of plane-wave reflection coefficients in anelastic media. *Geophys. J. Int.*, **175**, 94–102.

Slawinski, M.A. (2015). *Waves and Rays in Elastic Continua*. Singapore: World Scientific Publishing Co. Ltd.

Slawinski, R. and Krebes, E.S. (2002a). Finite-difference modeling of SH-wave propagation in nonwelded contact media. *Geophysics*, **67**, 1656–1663.

Slawinski, R. and Krebes, E.S. (2002b). The homogeneous finite-difference formulation of the P-SV-wave equation of motion. *Stud. Geophys. Geod.*, **46**, 731–751.

Spiegel, M.R. (1959). *Vector Analysis*. Schaum's Outline Series. New York: McGraw-Hill.

Spiegel, M.R. (1971). *Advanced Mathematics for Engineers and Scientists*. Schaum's Outline Series. New York: McGraw-Hill.

Stein, S. and Wysession, M. (2003). *An Introduction to Seismology, Earthquakes, and Earth Structure*. Oxford: Blackwell Publishing.

Stoffa, P.L., Buhl, P., Diebold, J.B., and Wenzel, F. (1981). Direct mapping of seismic data to the domain of intercept time and ray parameter – a plane wave decomposition. *Geophysics*, **46**, 255–267.

Stolt, R.H. (1978). Migration by Fourier transform. *Geophysics*, **43**, 23–48.

Stolt, R.H. and Benson, A.K. (1986). *Seismic Migration: Theory and Practice*. London: Geophysical Press.

Strick, E. (1970). A predicted pedestal effect for pulse propagation in constant-Q solids. *Geophysics*, **35**, 387–403.

Sun, Y. (2018). Solutions of the equation of motion with absorption for some common sources. M.Sc. thesis, Department of Geoscience, University of Calgary.

Taner, M.T., Cook, E.E., and Neidell, N.S. (1970). Limitations of the reflection seismic method; lessons from computer simulations. *Geophysics*, **35**, 551–573.

Taner, M.T. and Koehler, F. (1969). Velocity spectra – digital computer derivation and applications of velocity functions. *Geophysics*, **34**, 859–881.

Telford, W.M., Geldart, L.P., Sheriff, R.E., and Keys, D.A. (1976). *Applied Geophysics*. Cambridge: Cambridge University Press.

Thomas, M., Ball, V., Blangy, J.P., and Tenorio, L. (2016). Rock-physics relationships between inverted elastic reflectivities. *The Leading Edge*, **35**(5), 438–444.

Thomsen, L. (1986). Weak elastic anisotropy. *Geophysics*, **51**, 1954–1966.

Thomsen, L. (1988). Reflection seismology over azimuthally anisotropic media. *Geophysics*, **53**, 304–313.

Thomsen, L. (1993). Weak anisotropic reflections. In *Offset-Dependent Reflectivity – Theory and Practice of AVO Analysis*, 103-111, ed. J.P. Castagna and M.M. Backus. Tulsa: Society of Exploration Geophysicists.

Trorey, A.W. (1970). A simple theory for seismic diffractions. *Geophysics*, **35**, 762–784.

Udías, A. and Buforn, E. (2018). *Principles of Seismology*, 2nd edn. Cambridge: Cambridge University Press.

Ursin, B., Carcione, J.M., and Gei, D. (2017). A physical solution for plane SH waves in anelastic media. *Geophys. J. Int.*, **209**, 661–671.

Wang, Y. (2017). *Seismic Inversion: Theory and Applications*. Oxford: Wiley Blackwell.

Widess, M.B. (1973). How thin is a thin bed? *Geophysics*, **38**, 1176–1180.

Yilmaz, O. (1987). *Seismic Data Processing*. Tulsa: Society of Exploration Geophysicists.

Index